Western Civilization
in Biological Perspective

Western Civilization in Biological Perspective

Patterns in Biohistory

STEPHEN BOYDEN

Professorial Fellow in Human Ecology,
Centre for Resource and Environmental Studies,
Australian National University, Canberra

CLARENDON PRESS · OXFORD
1987

Oxford University Press, Walton Street, Oxford OX2 6DP
Oxford New York Toronto
Delhi Bombay Calcutta Madras Karachi
Petaling Jaya Singapore Hong Kong Tokyo
Nairobi Dar es Salaam Cape Town
Melbourne Auckland
and associated companies in
Beirut Berlin Ibadan Nicosia

Oxford is a trade mark of Oxford University Press

Published in the United States
by Oxford University Press, New York

© Stephen Boyden, 1987

British Library Cataloguing in Publication Data
Boyden, Stephen
Western civilization in biological perspective: patterns in biohistory.
1. Civilization, Occidental   2. Human biology—Social aspects
I. Title
909'.09821 CB245
ISBN 0–19–857672–2

Library of Congress Cataloging in Publication Data
Boyden, Stephen Vickers.
Western civilization in biological perspective.
1. Sociobiology 2. Human ecology   3. Human evolution.
I. Title
HM106.W48   1987   304.5   86–23575
ISBN 0–19–857672–2

Text processed by the Oxford Text System
Printed in Great Britain
at the University Printing House, Oxford
by David Stanford
Printer to the University

PREFACE

Broadly speaking, approaches to the study of human society can be said to be of two kinds. In one of these, scholars collect information about their own, or at least contemporary, societies, and this information is used to formulate laws and principles intended to explain what is taking place. In the other approach individuals gather information about human societies in the past, they attempt to recognize recurring or unfolding patterns, and they use the knowledge acquired as a source of insights into the nature of present-day society. This historical approach has been taken by Karl Marx and Arnold J. Toynbee, among many others.

In my view, the historical approach has the greater contribution to make towards our understanding of human society in the modern world. Nevertheless, it does have certain inherent hazards. One of the most obvious of these is the fact that it is unlikely to be useful if, as has often been the case, the information brought together about the past is inaccurate or unrepresentative. Of course, as more data become available through archaeological and historical research this potential pitfall may become somewhat less of a problem, although the temptation to make selective use of information to support a personal viewpoint still remains a risk.

I suggest that a much more serious flaw in the historical approach, at least as it has been applied up until now, lies in the fact that it takes far too little account of the biological components of the societal systems with which it is concerned. This is a weakness which it shares with the other conventional approaches to the study of society. Since, in fact, all human situations, past and present, involve a dynamic and continuous interplay between cultural and biological processes, it stands to reason that our understanding of society will always remain inadequate so long as one of these sets of processes, namely, the biological, is almost totally ignored.

It has been my view for many years that there is an urgent need for patterns of learning which reflect what really happened in the history both of life on earth and of civilization and which pay due attention to the interplay between biological and cultural processes which is so essential a feature of human situations. In essence, I am suggesting that the historical approach to the study of society would be a good deal more useful if it went back 4000 million years instead of a mere few thousand, and if it included study of the evolution of living organisms, ecosystems, and hominids, of the emergence of the human capacity for culture, and of the lifestyle and society of humans in the environment in which the species evolved. Such an approach would add an extra dimension to the understanding of human situations, providing appreciation of the sensitivities both of human beings and of the nature and scale of the changes which are taking place in the interrelationships between human society and biological systems.

I refer to this broadening of the conventional historical approach to the study of human situations as *biohistory*; and I suggest that the kind of understanding that it induces in the human mind has the potential to influence the way people perceive the world and human situations. I use the term *biosophy* to describe this kind of understanding and the general outlook or world view that it generates. I will not elaborate on this theme here, beyond stating the hypothesis that a pervasive biosophical perspective would influence societal values, attitudes, opinions, and aspirations in such a way that they would become more consistent than those which characterize our present society with the long-term survival of civilization and well-being of humankind.

It was thoughts such as these that led me to change direction academically some twenty years ago, in order to work in this area. It was a move that I have never regretted.

I regard this book, then, as a small contribution to the neglected but exceptionally important subject of biohistory. The book's theme, at an earlier stage of development, formed the basis of the Third Year course in the Human Sciences Programme at the Australian National University in 1974 and 1975. I would like particularly to express my heartfelt appreciation to the students and tutors in those two years—for their encouraging enthusiasm and for contributing so much to the development of the ideas which now find expression in this volume.

Apart from these people, I owe a great deal to many other individuals—students, colleagues, and friends—who have contributed immeasurably and in different ways to the whole process. There have been so many of them over the years that it would be impractical to try to mention each of them by name. I hope, however, that they all know how much I value their individual contributions. I will, however, mention the name of one stalwart colleague, because she has participated uninterruptedly in the work almost from the beginning: it is that of my secretary, Mrs Fay Goddard, whose help, patience, and perseverance have been very much appreciated.

I would also like to mention two other individuals who, although not involved in the work, have nevertheless played key roles. One of these is Professor Frank Fenner, whose understanding, encouragement, support, and advice have been invaluable throughout. The other is Sir Leonard Huxley, who, as Vice-Chancellor of the Australian National University at the time of my academic shift in direction, was sufficiently interested in and sympathetic towards my ideas to permit me to retain my position at this university in spite of the radical change in my field of study.

With respect to the book itself, I am aware that it has many deficiencies, and that certain biologically important aspects of human situations have received less attention than they deserve. For example, although, as I have indicated above, biohistory begins with the study of the history of life on earth and of certain important biological processes, it has been necessary, for reasons of space, to delete much of the part of the book which dealt with these aspects. The story is thus picked up with the emergence in evolution of early hominids. However, relevant evolutionary and ecological principles are discussed where appropriate in various places throughout the book. It also is likely that some readers will feel that human genetics has not been given the attention it warrants. Another omission lies in the fact that little attention is paid to children—the emphasis throughout is on adults of the human species. I hope to compensate to some extent for these and other deficiencies in later work.

Canberra   S. B.

CONTENTS

1

BIOHISTORICAL PERSPECTIVES

INTRODUCTION

The emergence in evolution of the human capacity for culture was a development of profound and far-reaching significance not only for humankind, but also, in the long term, for the whole of the living world. Human culture constituted a new force in the biosphere: while entirely dependent on the underlying inorganic and biotic processes, it was nevertheless different from them, and it interacted with them in many important ways. As, over the millennia, culture itself evolved and became more complex, so did the extent of its interplay with the processes of nature increase; and in recent times this interaction between culture and nature has intensified to such a degree that serious threats now exist to the survival of the human species and to the integrity of the life-supporting processes of the biosphere [1].

The word *biohistory* is used here to describe the approach taken in this book to the study of human situations in terms of the interplay between natural and cultural processes. This approach reflects the sequence of developments which took place in the history of life and of civilization on earth. Biohistory thus takes as its starting point our knowledge and understanding of the *biosphere*. It takes account of the fundamental processes on which all life is based, and it is concerned with the essential facts and principles of biological evolution and ecology. It moves on to consider the evolutionary history and biology of *humans*, and especially the appearance in evolution of the capacity for culture. Attention is then turned to human *society*, which, as a consequence of culture, eventually came to be very different in important ways from the societies of other animals and of early hominids [2]. Biohistory focuses particularly on the various impacts that society has had in the past, and is having today, on biological systems—especially on the ecosystems of the biosphere, on the biosphere as a whole, and on human organisms.

While some of the impacts that human society has had on biological systems may be perceived as desirable for human beings, such as increased yield of edible biomass in given areas, others may be seen as undesirable. Examples of the latter are declining bioproductivity in local ecosystems through erosion or salinization, contamination of the air with by-products of industrial processes, and the appearance in human populations of various 'diseases of civilization' such as cholera, plague, influenza, scurvy, beri-beri, cardiovascular disease, and lung cancer. In the case of these undesirable reactions of human beings to culturally induced environmental changes, biological adaptive mechanisms frequently come into play and contribute to the survival of some individuals or, in the long run, of populations. More important in the present context, however, is the fact that when the impacts of societal activities on natural systems are perceived by people to be undesirable, various *cultural* adaptive processes may be brought into action in attempts to overcome these difficulties, sometimes successfully and sometimes not. *Cultural adaptation* to culturally induced undesirable changes in natural systems is a recurring theme in biohistory.

Our interest in the cultural adaptive responses to culturally induced undesirable changes in natural systems raises a slight problem of terminology. 'Cultural adaptation' is an expression that has been widely used by cultural anthropologists who have been concerned especially with the extent to which cultural patterns in different societies can be regarded as adaptations to prevailing ecological conditions (as determined mainly by natural forces unmodified by human activities) [3]. The question therefore arises whether we should coin a new term, such as *cultural autoadaptation*, to describe the kind of adaptive response on which we are focusing our attention. I have decided against this course, and instead to make it clear that when the term 'cultural adaptation' is used in the pages that follow it will, unless otherwise stated, refer to *deliberate adaptive responses which are based on cultural processes, and which are aimed at overcoming undesirable culturally induced changes in biological systems.* This subject is of crucial significance in the modern world, in view of the fact that some of the culturally induced forces which at present threaten natural systems are of a magnitude unprecedented in the history of our species. The fate of civilization and, indeed, the very survival of humankind depend on whether or not the processes of cultural adaptation brought into play to counter these threats are successful.

It is self-evident that human culture, like the living components of the biosphere, has itself evolved and is still evolving. Indeed, much has been written on the subject of cultural, 'superorganic', or 'psychosocial' evolution [4], and there has been a good deal of debate about the extent to which the directions taken by culture in different populations are determined by the characteristics of the local environment, such as its agricultural potential and prevailing climate [5], and about the extent to which different societal and political arrangements are 'adaptive' to ecological circumstances [6]. In this book we will be concerned mainly with the *consequences* for biological systems of this evolution of culture, and we will not be dealing in detail with the actual process of cultural evolution itself. For the purposes of this volume, the following very brief summary of the situation will suffice.

Culture has evolved in different ways in different populations, and possibly the only common tendency has been for it to become progressively more complicated. Much of the recent increase in complexity has followed developments in that aspect of culture which we refer to as *technology*. In general, it is clear that a number of different factors affect the direction and quality of cultural change. These include environmental conditions, the previous cultural experience of a population, some phylogenetic predisposition of human beings to prefer one sort of cultural change to another [7], and various fortuitous factors—such as the chance appearance in a population of a charismatic individual with new ideas.

Mention should be made of the fact that some of the theoretical speculation about cultural evolution has involved attempts to apply biological concepts of evolution directly to cultural change. This approach is valid only so long as it is used as a basis for asking questions about cultural evolution. It cannot be used for drawing conclusions about the nature of cultural change—for the simple reason that cultural systems, although they contain biological components, are different from biological systems in many important ways. In this context it is worth stressing that the approach taken in this book *does not involve the use of biological analogy*. We shall be concerned strictly with the interrelationships between the biological and cultural aspects of human situations, and will not be assuming, as was the case for example in Social Darwinism, that biological principles and concepts apply to cultural processes [8].

I would like to emphasize the fact that the study of biohistory requires the construction of a comprehensive conceptual framework to facilitate thinking and communication about the interrelationships between the disparate aspects, inorganic, biological, and cultural, of human situations. While the paradigms of biology, sociology, economics, and anthropology may well be relevant in one context or another, none of them provides such a framework. Recognizing this need, my colleagues and I have devoted some effort over the years to the development of such a framework to assist us in our deliberations about the dynamic interactions between cultural and natural processes. It is based initially on the simple model introduced at the beginning of this chapter—that is, the recognition of three broad clusters of variables relating, respectively, to the *biosphere*, *humans*, and *society* [9]. It is not my intention to impose this framework too forcibly on the readers of this book. However some of the conceptual schemes are presented in Appendix 1 for those who may be interested.

The arrangement of this book reflects the biohistorical sequence outlined above, except in so far as it assumes that the reader has some basic knowledge of evolutionary history and the diversity of life on earth before the appearance of humankind, of the history of the atmosphere and the soil, and of the fundamental principles of ecology. Thus, after these introductory comments we take up the story in the second part of this chapter with a brief summary of the state of knowledge about the relatively recent evolutionary background of the human species. This is followed by two theoretical chapters, one of which deals with the different kinds of adaptation, biological and cultural, available to humankind. The other considers the contribution of our knowledge of evolutionary processes to the understanding of human health needs and human behaviour.

The rest of the book is concerned with the impacts, throughout history, of human society on biological systems—in particular, on the ecosystems of the biosphere and on human organisms. The organization of this, the main part of the book, will be based on the fact that we can recognize four distinct ecological phases or modes in human history. Each of these differs from the others in important ways, in terms both of the ecological relationships between society and the biosphere and of the actual biological conditions of life and the patterns of health and disease of human beings. These four phases will be referred to as follows:

Phase one—the primeval (or hunter-gatherer) phase. This is by far the longest of the ecological phases. It was the only way of life of our ancestors for the tens of thousands of generations.

Phase two—the early farming phase. This ecological phase began independently in several regions of the world around four or five hundred generations ago.

Phase three—the early urban phase. This phase goes back about two or three hundred generations to the formation of the first cities in the Near East.

Phase four—the modern high-energy phase. This most recent ecological phase began in Europe and North America seven or eight generations ago. As a consequence of technological developments, it has been associated with a 'population explosion', a massive increase in the destructive power of weapons designed for killing other members of the human species and an enormous growth in the per capita use of extrasomatic energy and non-renewable resources [10].

It must be emphasized that the use of the term 'phase' is not intended to imply that these are inevitable steps in cultural evolution, nor that they are irreversible in all respects. Nevertheless, we may note that the modern high-energy phase could not have come into being if there had not been an early urban phase, and the latter could not have existed if there had not been an early farming phase, which in turn could not have come about if there had been no primeval phase. It is also clear that the appearance of each new phase did not result in the disappearance of all populations with economies characteristic of earlier phases. Societies based on these different ecological phases can, and do, coexist [11].

From the macroecological standpoint, it is significant that the outstanding characteristics of ecological *phase four*, the high-energy phase, are such that it cannot last indefinitely. Indeed, the threats to natural systems inherent in the build-up of weapons of mass destruction and in the progressive intensification of resource extraction, energy use, and waste production are on a scale and of a kind such that the limits of tolerance of the biosphere as a whole may soon be reached. If civilization is to survive, the fourth ecological phase must at some time give way to a *phase five*—a phase characterized by restoration of ecological balance between human society and the biosphere and by the eradication of the weapons of mass destruction which could so easily bring an end to civilization and humanity.

The four ecological phases will be discussed in order in chapters 4 to 10. In each case we will consider: (a) the *interrelationships between human society and the biosphere*, (b) aspects of *human society* itself which have a bearing on these interrelationships, and (c) *human organisms* as they are affected by cultural and societal developments.

The final chapter will consist of a discussion of some of the implications of the biohistorical perspective for the future of humanity.

Reference must be made in this introductory chapter to an especially important ecological consequence of cultural processes. It is the fact that culture has added an extra dimension to the inputs and outputs—or the metabolism—of human populations. In addition to *biometabolism*—that is, the inputs into human organisms of food, water, and oxygen and the discharge of organic wastes—culture has resulted in extra inputs and outputs of resources and energy which are used, not in biological metabolism, but in various kinds of technological processes. This additional set of inputs and outputs of society is referred to as *technometabolism*.

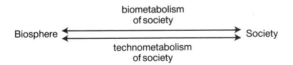

Technometabolism has been a feature of human society since fire first came to be used on a regular basis by our ancestors. The introduction of metallurgy was another particularly important development in this regard. It is in the modern high-energy phase, however, that a massive intensification of technometabolism has occurred, involving increases as great as fiftyfold in the per capita inputs and outputs of energy and materials in some Western communities.

Another extremely important biological consequence of cultural developments is the fact that, in the words of Tinbergen: 'We are creating a habitat that diverges more and more and with increasing speed from that to which genetic evolution has adapted us' [12]. This process began, of course, with the domestic transition and the introduction of farming. Its wide-ranging implications for human health and well-being as well as for the interrelationships between human society and the ecosystems of the biosphere will be one of the main themes of this book.

One further comment is necessary on the aims of this volume. An important aspect of the approach taken is that it recognizes that many important variables in human situations are of a kind that are difficult, if not impossible, to quantify, and many are also hard to describe precisely. There can be no more potent ecological force in the modern world, for instance, than certain components of the dominant value system of Western society; and on the level of human experience cultural developments can markedly influence such intangible but important aspects of human experience as a sense both of personal involvement and of frustration. To ignore such variables in our analysis of any particular issue or problem simply because they cannot be measured or counted would be unscientific, and would lead to an incomplete, unbalanced, and misleading picture. Consequently, where such intangible aspects of situations are considered important, a deliberate effort will be made to take them into account and to deal with them as systematically as possible.

HUMAN EVOLUTION

After life on earth had been in existence for some 4000 million years there emerged, around 65 million years ago, during the last part of the dinosaur era, a small group of mammals which are classified as *primates*. Among them were the ancestors of humankind. They were small, shrew-like creatures which had taken to living in trees, where most primates have remained to the present day. In fact, the arboreal environment exerted selection pressures which had important influences on primate evolution and which left their mark on the forms which are now completely terrestrial. The important evolutionary consequences of tree-living in the primates included the development of the limbs as grasping organs, the replacement of claws by flattened nails, the development of the forelimbs as exploratory organs, a reduction in the olfactory sense and a compensatory development of visual acuity, and, eventually, stereoscopic vision. Changes in the skull included a downward movement of the *foramen magnum*, reduction in the size of the snout, and enlargement of the eyes, which migrated to the front of the face. In the brain itself there was a progressive development of the parts of the cerebral cortex concerned with sensory representation. In addition, areas of the brain associated with fine co-ordination of movement and balance, namely, the cerebellum and parts of the cortex, became elaborated. The end result was the development of a relatively large brain [13].

By some time before 5 million years ago, and probably before 15 million years ago, the ancestors of the human species had come down from the trees and were living mainly in open grassland [14], and there is good evidence that by 4 or 5 million years ago there were terrestrial primates, walking upright, in the East African savannah. One fossil that has given rise to much interest is that of a young female hominid, referred to as 'Lucy', found at Hadar in Ethopia and dated about 3 million years before the present [15]. She was designated by her discoverers as belonging to the species *Australopithecus afarensis*. This remarkably complete specimen had a skull and brain case very like that of a chimpanzee, but she walked erect, more or less in the human manner. Recently a jaw-bone similar to Lucy's and dated about 5 million BP has been found in East Africa [16].

Another remarkable discovery was made at a place called Laetoli, near Lake Eyasi in Tanzania. It consists of a set of fossilized footprints of three upright-walking individuals—two adult and one young—believed to date back about 3.5 million years [17].

By 2 million years ago there was a hominid in East Africa which has been ascribed to the genus *Homo*, and named *Homo habilis*. One skull of a member of this group found at Koobi Fora near Lake Turkana in northern Kenya contained a brain with a volume of 800 cc (that is, about 300 cc bigger than that of the average chimpanzee) [18]. *Homo habilis* was an omnivorous species, consuming both plant and animal food.

There were at least two other species of hominid also living in Africa at this time. They have been ascribed to the genus *Australopithecus*, and they differ from *Homo habilis* in having smaller brains and bigger and more protruding faces. The larger of these has been called *Australopithecus robustus*; it was up to 5 feet tall and had a brain size of up to 550 cc, and its skull had a distinct sagittal crest. Its teeth contrast with those of all other hominids in that the molars and premolars were enormous and the incisors and canines quite small. This has been interpreted as indicating that this species was almost entirely herbivorous, while other hominids were omnivorous. The smaller and more 'gracile' form of hominid which coexisted in Africa with *Homo habilis* has been called *Australopithecus africanus*. It stood about four feet tall and had a cranial capacity of around 450 cc [19].

By around 1.5 million years ago there were hominids in Africa which were more like modern human beings. Their general physique was similar, although they had a somewhat stockier build. The skull was characterized by a strongly developed brow ridge extending above the orbits of the eyes as an uninterrupted bar of bone. They have been described as early examples of *Homo erectus*, a species which is considered to have persisted until around 300 000 years ago. The brain size of *Homo erectus* ranged from about 900 cc to 1200 cc in later specimens. In its early days *Homo erectus* shared the African savannah with the two australopithecine species, which eventually died out about a million years ago.

The oldest stone tools found in East Africa consist of choppers, scrapers, and flakes. These date back at least 2.5 million years and are generally considered to have been made and used by members of the species *Homo habilis* [20]. Gradually, over hundreds of thousands of years the tool kit became more varied and complex. About 1.5 million years ago there appeared a new stone industry, known as Acheulian, which was characterized by its 'hand axes'. These are pear-shaped implements, some of them of a size that can be comfortably held in the palm of a hand, but others much larger. They were presumably made by members of the species *Homo erectus*. Acheulian hand axes have also been found in Europe, where they were used at least until 200 000 years ago.

The very few human remains that have been found dating back to between 200 000 and 300 000 years ago suggest that at this time evolutionary change in the hominid line was proceeding more rapidly. The people of this period, as represented by the remains of skulls found at Steinheim in Germany and at Swanscome in Britain, are sufficiently like ourselves for some authorities to classify them as members of our own species, *Homo sapiens*. They have variously been described as *Homo sapiens presapiens* and as *Homo sapiens steinheimensis* [21]. Their brow ridges were less pronounced than those of *Homo erectus*, and their brain size was 1200–1300 cc.

It is likely that the people of this period who were living in Europe took to wearing clothes for the purpose of keeping warm—a practice which would have stood them in good stead for the approaching Riss glaciation; and they also developed a new sophistication in tool manufacture, involving the striking of the stone core with a softer material such as wood—a method known as the Levallois technique. They were also using fire, as their *Homo erectus* ancestors had done some two or three hundred thousand years before them. Indeed, it has recently been sugggested that hominids might have been using fire as long ago as 1.5 million years, although this view is not universally accepted [22].

The course of human evolution over the following 200 000 years is far from clear. It has recently been suggested on the basis of pollen analysis that human beings using fire arrived in Australia about 150 000 years ago, but at this stage the suggestion must be treated with caution [23]. It is known, however, that about 100 000 years ago, and during most of the first part of the fourth or Würm glaciation (70 000–40 000 BP), western Europe was occupied by a very distinctive form of humanity which has been classified scientifically as *Homo sapiens neanderthalensis*. The classic 'Neanderthals' were short people of a generally stocky build, the men being on average a little over five feet tall and the women a little less. The skull was rather flattish on top and noticeably rounded at the back, and a characteristic feature was a pronounced brow ridge reminiscent of *Homo erectus*. The jaws were massive. The size of the brain ranged from 1450 cc to 1650 cc, and the average brain size exceeded that of modern humankind. These people were apparently very strong, their bones being especially rugged and showing signs of strong muscle attachment. They were well acquainted with the use of fire, they hunted big game, and they dressed themselves in animal skins. Some of them buried their dead, and they made use of paints for decorating their bodies [24].

By around 40 000 years ago another type of human being had appeared in Europe. If it were possible to bring some of these individuals back to life, to dress them appropriately, and then to set them loose on a city street, we would not be able to distinguish them from the better physical specimens of modern humanity. These people, typified by skeletal remains found at Cro-Magnon, were tall and had steep foreheads and rounded skulls, with a cranial capacity of around 1400 cc. Their brow ridges were only moderately developed and were not continuous from one side to the other, and they had well developed chins. They are classified as *Homo sapiens sapiens*, the same species and sub species as ourselves.

It is apparent that by this time, at the end of the first phase of the fourth or Würm ice-age, modern humankind had replaced the classic Neanderthal people in Europe. Just how this change came about, and where the modern people came from, is not known [25]. Their implements were distinctly superior to those of the Mousterian flint industry of the Neanderthals and it has been suggested, rather oddly perhaps, that this is an indication that they possessed a more complex social organization [26]. It is not clear whether people of this kind, much more like ourselves than like the classic Neanderthals, were in existence outside Europe during the Neanderthal era, although this seems likely. It is possible that around sixty to one hundred thousand years ago, when the Neanderthals were still occupying Europe, there were humans living in Africa very similar in their general appearance to modern human beings—people who would be classified by biologists as belonging to the sub species *Homo sapiens sapiens*.

It was people of this modern type who were responsible, over the following 30 000 years, for a marked diversification and sophistication of culture, at least as reflected in the tangible evidence that they left behind in the form of artefacts of many kinds, including scraping tools, knives, burins, awls, needles, spatulae, and various kinds of weapons. When we look at many of the implements of this period, it is hard to avoid the impression that a great deal more work went into shaping them than was really necessary to render them functional; and apart from these utilitarian objects, people were finding it worth their while to make pendants, necklaces, arm bands, musical instruments, statuettes, and paintings [27].

By 40 000 years ago, perhaps much earlier, humans had penetrated to all continents of the world, with the possible exception of North and South America, where they may not have arrived until around 20 000 years ago [28]. The spread of the species was associated with some genetic divergence, giving rise to observable differences between people in different parts of the world with respect, for example, to stature, skin and hair colour, and facial features [29]. Other genetic differences, such as differences in the distribution of blood-group antigens between populations from different regions, are detectable by various scientific procedures. Presumably these differences were brought about by a combination of genetic drift and natural selection [30]. It has been speculated, for example, that a light skin is of selective advantage in areas of the world where sunlight is weak, because it allows the formation of vitamin D below the skin, whereas in tropical areas a dark skin would prevent the synthesis of large amounts of the vitamin, which would be harmful. Similarly, the suggestion has been made that pathological changes which occur in the skin of white people, especially children, in the tropics might be the basis of a negative selection for white skin in these areas. It is impossible, however, to determine the relative roles played by genetic drift and by the various postulated selective factors in determining the genetic differences between populations in different regions of the world.

The main genetic types or 'races' of humankind have conventionally been classified as Negroids, Caucasoids, Mongoloids, and Australoids; but Dobzhansky, for instance, defined 14 different races, and Garn recognized 32 [31]. However, partly because of the frequency of 'hybridization', but also for other reasons, attempts to classify the different human groups in this way are of dubious value.

THE EMERGENCE OF HUMAN CULTURE

There has been much speculation among scholars about the origins in our evolutionary history of the capacity for culture [32]. Broadly, it can be seen as an outgrowth of the ability, so well developed in primates, to learn new ways of doing things through experimentation. Another primate characteristic which certainly played a part was the tendency to imitate, so that learned skills could be passed on more effectively among the members of a population and from one generation to the next. The relative manual dexterity of the higher primates was also a prerequisite for the development of the capacity to make tools.

At some stage in the evolutionary process experimentation must have taken the form of making and contriving new sounds, distinct from the various innate expressions of emotions [33]; and the idea must have occurred of using these new sounds to convey definite meaning. We see in this critical step—the invention, learning, and use of symbols—the most essential element of culture, as reflected in most of the numerous definitions of this term [34]. In time our ancestors developed the capacity to use several symbolic sounds, or 'words', together in order to communicate more complex thoughts or observations, leading ultimately to language as it exists today. This development of the capacity for speech and, eventually, of language itself was surely the single most significant factor in the evolution of the human species. In the words of Wilson:

All of man's unique behavior pivots on his use of language, which is itself unique ... . The development of human speech represents a quantum jump in evolution comparable to the assembly of the eucaryotic cell [35].

At what particular stage in hominid evolution our ancestors developed this ability to communicate through the spoken word is a matter for conjecture. Although the suggestion has been made that it did not occur until after the Neanderthal era [36], a more common view is that the capacity for speech, if less developed than in modern humankind, goes back to *Homo erectus* [37]. In my view, it may well go back further than this, since it seems possible that it was this emerging capacity for inventing and using language which played the decisive role in the almost explosive (on the biological time-scale) increase in the size of the brain in the hominid line. This enlargement of the brain was well underway by the time that *Homo erectus* was on the scene [38].

The word *culture* itself has been given many different meanings. In this book it is used for that set of phenomena and processes which are characteristic of human societies and which involve the acquisition and accumulation of information and its transmission by non-genetic means, mainly through the use of learned symbols, from one human to another, from one society to another, and from one generation to another [39]. Culture includes language, beliefs, knowledge, ideas, art, as well as learned patterns and ways of doing things; it thus includes technology—that is, knowledge of techniques for modifying the environment [40]. The word *nature* is used here for all other aspects of reality—aspects, that is, which are not covered by this definition of culture. *Natural processes* are processes (inorganic and biotic) of a kind which existed on earth before the advent of human culture. *Cultural processes* are processes which depend on and involve human culture [41].

It makes no sense to try to decide precisely at what point in hominid evolution culture began. We know that it has been in existence for tens of thousands of years, and that it was presumably not in existence 5 million years ago. All the evidence, as well as commonsense, tell us that the human capacity for culture did not appear with a 'big bang', but that it developed gradually over a period of at least several million years.

CULTURE–NATURE INTERPLAY BEFORE THE DOMESTIC TRANSITION

As soon as culture came into existence, even in rudimentary form, it began to have impacts on natural processes [42]; and in turn, some of the changes that it brought about in biological systems influenced the processes and patterns of culture. In the early stages of the evolutionary development of the potential for culture, by far the most significant influence was the introduction of new selection pressures acting on the early hominid populations bearing culture. Indeed, as many authors have pointed out, culture added an extra dimension to the adaptive opportunities of its bearers, and this applied no doubt even to the embryonic 'protoculture' of several million years ago. Thus, once an incipient culture had come into being there existed a strong selective advantage for genetic variants that had a greater capacity for culture than others. It is a feature of biological evolution that, once a structure comes to acquire a new function of adaptive value, the rate of evolutionary change through natural selection, involving improvement and elaboration of the structure, is likely to be rapid [43]. This is apparently what occurred in the case of the brain of hominids between the time of the relatively small-brained australopithecines (with a cranial capacity about 500 cc) of around four million years ago and the appearance of the Cro-Magnon people about forty thousand years ago (with an average cranial capacity of about 1400 cc).

While it is abundantly clear that the capacity for culture must have been of selective advantage in human populations living in the conditions in which this capacity evolved, it is reasonable to question whether this capacity will always, under entirely different conditions, continue to be an advantage, or whether it might ultimately prove maladaptive. We will return later to this point.

As we shall discuss later, culture has continued to have some influence on the genetic structure of human populations in more recent times; and, conversely, it is obvious that the genetic attributes of the human species have, in many ways, influenced the direction and nature of cultural change. This *coevolution* of cultural and natural processes has proved a source of much discussion and speculation among anthropologists and biologists in recent years [44].

The impact of human culture on other species and on the world's ecosystems during the primeval phase of human existence was slight, as compared with the situation which followed the domestic and industrial transitions. Nevertheless, there must certainly have been some local effects on the populations of various animals and plants, and there is evidence of two major and relatively widespread influences. One of these general effects was the razing by fire of large areas of woodland which came to be replaced by grassland. To what extent this action was deliberate is unknown, but there is evidence of far-ranging fires in the forests of continental Europe as long ago as 200 000 years. The replacement of forest by grassland had important ecological impacts on animal populations and is thought to have resulted in big increases in the herds of grazing animals and consequently in the supply of animal protein for human beings. Fires resulting from human activities also had a major impact on the vegetation in parts of Australia long before the invasion of the continent by Europeans [45].

The other widespread ecological change which some authorities believe to have been the result of human activities was the phenomenon that has come to be known as the Great Pleistocene Extinction. The fossil record shows that late in the Pleistocene there occurred, in every continent, a wave of extinction of large animals affecting at least two hundred genera— consisting mainly, but not entirely, of herbivores. There is some disagreement whether humankind was responsible for these extinctions, but the evidence put forward to support the hypothesis seems rather convincing [46].

So, on the eve of the domestic transition, and as the ice sheets were finally receding at the end of the last wave of glaciation, we find the human species, with its big brain and unique capacity for culture, firmly established on all continents of the world. By this time, which was only about 400 generations ago, our species had been in existence for tens of thousands of generations, and the sub species, *Homo sapiens sapiens*, for at least 2000 generations.

NOTES

1. In this book the word *culture* is used to represent that set of phenomena and processes which are characteristic of human societies and which involve the acquisition and accumulation of information and its transmission by non-genetic means, mainly through the use of learned symbols, from one human to another, from one society to another, and from one generation to another. 'Culture' is therefore used here in the sense of an 'ideational system' (see Keesing 1976, p. 139). It is thus distinct from its *products*— which may be material objects, like buildings, machines, and paintings, or arrangements, like economic systems and organizations (see also notes 34 and 40).
2. Here I use the word *society* to include all the various products of human culture on the population level. It includes societal institutions and organizations, artefacts or products of labour (e.g. buildings, machines, manufactured commodities), societal arrangements (e.g. economic systems), and societal activities (e.g. farming, industrial production, transportation, mining, manufacturing and use of weapons, and education). The term also includes those products of culture which comprise culture itself—such as knowledge, technology, belief systems, value systems, and societal aspirations.
3. For a useful introduction to the debates on this area, see Keesing 1976, chapter 12.
4. See Alland 1970, chapter 2; Phillips 1971; de Beer 1975, p. 417; Durham 1979.
5. See Meggers 1954; Keesing 1976, chapter 9.
6. Durham 1978; Chagnon and Irons 1979; Reynolds 1980.
7. For speculative discussion on phylogenetic predisposition to certain kinds of cultural change, see Lumsden and Wilson 1981.
8. For discussions on Social Darwinism, see Banton 1961; Peel 1971; Jones 1980.

9. For more detailed discussion, see Boyden and Millar 1978; Boyden 1979; Boyden, Millar, Newcombe, and O'Neill 1981.
10. *Extrasomatic energy*, in contrast to *somatic energy*, is energy which does not flow through living organisms.
11. For this reason the word 'mode' might seem more appropriate than 'phase'. However, in this book the term 'phase' will be used throughout in the sense of a 'stage of change or development' (*The Concise Oxford Dictionary*).
12. Tinbergen 1972, p. 386.
13. For a fuller description see Harrison, Weiner, Tanner, Barnicot, and Reynolds 1977, chapter 2, which is the main source of the information presented in these paragraphs.
14. The suggestion has been made that the evolutionary ancestors of humankind of around 10 million years ago were aquatic and that this explains certain characteristics of the human species, such as relative hairlessness and subcutaneous fat (see Morgan 1982, 1984).
15. Johanson and Edey 1981.
16. There has been a great deal of discussion and debate in recent years about the significance of and relationships between the various fossil remains of early hominid and hominid-like individuals which have been uncovered—mainly in Africa. For useful accounts of the state of the debate, see Tobias 1975; Wood 1978; Johanson and Edey 1981; Leakey 1981; Reader 1981; Cherfas 1983; Bunney 1984.
17. Hay and Leakey 1982.
18. Leakey 1973. For a general description of *Homo habilis*, including the first specimens found at Olduvai, see Leakey 1981.
19. Many different scientific names have been given to the remains of hominids found in southern and eastern Africa dating from about 1 million to 2 million years ago. However, the prevailing view is that no more than three species existed: *Homo habilis*—transforming into *Homo erectus*, *Australopithecus africanus*, and *Australopithecus robustus*. For discussion on the topic, see Harrison *et al.* 1977, chapter 4 and Stringer 1984.
20. Whether or not the australopithecine species also made tools is at present a matter for conjecture (see Isaac 1978).
21. Dobzhansky 1962; Young 1971.
22. Gowlett, Harris, Walton, and Wood 1981; see *New Scientist* 1981, **92** (1280), p. 505.
23. Singh and Geissler 1985.
24. Young 1971, chapter 37.
25. For a discussion on this point, see Wood 1978, pp. 72, 73.
26. See Harrison *et al.* 1977, p. P5.
27. Edwards 1978; Leakey and Lewin 1979; Paddayya 1979.
28. Adovasio and Carlisle 1984.
29. Coon 1962, who recognizes five human 'subspecies'—Mongoloid, Australoid, Negroid, Capoid, and Caucasoid—takes the view that the genetic divergence which led to the existence of distinct races began in the time of *Homo erectus*, different groups of which evolved independently into *Homo sapiens*. For discussion on this view, see Dobzhansky 1963 and Cavalli-Sforza and Bodmer 1971, chapter 11.

30. On genetic drift, see chapter 2, p. 20 in this volume. 31. Garn 1961; Dobzhansky 1962.
32. See, for example, Spuhler 1959; Garn 1964; Foley 1984. 33. By innate sounds I mean, for instance, the sound of crying, laughter, wailing in pain, gasping in fright (or their equivalents in prehuman primates). The emission of such sounds is, I assume, a phylogenetic characteristic of the species, and does not have to be learned. Whether the *understanding* of this meaning is learned or innate is debatable.
34. Already in 1952 Kroeber and Kluckhohn were able to pick out 164 definitions of culture from the literature. See also Phillips 1971, p. 360 and Keesing 1976, pp. 138, 172.
35. Wilson 1975, pp. 555-6.
36. Lieberman, Crelin, and Klatt 1972.
37. Falk 1975; Harrison *et al.* 1977.
38. There has been much discussion about the relationships in evolution between the development of the capacity of speech and the development of the capacity for making tools; see, for example, Washburn 1960; Lenneberg 1967; Hewes 1973; Leakey and Lewin 1979.
39. The use of symbols is by no means restricted to hominids. Numerous species emit sounds as a means of communication which are symbolic indicators of some important fact or state (e.g. to indicate the presence of a predator or that the hunt is on, or to invite sexual attention). However, the emitting of such sounds is determined by the genotype; it is not learned. The symbols are not invented by individual animals.

    Nevertheless, many animals are able to *understand* the symbolic meaning of sounds which have no phylogenetic basis. Thus Pavlov's dogs learned the meaning of the sound of a bell, and dogs also readily catch on to the meaning of human words, such as 'walk' and 'sit', and, in the case of sheep-dogs, of different kinds of whistles. This capacity is also important in nature. Moreover, the meaning of sounds and scents originating from other species may even be passed on, through teaching, from one generation to the next. It is said, for instance, that lionesses, through their own behavioural responses, communicate to their cubs the implications of environmental stimuli. Chimpanzees and gorillas in captivity have been taught the meaning of symbols invented and communicated by human beings, and in the case of sign language have shown some capacity to use these symbols themselves to express certain feelings and concepts. To my knowledge, however, no species other than human beings have shown a capacity to *invent* new symbols and to communicate their meanings to others.
40. These comments reflect the fact that the concept of *culture* as generally used has two distinct facets. One of these relates to the ability to learn and imitate new *techniques*. Monkeys have this ability, and we can be sure that it was very well developed in our early hominid ancestors. The other facet is the ability to invent, learn, and use symbols for purposes of communication (and later in cultural evolution for storing information).
41. The distinction made here between cultural processes and natural processes is sometimes misunderstood, and people react to it by pointing out that there is nothing 'unnatural' about culture. This, of course, is true. As I have indicated

the capacity for culture, and hence culture itself, is indeed the product of the processes of nature; but just as we have terms to differentiate between biotic processes and inorganic processes, so do we need terms to differentiate the processes of culture from the underlying inorganic and biotic processes. We could, indeed, avoid using the expression 'natural processes' by referring instead to 'inorganic and biotic processes', but this would be unnecessarily clumsy. We could also refer collectively to the latter processes as 'non-cultural processes', but this term also has its disadvantages.

42. As discussed on p. 6, a number of characteristics of primates 'pre-adapted' our prehuman ancestors for the capacity for culture. In light of this fact, and of the obvious Darwinian advantage of culture, the question arises why only a single species exists with this capacity. In view of the clear adaptive advantage value of culture (at least up until the present time), and from our knowledge of the principle of evolutionary opportunism (Simpson 1950), we might expect that an almost explosive adaptive radiation occurred of different culture-learning species emerging in the Pliocene and Pleistocene. The fact that this did not happen might be due to chance. On the other hand, it may be that the capacity for culture puts its bearers into an ecological niche in which there is room only for a single species. One could argue that the palaentological evidence is consistent with this view, since it seems likely that there were at least two and perhaps three or more species of bipedal primates in Africa at the 'protoculture' stage of development around two million years ago. Tobias (1969), for example, interprets palaentological findings as indicating that *Australopithecus africanus* had a lifestyle not very different from that of *Homo habilis*. Another contender for a rudimentary capacity for culture might be *Australopithicus robustus*. By 500 000 years ago, however, and probably much earlier, only a single species of hominid was in existence.

43. Smith 1975, p. 256.
44. Chagnon and Irons 1979.
45. Jones 1975. See also note 23.
46. Martin 1967; Webster 1981.

2

ADAPTATION

We often hear it said that man is the most adaptable of species. This book does not deny this proposition, but it does suggest that human adaptability is a concept which needs to be examined very critically. The topic is an important one, for the fate of our species over the coming decades depends on the success or otherwise of human adaptive processes in dealing with challenges of a magnitude and of a kind not previously encountered by human populations. Improved understanding of the different processes of adaptation available to humankind may well be one of the prerequisites for a successful transition to an ecologically viable fifth ecological phase of human existence.

Before proceeding further, it is necessary to discuss the way in which the word *adaptation* is used in this book. One of the definitions of this word given in the *Shorter Oxford Dictionary* is *the process of modifying to suit new conditions* [1]; and it is in this sense that the word is used here. Nevertheless, we must appreciate that *new* conditions are not an essential requirement for adaptation in the broader meaning of the term. Thus, even in a relatively unchanging environment an animal population may, through natural selection, continue to undergo progressive adaptive changes which render its members more likely to survive and reproduce under the prevailing conditions [2]. However, we shall be mainly concerned in this volume with adaptation of a kind which is consistent with the above definition, since we shall be paying special attention to adaptive responses of human beings to the undesirable impacts on biological systems of culturally induced, and hence biologically 'new', conditions.

Expanding on the above definition, *adaptation* will be used in this book to mean: *a modification in an organism or a population which occurs as a consequence of the introduction of an environmental threat and which renders this organism or population better able to cope in the new conditions* [3].

Certainly, the human species has shown, overall, signs of being remarkably adaptive, at least if population size is taken as the criterion of successful adaptation. Much of this success has been due to the fact mentioned in chapter 1 that humankind has at its disposal, in addition to those forms of adaptation which it shares with other animal species, an extra set of adaptive capabilities—the processes of cultural adaptation [4]. It is necessary, however, to maintain a sense of perspective. First, it is important to appreciate the fundamental, if obvious, principle that all kinds of adaptation have their limits, including, as history shows us, the processes of cultural adaption. Second, many groups of people in the past have *not* adapted

17

successfully to environmental changes and have been almost, or completely, wiped out. We are descendants of members of groups, perhaps the minority, who chanced to be more successful. In this connection, it is significant that for the first time in human history the whole population can now be regarded ecologically as a single group, in that some of the threats to the well-being and survival of humankind are on a global rather than on a local scale. On the one hand, our knowledge and understanding of natural and societal processes today is greater than at any time in the past, better equipping us to develop successful cultural adaptive responses and to put them into effect. On the other hand, the culturally induced threats to the survival of humankind are also much greater than at any earlier time.

It is worth drawing attention in these introductory comments to the distinction which exists between adaptive responses which are effective in the *long term*, and those which have only a transient or *short-term* value. A related distinction is that between *dead-end* adaptation and *open-road* adaptation. Dead-end adaptive responses are those which help individuals or populations to cope with threats in the short run, but which actually render them less well able to cope with or adapt to subsequent environmental changes. *Open-road* adaptation occurs when adaptive responses which help an organism or population to cope in the short term offer no disadvantage, or perhaps even offer advantage, in a changing environment in the long term. Examples of each kind will be discussed below in relation to both biological and cultural adaptive processes.

This chapter outlines the fundamental kinds of adaptation which have played a part in the adjustment of human populations and of individuals to changing environmental conditions.

The different classes of adaptation will be considered under four headings: evolutionary adaptation; innate adaptation; adaptation through learning; cultural adaptation.

EVOLUTIONARY ADAPTATION

The term *evolutionary adaptation* is used to describe the set of processes by which different species of animals and plants and other organisms have come into existence and by which changes take place over the generations in the phylogenetic characteristics of populations, with the result that individual members of the populations are better suited to their environment.

Evolutionary adaptation, according to the Darwinian explanation, comes about through natural selection. At any given time there will be a degree of variability in genetic make-up among the members of a population of living organisms. This genetic variability is in part due to the fact that spontaneous changes or mutations occur from time to time in a random manner in the genetic material; that is, in the DNA of the genes. It is also partly the consequence of the fact that the process of inheritance may result in genes being transmitted to offspring in different combinations than those which exist in the chromosomes of the parents. Because the individuals in a population at a given time are not genetically identical, some are likely to be better suited in their genetically determined characteristics than others to the prevailing conditions, and these better suited individuals will tend to be more successful in surviving and reproducing. Consequently they are likely, through reproduction, to contribute a higher proportion of individuals to the next generation than will the less well suited members of the population; and the progeny of these better suited individuals will carry the genes or gene combinations which rendered their parents at an advantage in the prevailing conditions. Thus, through this process of genetic selection, a population may, generation by generation, become increasingly well suited to the environment in which it lives [5].

Similarly, when a change occurs in the environment and persists, some individuals are likely, because of the genetic variability, to be better suited in their genetic characteristics to the new conditions than are others—so that subsequent generations will contain a higher proportion of individuals which are better suited to these new conditions. Not all populations adapt successfully in this way to environmental change; in fact, the majority of species which have existed in evolutionary history have failed to adapt to new environmental conditions and have become extinct, without leaving descendants [6].

This fact draws attention to an interesting paradox: for most species evolutionary adaptation is important because it has given rise to them in the first place; but it will not be sufficiently effective to protect them in the long run from the environmental changes which lie ahead. Just as they are the product of natural selection, so will they eventually be eliminated by it.

Although the rate of evolutionary adaptation will vary somewhat according to the kind and degree of genetic change which would be required to render genotypes more resistant to the change, it is always a slow process and it is unlikely to be effective except in cases where the new environmental conditions deviate only slightly from those to which the population had already become adapted through natural selection. The mutation rate for individual genes is estimated to be around one mutation per 100 000 spermatozoa or ova, and most mutations are likely to be harmful rather than beneficial. The chances of a suitable or helpful mutation arising in an appropriate gene in a small population suddenly exposed to a new detrimental environmental condition are therefore negligible. In the long run, however, all evolution depends on the introduction of new genetic characteristics through random mutation.

It is necessary to emphasize two important ways in which evolutionary adaptation differs from the other classes of adaptive process that we shall consider. First, although it is the result of the differential reproduction of individuals, evolutionary adaptation is nevertheless a process that takes place in populations, not in individuals, and the adaptive changes in populations undergoing evolutionary adaptation are evident only in subsequent generations. Second, evolutionary adaptation is *not a response*, in that no new processes are brought into action as a result of the environmental threats. It is, in fact, a consequence of the differential *elimination* of less resistant genotypes (through differential mortality and differential reproduction). Thus, not a single individual in the population first exposed to a new detrimental influence derives the least benefit from the process of evolutionary adaptation.

One reason why most biological species have become extinct is 'because selection promotes what is immediately useful, even if the change may be fatal in the long run' [7]. Thus, while evolutionary adaptation may produce a population better able to cope with the existing environment, it may also result in a loss of adaptive potential should the conditions change. In the words of Gould:

Darwinian evolution decrees that no animal shall actively develop a harmful structure, but it offers no guarantee that useful structures will continue to be adaptive in changed circumstances [8].

The case of the sabre-tooth tiger is often mentioned as an example of such *dead-end adaptation*. Through natural selection, this animal developed extraordinarily large canine teeth. Clearly, these structures must have been of selective advantage during the evolutionary history of the species; but in the long run, when environmental conditions deviated from those which prevailed when they were evolving, they may well have been maladaptive, contributing to the extinction of this species. In fact, *overspecialization* in genetically determined characteristics appears to have been a major cause of extinction of animal species throughout the history of life on earth (9).

The converse situation, *open-road adaptation*, occurs when genetic changes resulting from selection pressures operating in one set of environmental conditions render organisms better able, at a later date, to cope in, or further adapt to, new conditions. For example, as discussed in chapter 1, many of the characterisitics which evolved in primates as a consequence of the selection pressures operating in the arboreal environment, such as the development of the hands as grasping organs, proved an advantage in the very different ecological niche of the early hominids.

As mentioned in the last chapter, it is a valid question whether the human capacity for culture is an example of dead-end or open-road adaptation. Certainly, as a consequence of cultural processes themselves, this phylogenetic capacity is now operating in an environment very different from that in which it evolved.

A phenomenon which has played an important role in evolution is *genetic drift*. Groups within populations sometimes become split geographically as a result, for example, of migration. Although the distribution of genotypes among the two such separating groups may be random, if the groups are small one group may, by chance, contain a higher proportion of particular genes than the other. Then if both populations increase in size and at the same rate, there will ultimately be two large populations, each with its own distinctive pattern of gene distribution. Another possibility is that, at the time of separation of the small groups, one group may contain a higher proportion of individuals carrying a gene or genes which favour the survival of the group as a whole. This group will thus tend to survive and perhaps to increase in numbers— whereas groups containing a lower proportion of such genes are less likely to do well and may decline in numbers or die out. Thus, although separated, and not in direct competition with each other, some groups may do much better, in the Darwinian sense, than others [10].

There is one very important, if rather self-evident, consequence of the evolutionary process that has special implications for human beings living under the conditions of civilization. As we have noted already, natural selection produces animal populations in which the majority of individuals are well suited in their genetic characteristics to the set of conditions prevailing in the ecological niche in which the evolutionary forces are acting. These forces are selecting for optimum performance in the given environment. It follows, therefore, that if the conditions of life suddenly deviate from those of the natural habitat (to which the species has become adapted through natural selection), then it is likely that the individual animals will be less well suited, in either physiological or behavioural characteristics, to the changed conditions. Consequently some signs of maladjustment, physiological or behavioural, may be evident. In this book this fundamental concept is referred to as the *principle of evodeviation*, and the adjective *evodeviant* is used to describe life conditions which are different from those which prevail in the natural habitat of a species [11]. Disturbances in physiology or behaviour which result from evodeviations are referred to as examples of *phylogenetic maladjustment*, because they are essentially responses typical of the species and are due to the fact that the phylogenetic characteristics of the species are not suited to the new conditions.

This principle does not infer that every conceivable evodeviation in the conditions of life of a species will *necessarily* give rise to maladjustment. It is suggested, however, that any definite deviation in conditions from those that prevail in the natural habitat of an animal should be regarded as a potential cause of maladjustment until proven otherwise.

INNATE ADAPTATION

The term *innate adaptation* is used here to include all forms of physiological adaptive responses which occur in individual animals, as well as all adaptive behavioural responses which do not depend on learning.

The innate adaptive responses are phylogenetic characteristics and are the outcome of evolutionary adaptation. From this we can deduce that the innate adaptive mechanisms of any species are the product of natural selection, and as such they are evolutionary consequences of threats experienced in the environment in which that species evolved. We cannot expect, therefore, that the innate adaptive responses of the human species will necessarily be effective in assisting individuals to cope with environmental hazards of an entirely new kind experienced in the habitats of civilization. Nor can we expect innate adaptive responses to deal adequately with threats of a kind which, although they may have existed in the evolutionary history of the species, occur in greater intensity under new conditions.

Innate adaptation includes a wide range of different physiological adaptive responses. Some of these come into play almost instantaneously when an acute threat develops or is perceived. An example is the increase in heartbeat and the outflow of hormones from the adrenal gland which occur within seconds of the sudden appearance of a threat and which favour vigorous muscular activity in flight or in defense. Others may take weeks to become fully effective, as in the case of the increase in the concentration of haemoglobin in the blood which occurs when people move from low to high altitudes and which increases the blood's oxygen-carrying capacity. The processes of wound-healing provide a readily observable example of innate adaptation, and they include both immediate responses, such as the clotting of blood, as well as the slower cellular reactions which result in the removal of dead tissue and the growth of new skin.

From among innate behavioural adaptive responses we may cite as an example the spontaneous 'crouching' reaction of young birds of various species when they observe a novel shape to which they are unaccustomed. The birds have to learn *not* to respond to objects which do not pose a threat; that is, they react to *any* shape until they learn otherwise [12]. Another example is the rapid response of a human being who inadvertently sits on an upturned pin [13].

An important kind of innate adaptation is seen in the decline, after repeated stimulus of a given innate adaptive response. The case of the young birds mentioned in the last paragraph is an example. This kind of adaptation is called *habituation*, which has been defined as the loss or waning of a response as a result of repeated stimulation [14]. Later we shall consider habituation as an important aspect of adaptation in human beings.

ADAPTATION THROUGH LEARNING

This third set of adaptive processes depends on memory, and examples of adaptation through learning are seen when individual animals in a given situation behave in a certain way as a consequence of having experienced similar situations in the past. We can recognize two forms of adaptation through learning. First, individual animals can learn, through experience, the significance to themselves of various aspects of the environment, and hence whether these are to be sought after, ignored, or avoided. Second, animals can learn, through trial and error or through mimicry, behavioural techniques which enable them to deal appropriately with components of their habitat and to cope with environmental threats.

Adaptation through learning, although seen in such relatively simple creatures as planaria and earthworms, becomes increasingly important in animals higher in the evolutionary scale. It is of supreme importance in *Homo sapiens*, in which species the phenomenon of culture adds an extra dimension to process.

CULTURAL ADAPTATION

As indicated in the last chapter, the emphasis in this book is on the contribution of cultural adaptive processes to the adaptation of individuals and of populations to undesirable effects of culturally induced changes in biological systems. This kind of cultural adaptation and its role in the history of humankind and in modern society surely provides one of the most significant and, in my view, interesting subjects for serious study. Yet, while volumes have been written on evolutionary adaptation and on innate adaptive mechanisms, cultural adaptation to undesirable impacts of cultural processes on biological systems has received, as a theme in its own right, only scant attention from human biologists and social scientists [15].

A simple example of cultural adaptation is provided by the sequence of events that followed the introduction of refined carbohydrates into the diets of human populations. This change was culturally induced, being the consequence of technological developments and economic conditions. It represents a significant deviation from the evolutionary conditions of life of the human species and it has been blamed by various authors for a whole series of different forms of phylogenetic maladjustment (see chapter 10). The formation of cavities in the teeth is perhaps the most widely accepted consequence of this dietary change, and society has responded to this particularly painful disorder by establishing a profession of men and women who are trained in the skill of filling these dental cavities. Other cultural adaptive responses to this form of maladjustment include the addition of sodium fluoride to drinking water, the daily cleaning of the teeth with special brushes and chemical cleansing agents, and, occasionally, reversion to a diet free of refined carbohydrate.

By definition, cultural adaptation involves culture and it is consequently dependent on, and always involves, societal processes. That is to say, the adaptive responses are based on information communicated between the members of society and sometimes on societal activities, such as in the manufacture of aspirins or antibiotics. In some cases all aspects of the cultural adaptive responses are implemented on the level of society as a whole, as in the case, for instance, of vaccination programmes [16]. Such responses are examples of *societal cultural adaptation*. In others, the final stage in the process may be on the level of the individual, as, for instance, when someone decides to take an aspirin; this is an example of *individual cultural adaptation*.

A few important broad distinctions among the various forms of cultural adaptation will be discussed in this chapter. Examples of these different forms will be encountered in subsequent chapters. First, there is the important distinction between *corrective* and *antidotal* cultural adaptation. Corrective adaptation occurs when the adaptive process involves correcting the underlying cause of maladjustment or disharmony. An example is provided by the restoration of vitamin C to the diet of a population suffering from scurvy [17]. In antidotal adaptation the unsatisfactory conditions which are the fundamental cause of disturbance are not modified, and the adaptive response is aimed rather at alleviating the symptoms or at an intermediate factor. Most, but not all, of the work of the medical profession in present day Western society is antidotal in nature, as is that of the police force.

Another important distinction is that between short-term and long-term cultural adaptation. As in the case of evolutionary adaptation, cultural adaptation tends to promote 'what is immediately useful'. However, this response may on occasion result in a society being less well able than it would otherwise have been to cope with a subsequent threat. Examples of such dead-end cultural adaptation occur especially when societies respond to environmental pressure by introducing new technologies on which they eventually become totally dependent, thus reducing their adaptive possibilities for any future threats. It also occurs when adaptive responses, to population pressure, for example, result in the use of new techniques of farming which, although initially increasing food production, in the end bring about degradation of the ecosystems so that bioproductivity falls permanently below even the original level.

Prerequisites for successful cultural adaptation

Successful cultural adaptation depends on the satisfaction of four sets of conditions. First, there must be *recognition* by the individual, or by society, *that an undesirable state exists*—that something is wrong. In some situations, such as when large numbers of people are dying in an outbreak of cholera, this prerequisite is readily satisfied. In other cases it requires some special knowledge or understanding. Second, some *knowledge* is necessary *of the cause or causes of the undesirable state or* failing that, at least some knowledge *of ways and means of overcoming it*. Third, the individual or society must possess the *means to deal with the undesirable state* in terms of the financial, technical, or human resources necessary to put a programme of adaptation into effect. Fourth, there must be *motivation to take appropriate action on the part of those in society who make the relevant decisions*. Some discussion on each of these four prerequisites will draw attention to some further important characteristics of cultural adaptation.

Recognition that an undesirable state exists

The fact that cultural adaptation requires that a state of maladjustment be recognized to exist may seem so obvious as to be hardly worth discussing. However, the full significance of this requirement is often not appreciated.

With respect to threats to human well-being, no problem arises in this regard in the case of such conditions as plague, appendicitis, severe malnutrition, or highly aberrant and dangerous behaviour. But if the onset of a relatively undramatic, chronic form of maladjustment is gradual during the course of an individual's lifetime, or if its increasing prevalence in a community occurs slowly over several generations, the possibility exists that it might pass unnoticed, and that its manifestations might come to be regarded by the individual, or by society, as 'normal' (which, of course, it would be, under the new conditions), and that it might not be recognized as a departure from the optimal state. This possibility reminds us of the case of the boiling frog. If a frog is placed in hot water it will make frantic efforts to escape; it is said, however, that if the animal is put into cold water, which is then slowly heated, it may, after passing through various stages of phylogenetic maladjustment, be boiled to death without so much as a struggle.

Does this *boiling frog principle* apply to the human species in civilization? The evidence shows that it does, on the level both of the individual and of society, and a few examples on the societal level serve to illustrate the significance of this first prerequisite for successful cultural adaptation. For instance, for many years textbooks of medicine described progressively increasing blood pressure after the age of twenty years to be a normal characteristic of human beings. More recently, it has been shown that people living in certain relatively simple 'non-westernized' societies show no such increase in blood pressure with advancing age. This finding strongly suggests that the almost universal increase in blood pressure which accompanies increasing age in the high-energy societies is, in fact, a pathological response of the human organism to adverse environmental conditions.

The boiling frog principle may be especially important in the case of behavioural and mental disorders. Suppose, for example, that a slowly introduced environmental change were to produce in most people an increase in irritability, fatigue, aggressiveness, a general deterioration in the quality of personal relationships, and perhaps some interference with the capacity to make rapid and wise decisions. The strong possibility exists that in such a situation these signs of maladjustment would permeate society unrecognized as deviations from the normal or healthy state. In terms of the well-being of people, however, a change of this kind, affecting the mood of the whole population, would be a serious matter. If the effect of thalidomide on the foetus had been to interfere with development in the central nervous system, so that when the victims grew up they had significantly lowered IQs, or were unusually aggressive, the product would probably still be in use today.

Even severe forms of maladjustment can sometimes be subject to the boiling frog principle. In 19th century Europe, tuberculosis dwarfed the other contagious diseases, such as typhus, typhoid, and cholera, as a killer of humankind, especially young men and women. But the disease caused little alarm; it was mutely accepted as a fact of life. It had a reputation for running in families, and when it occurred it seems to have been regarded almost as an inevitable hereditary misfortune. When reformers called for improved living conditions for the inhabitants of the industrial cities—a call which eventually led to the Public Health Movement and greater government responsibility for community health—they were motivated by concern about typhus and cholera, and not about tuberculosis. It has even been argued that the ubiquity and inexorability of tuberculosis affected aesthetic standards during the 19th century. Dubos and Dubos have written:

In France, the boisterous female type of the Revolutionary era had been displaced in popularity shortly after 1800 by the languishing beauty dressed in vaporous muslins... . When languid pallor became a fashionable attribute of women, the use of rouge was abandoned in favor of whitening powders... . The impact of consumption on nineteenth-century aesthetics appears in the elongated 'women with cadaverous bodies and sensual mouths' of the Pre-Raphaelite paintings... . The fragile silhouette, with long limbs, long fingers, long throat, the tired head leaning on a pillow, with prominent eyes and twisted sensual mouth, became the unhealthy, perverted symbol of Romanticism [18].

So long as signs of maladjustment are accepted as normal, or at least as inevitable, it is unlikely that effective processes of cultural adaptation will get underway. In the case of insidious and chronic interference with human health or, for example, with the bioproductivity of ecosystems, recognition on the societal level that an undesirable state of maladjustment exists often awaits the vigorous activities of a relatively small number of individuals who are alert to the nature of the situation and to the need for action. I refer to them as *first-order reformers*. It is predictable that whenever such individuals raise their voices there will emerge others, the *anti-reformers*, usually representing vested interests, to counter their arguments and to describe them as alarmists, fanatics, and extremists. The cause of the first-order reformers is later taken up by others—the *second-order reformers*—and the societal adaptive process can then be said to be properly underway. Examples of this general pattern will be given in later chapters. Clearly, effective cultural adaptation on the societal level is unlikely to be achieved before the recognition of the undesirable state reaches those who hold power in society, and even then, as we shall see, it may not come about due to lack of satisfaction of one or more of the other three prerequisites.

Knowlege of the cause of the threat or of ways and means of overcoming it

A human population is in a much better position to respond efffectively to an environmentally or culturally induced disturbance if it possesses knowledge of the true nature of the disorder in question and of all the contributing factors. However, lack of such information does not entirely preclude the possibility of successful adaptation. In the complete absence of any knowledge of the origin of a sign of maladjustment, a period of fumbling counter-activity usually ensues, based partly on incorrect interpretations of the situation and partly on trial-and-error tactics. In the case of ill health, people are generally convinced that there does exist a 'cause' of their unwanted afflictions, and when their knowledge of the material world fails to provide them with a ready explanation they may seek one in terms of spiritual intervention. The concept of disease as a divine sanction reflecting the displeasure of a deity has been almost universal in the past. Prayers and rituals of appeasement addressed to a god or gods have been commonplace—often accompanied by magic rites performed on the sick persons aimed at countering an evil influence or driving it out of their bodies. This form of cultural adaptive response is mainly *post hoc* and hence therapeutic in intention; but this is not always the case—many people in simple societies, and some in the complex ones, wear magic objects, amulets, or talismans to protect themselves against disease.

The trial-and-error approach has often been successful in providing effective procedures for dealing with forms of maladjustment in the absence of knowledge concerning their true nature. There are many stories of the efficacy of various concoctions prepared from herbs by simple societies for the treatment of diverse ailments. One of the most interesting of these is cinchona bark, which was discovered by the indigenous inhabitants of Peru to be effective in the treatment of malaria. It was brought to Europe by the Jesuits in 1632, and the revolution it brought about in the history of European medicine has been described as comparable only to the introduction of gunpowder in the art of war [19]. The active principle of cinchona bark is quinine, which persisted as the specific treatment for malaria until quite recently.

It was partly the effectiveness of cinchona in the treatment of malaria that led Edward Stone, who lived in East Anglia in the mid-18th century, to investigate the medicinal properties of extract of willow (*Salix*) bark, which has a bitter taste similar to that of cinchona bark. He was also motivated by the ancient doctrine that the cures of maladies often occurred naturally in the same regions as their causes. In 1763, Stone presented his paper entitled 'An account of the success of the bark of the willow in the cure of agues' to the Royal Society in London [20]. One of the constituents of extract of willow bark is salicin, which was later found to be very effective in pure form in the treatment of febrile and rheumatic conditions and also as a mild analgesic. Today the famous derivative of salicin, acetylsalicylic acid or aspirin, is probably the most widely used chemical compound in the efforts of human beings to overcome biological maladjustment. In the USA alone some 6000-7000 tonnes of this substance are consumed annually [21].

Successful adaptive procedures arising from a purely empirical approach are usually, although not always, of the antidotal variety. An example of the trial-and-error method leading to successful corrective adaptation is seen in the discovery in the 17th and 18th centuries that scurvy could be successfully treated, and also prevented, by the consumption of fresh fruit and vegetables.

In general, however, attempts at cultural adaptation are not likely to be successful when society remains in total ignorance of the nature of the disorder that it is trying to overcome. Nevertheless, as time goes on and new facts come to light, a stage of partial understanding may be achieved; and some notable cases of successful cultural adaptation have been based on such partial understanding. One of the best examples is the public health movement of the last century in Britain, directed mainly against such endemic and epidemic diseases as typhus, dysentry, and typhoid which were rife in working-class populations in British towns (see chapter 7). The prevailing view was that these diseases were due to the 'foul odours' which were characteristic of the urban environment at the time and which were given off by decomposing organic matter.

The dramatic discoveries of Louis Pasteur and other bacteriologists late in the 19th century provided a better understanding of the causes of certain diseases: cholera was due to the cholera vibrio, consumption was due to the tubercle bacillus, and anthrax was due to the anthrax bacillus. In fact, science seemed to be demonstrating that each disease was caused by a specific pathogenic agent—an assumption which had a major impact on the approach of Western medicine. The main task of medical science was seen to be to discover compounds which would specifically destroy the pathogenic organisms or else to prepare vaccines which would provide complete protection against them. More recently it has been argued that this microbial theory of disease has served, in fact, to obscure the full picture. Certainly, in the last decade or so there has been a tendency in some sections of medical science to take a more comprehensive view of patterns of health and disease. We read of the 'multifactorial causation' of diseases. In the case, for example, of cardiovascular disease, statistical studies show that correlations exist between the incidence of this form of ill health and a whole range of different biosocial variables, including calorie intake, consumption of saturated fats, consumption of refined carbohydrates, levels or kind of physical exercise, the drinking of coffee, the quality of drinking water, cigarette smoking, and personality type. In such cases, it may be that the multifactorial approach is also tending to obscure the essential relationships (see chapter 10).

In summary, knowledge about the various causes of, or at least remedies for, undesirable changes in natural systems, human or otherwise, clearly facilitates successful cultural adaptation. It was because of the lack of such knowledge that infectious disease remained a major cause of ill health and death throughout the early urban phase of human history and that the ecosystems of Sumer and Akkad and of many other regions of the world eventually became incapable of producing sufficient food for the local populations (see chapters 5 and 6). It is not the reason, however, why today there continues to be a high incidence of infectious disease in some developing regions of the world, or that progressive degradation of agricultural land proceeds apace in, for example, large areas of Australia. For explanations of these facts we must look to the third and fourth prerequisites for successful cultural adaptation.

The means to deal with the undesirable state

In many developing countries of the world today the main impediment to successful cultural adaptation against undesirable states such as infectious disease or malnutrition is not lack of knowledge, but rather lack of the economic, technological, or human resources necessary to overcome the problem. In some situations such resources have been provided, to a greater or lesser extent, by rich high-energy societies in various international aid programmes. However, the financial, technological, and human commitment of the more affluent nations to such assistance is equivalent to only a minute fraction of their commitment to preparations for military conflict with each other.

Motivation to take appropriate action

The degree of motivation to take adaptive measures is a most important variable affecting whether or not cultural adaptive processes are brought effectively into play, both on the level of the individual and on the level of society. With respect to individuals, the motivation to take steps to overcome threats to health and well-being may be influenced considerably by the value system of the subculture to which they belong. In certain socio-economic groups in high-energy societies, success of bread-winners at work may be considered, by the bread-winners themselves as well as by their families, to be more important than their physical health.

On the societal level, an important variable in the modern world relevant to cultural adaptation is the degree of motivation of the large corporate organizations in society—or, in the case of totalitarian regimes, of individual dictators—or the small bands of individuals who wield political and economic power. Lack of motivation, even active resistance, on the part of the corporate organizations which hold power in society can effectively block useful cultural adaptive responses.

A typical chain of events in a cultural adaption response to an undesirable impact of a societal activity on a biological system is summarized in Fig. 2.1.

It is worth stressing again the key importance in cultural adaptation of knowledge and understanding in the community at large and of the

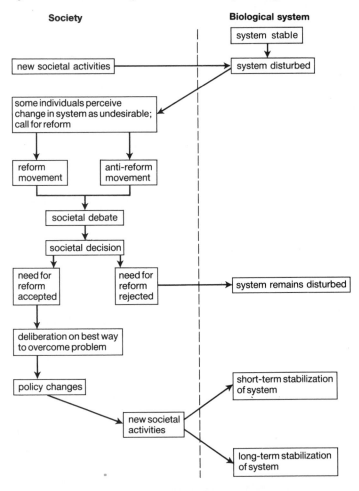

Fig. 2.1. Typical patterns of cultural adaptive response to undesirable impacts of societal activities on biological systems.

widespread dissemination of relevant information. Ignorance is so often the major factor blocking desirable and effective cultural adaptive responses at the societal level. Some other impediments to successful cultural adaptation will be discussed in later chapters.

NOTES

1. Some authors make the point that 'adaptation' should be distinguished from 'adjustment', which is seen as 'merely the removal of tension and stimuli without effective solution to the problems and true satisfaction of basic needs' (Rotondo 1965, p.38).
2. In the main, however, evolutionary forces are conservative and work to maintain the existing phylogenetic characteristics of species rather than to favour further genetic change.
3. There are many other definitions of the word adaptation as used by anthropologists (see, for example, Bennett 1976, pp. 245–52) and biologists (see, for example, Prosser 1958, pp. 167–80). There has been a tendency, especially in medical science, to apply the term *adaptation* to *any* response of an individual to new environmental influences, regardless of whether the response renders the individual better able to cope with these influences. For example, it has been used to describe the pathological changes which occur in response to exposure to ionizing radiation and for the chronic bronchitis which occurs in a polluted atmosphere.
4. Cultural adaptation is here defined as adaptation which is the result of, and which depends on, cultural processes.
5. Some authors prefer to see biological evolution in terms of the survival of genes rather than of whole organisms—the 'selfish' gene' interpretation (see Dawkins 1976). For a critical discussion of this theme, see Midgley 1978.
6. It has been estimated by Romer that perhaps not more than one per cent of the tetrapods of the Middle Mesozoic have left living descendants (see Simpson 1950, p. 80).
7. Dobzhansky 1958, p. 1091.
8. Gould 1977, p. 90.
9. Simpson 1950, p. 205.
10. The question of 'group selection' is one which has given rise to much debate (see Borgia 1980, p. 166). According to Irons (1979, p. 12): '... very few theoretical biologists presently support the proposition that group selection can be expected, under any but very rare circumstances, to overcome the effects of individual selection and produce group level adaptations'. However, the matter is far from resolved. In my view, evolutionary theorists pay too little attention to the implications of the constant splitting up of human populations into small relatively independent groups—groups which, because of random gene distribution, differ from each other with regard to their respective content of genes favouring group survival and growth.
11. Boyden 1973; Boyden, Millar, Newcombe, and O'Neill 1981, p. 98.
12. Tinbergen 1966, p. 130.

13. This reaction is a mixture of innate and learned responses. Humans have to learn how to stand up.
14. A narrower definition is as follows: a stimulus-specific decline in a response resulting from the effect on the central nervous system of repeated stimulation; see Barnett 1981, pp.129-30.
15. For two exceptions to this statement, see Alland 1970 and Bennett 1976.
16. The examples of cultural adaptation given in this chapter mainly relate to responses aimed at overcoming environmentally induced disturbances in physical health, because these illustrate most clearly the pertinent principles. However, the generalizations which are made apply equally to cultural adaptive responses aimed at countering mental ill health, societal disharmonies, as well as undesirable disturbances in natural or agricultural ecosystems.
17. Corrective adaptation can be seen as a form of negative feed back—a return to a former state. It has been pointed out that 'negative feedback is rare in human systems, or at least is not the usual or most important way humans cope' (Bennett 1976, p. 57). Evolutionary adaptation, when successful, is mainly a form of positive feedback. Adaptation by innovation, that is, positive feedback, is more common.
18. Dubos and Dubos 1953, pp. 54-7.
19. Garrison 1929, p. 290.
20. Stone 1763.
21. Hughes and Brewin 1979.

3

HUMAN NEEDS AND HUMAN BEHAVIOUR

CONCEPTUAL SCHEMES

A feature of the conceptual framework on which the organization in this book is based is the fact that, while it ensures that due consideration is given to dynamic interrelationships between human society and the biosphere, it also recognizes that there exists, within this *total environment*, a large number of particularly interesting subsystems in the form of individual human beings. Each of these individuals lives in his or her own personal or *immediate environment*, which is only a small part of the total environment; and within that immediate environment each of them behaves in a certain way and exists in a certain state of body and mind.

The framework thus recognizes three sets of variables which relate intimately and directly to the life experience of individual human beings (see Fig. 3.1). The term *biopsychic state* is used for the set of variables relating to the individual's state of body and mind [1]. This term, which is short for biological and psychological state, incorporates everything that a person is at any given time. It thus includes such measurable aspects of the individual as stature, weight, hair colour, and blood pressure, as well as relatively intangible factors, such as feelings, knowledge, and values. The concept of the biopsychic state thus incorporates not only those aspects of individuals which would ordinarily be regarded as related to health and disease, but also others, like skin colour or hairiness, which may have no special significance for health.

The second set of variables describes the individual's *behaviour pattern*, a term which applies to everything that an individual actually *does*. It is reserved for physical behaviour (e.g. running, dancing, making things, eating) and does not include mental behaviour (e.g. developing attitudes,

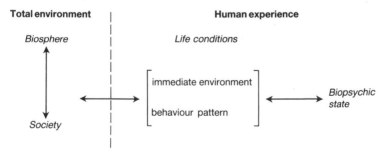

Fig. 3.1. The relationship between the three aspects of human experience and the total environment.

33

changing an opinion, doing mathematical calculations), which is a dynamic aspect of the biopsychic state.

The third set of variables is referred to as the *immediate environment*. This is that part of the total environment which impinges directly on the individual. It includes variables ranging from material or physical components, such as the quality of air and levels of environmental noise, to such intangible factors as the size and nature of social support networks and the incentives and opportunities offered by the social environment for co-operative small-group interaction or creative behaviour.

For some purposes it is convenient to refer to the individual's immediate environment and behaviour pattern together as his or her *life conditions*.

The relationship between the immediate environment and the total environment is not a simple one, and different individuals sharing the same total environment (e.g. a city) may have very different immediate environments. Each individual can be said to be separated from the total environment by a series of *filters* which determine what aspects of the total environment become a part of the individual's immediate environment. These filters include economic, cultural, and geographical factors.

About 2400 years ago Hippocrates and his colleagues recognized that the state of health of individuals is greatly affected by the properties of the environment in which they live, as well as by their pattern of behaviour in that environment. This simple but fundamental principle will be referred to as the *Hippocratic postulate*. In terms of the conceptual scheme discussed above, the Hippocratic postulate may be restated as follows: the biopsychic states of human beings are much affected by their life conditions—that is, by their immediate environments and behaviour patterns.

The biopsychic state of individuals, as well as the biopsychic responses to life conditions at any given time, are influenced not only by current life conditions but also by *previous life conditions*—such as diet in infancy and socialization experience. This refinement of the Hippocratic postulate, as well as the essential contribution of the genotype to the biopsychic state, are depicted in Fig. 3.2. This figure also draws attention to another important feature of the conceptual scheme—the fact that the biopsychic reaction to any given life conditions often depends on the way individuals *perceive* aspects of their life conditions.

HUMAN HEALTH NEEDS

Turning to the subject of human needs, there is an extensive literature on this topic, but it is complicated and confusing [2]. This is unfortunate, since

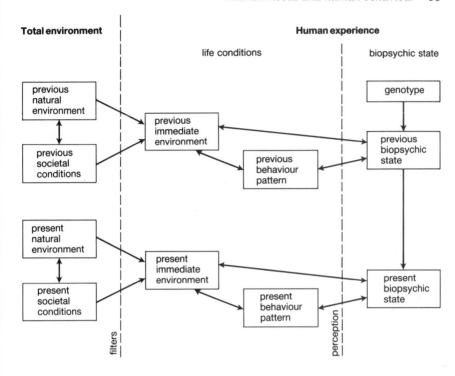

Fig. 3.2. Interrelationships between genotype, present and past experience, and biopsychic state.

there can surely be no subject of greater importance to humankind or of greater relevance to our efforts to work towards an ecologically sustainable and humanly acceptable society for the future. The confusion is partly due to difficulties in defining some of the key terms used, and partly to the intangible nature of some of the relevant concepts and variables. Words like 'needs', 'wants', 'drives', 'motivation', 'preference', 'choices', 'instincts', and 'desires', abound in the literature and are used in different ways by different authors, and often synonymously with each other.

The approach taken in this volume is based on appreciation of the importance of the difference between three connected but nevertheless quite distinct aspects of the problem, for which the words *needs*, *wants*, and *behaviour* will be used respectively. The word *need* will be used strictly in the sense of *life conditions which are necessary for the attainment of a given state*. The word *want* is used exclusively for a *state of mind* and it is synonymous with *desire*. The word *behaviour* is the word used for actions that people perform, sometimes in response to a want and sometimes with the effect of satisfying a need.

Thus, in our terminology, a *human need* is a condition which is necessary in order that a human being attains a certain state [3]. However, the expression still has little meaning unless we make it clear what particular state we have in mind. It is necessary to clarify, for example, whether we are speaking of conditions which are necessary for survival, for longevity, for health, for contentment or happiness, for spiritual salvation, or for some other state perceived to be desirable.

In this book the emphasis will be on *human health needs*—that is, on those conditions which are necessary for human beings to live in a state of health and well-being. The problem of definition of health and well-being will be discussed later; at this stage let us simply note that health and well-being is an aspect of the biopsychic state. Conditions which are necessary for human survival are referred to as *human survival needs*.

The terms 'health needs' and 'survival needs' as used here will apply only to variables on the level of life conditions (see Fig. 3.1): they are those aspects of life conditions which are necessary for the attainment of the desired biopsychic state. They will not be used for aspects of the biopsychic state itself. Consequently, although it is possible to argue that blood pressure between certain levels is necessary for health and well-being and is therefore a human health need, in the present text satisfactory blood pressure will be described as an aspect or component of health and well-being rather than as a health need. Similarly, when later in the book we discuss the importance of a sense of personal involvement, it will be described as an aspect of health and well-being; the corresponding health need would be life conditions which are conducive to a sense of personal involvement.

One of the most important distinctions, conceptually and practically, among different kinds of human health and survival needs, is that which exists between those species characteristics which apply to all human beings in all societies, and those which are determined by particular life conditions and which must be met under these conditions (but not necessarily under other conditions) in order that the more basic needs are satisfied. The former are here referred to as *universal needs* and the latter as *contingent needs* [4]. Universal health needs include, for example, water for drinking, oxygen for breathing, and certain chemical compounds in the diet, such as vitamins A and C; and they embrace, of course, all those factors necessary for survival—although the amount necessary for health and well-being may be different from that necessary for survival.

The importance of universal health and survival needs lies in the fact that they are, as phylogenetic characteristics, fixed. No amount of technological advance can modify, substitute, or eliminate them. This is in contrast to the contingent needs, which are not inevitable and which are nowadays to a large extent culturally determined. This applies especially to some of the less tangible health needs. For example, the specific way in which an effective emotional support network is provided might well vary from one cultural setting to another.

In the literature the word *need* is sometimes used, synonymously with 'drive' or 'instinct', for an innate urge which leads to behaviour of a kind which is necessary for the attainment of a state of well-being. Thus the urge to eat when hungry or to engage in sexual intercourse would be regarded, in this sense, as a need. Although the word is not used in this way here, we should note that a particular form of behaviour (as distinct from an urge which is an aspect of the biopsychic state) may well be a health need. For example, there is a strong body of biomedical opinion today which holds that a certain level of physical exercise is necessary for the attainment or maintenance of health and well-being. Another link with behaviour is apparent when a particular action is directed at, or results in, the satisfaction of a health need. Many innate drives in animals have this function (while others may have the function, for example, of promoting reproduction).

Attention must be drawn to the important, if self-evident, *principle of optimum range*, which states that for many human health needs there is an optimum range of concentration, intensity, or quantity. For example, a certain amount of vitamin A in the diet is essential for health, but too much can cause sickness and even death. The principle of optimum range applies also to various intangible aspects of life conditions, such as variety in daily experience, too little or too much of which may be detrimental to health and well-being.

One further concept relevant to the relationship between life conditions and health and well-being deserves mention. There is an important class of experiences of a psychosocial and intangible nature which, although they may not be absolutely necessary for health and well-being, certainly tend to promote this state if they are present and to counter the potential ill-effects of various environmental or behavioural stressors. The word *melior* will be applied in this text to experiences of this kind. For instance, co-operative small-group interaction, creative behaviour, or life conditions which promote a sense of belonging may be postulated to act as meliors. Although specific meliors such as these may not each be a health need, in that a state of health and well-being may be possible without them, it is reasonable to assume that a certain overall level of melioric experience is necessary for optimum health. The level of meliors required for health in a given situation will depend on, among other things, the level or intensity of stressors in the individual's life experience [5]. Indeed the balance between meliors on the one hand and experiences which induce a state of distress on the other is surely one of the most important influences on human health and well-being.

As indicated above, the word *want* as used in this book refers strictly to a state of mind. It is therefore an aspect of the biopsychic state, unlike health or survival needs which, according to the definition adopted here, are on the level of the life conditions [6]. The concept of wants, as distinct from needs, emphasizes the importance of *perception* as an influence on the biopsychic response to given life conditions. In many cases the relationship between wants and needs is direct and obvious. A woman, when thirsty, may want water to drink, and water is also a health and survival need; on the other hand, she may want vitamin tablets because, as a consequence of cultural influences, she perceives additional vitamins in her diet to be necessary for her health. The perceived need may not, however, be a real health need, since she may be consuming plenty of vitamins in her normal diet.

The subject is complicated by the fact that there is much to be said for the view that the satisfaction of all wants is not necessarily a desirable state of affairs, and that optimum health and well-being is associated with a balance between satisfaction and non-satisfaction of wants. Life may be more enjoyable and richer if there is something to look forward to and something to strive for: the principle of optimum range thus comes into play. Nevertheless, a chronic feeling that important wants are unsatisfied, whether they be directly related to universal health needs or purely culturally determined, is likely to lead to a sense of deprivation and consequently to interfere with health and well-being.

Identifying human health needs

Something must now be said about the expression 'health and well-being'. The word 'health' itself has proven to be notoriously difficult to define, and countless attempts at definitions have been put forward in the literature [7]. It would not be useful to try to discuss here the pros and cons of the various shades of meaning implied in the different definitions. In fact, it is hardly necessary to define the term as it is used in this book, since most readers have their own mental picture of what 'health' means to them, and I suspect that in most cases this meaning would be entirely consistent with my own use of the word. The expression 'health and well-being' is used in this text synonymously with 'health'; 'well-being' is added as a reminder that we are thinking of something broader than mere physiological health.

For the sake of completeness, however, I will provide two definitions of the word 'health'; the first a biological one, and the second more subjective. According to the biological definition, health is a state of body and mind similar to that which would be most likely to result in survival and successful reproduction in the evolutionary environment. According to the second definition, health is a state of body and mind conducive to and associated with full enjoyment of life. I strongly suspect that these two states are one and the same thing. Ill health is the converse of health, and could be defined appropriately in terms of the two definitions given above.

According to the Hippocratic postulate, the state of health and well-being of an individual is largely a function of his or her lifestyle. This postulate does not tell us, however, what sort of life conditions are likely to be conducive to, or detrimental for, health. For this kind of information we have to look to the results of biomedical research, to personal experience, or to evolutionary concepts. Let us first consider the contribution of this last factor to the understanding of human health needs.

Implications of the evolutionary background of the human species

All the innate characteristics of an animal species, including its phylogenetic health needs, are the product of selection pressures operating in its evolutionary history. The effect of the processes of evolution through natural selection is to produce populations with genetic characteristics such that the individuals perform optimally, in terms of survival and reproduction, in the ecological niche in which they are evolving. As pointed out in the last chapter, one of the consequences of such evolutionary adaptation is that if an animal is exposed to new conditions which are significantly different from those to which the species had become adapted through natural selection, the likelihood is that it will be less suited to the new conditions and, as a result, it is likely to exhibit signs of phylogenetic maladjustment. People tend to take this fundamental biological principle for granted in the case of animal species other than *Homo sapiens*. If a wild creature is captured and placed in a zoo and then begins to show some signs of maladjustment, the first question that the zoo authorities ask is 'what are we doing wrong?'. They look to the life conditions of the animal in its natural environment to check, for example, whether they may be feeding it the wrong food, or whether perhaps it needs water to wallow in, or branches to climb.

This *principle of evodeviation* applies as much to humans as it does to any other species. Indeed, in my view, it represents one of the most important of the implications of Darwinian theory for human society. The phylogenetic characteristics of our species, including its universal health and survival needs, were determined by the selection pressures operating before and during the long hunter-gatherer phase of human existence. Cultural developments over the past ten thousand years have brought about many changes in the conditions of life of most sections of the human population. As a consequence of these culturally induced changes in human life conditions, there have occurred countless examples of phylogenetic maladjustment—including numerous forms of nutritional, infectious, and organic diseases.

With respect in particular to the conditions of life characteristic of modern urban society, there are several reasons why we can rule out the possibility of any major evolutionary adaptation of human populations to the new situation. In the first place, the earliest cities came into existence only about two hundred generations ago, and since that time only a small minority of the human population has actually lived in an urban environment. Moreover, many of the significant evodeviations associated with modern urban society today have been introduced within the last one or two generations. Even over the longer period, significant change in the genetic constitution of urban populations in response to new environmental conditions would be anticipated only in cases in which the selective advantage of a given genetic variant was especially strong. It has been suggested that certain infectious diseases associated with crowding in cities, such as tuberculosis, have exerted selection pressures sufficiently powerful to render later populations more resistant to them [8].

Nevertheless, the number of human beings in the world has not declined since the time of the domestic transition; in fact it has increased about a thousandfold. There are two reasons for this. First, the people in villages and cities who experience ill health are at less of a survival disadvantage than they would be in the more hazardous primeval situation. They are not required to be nomadic, they have their food and water brought to them, they are protected from unfavourable weather conditions by permanent shelter, and they are unlikely to be attacked by predators. They are therefore more likely to be kept alive long enough to recover; and even if they do not recover, they may continue to live and may still be able to reproduce. This effect is referred to here as the *molly-coddling effect* of civilization. The second reason, which applies especially to the last half century, lies in the processes of cultural adaptation, which take the form of various specific measures, such as vaccination and the use of pharmaceutical drugs.

We can safely conclude, then, that modern cities are not inhabited by a new breed of the human species whose members are better adapted in their genetic characteristics than their palaeolithic ancestors to the prevailing urban conditions.

The principle of evodeviation applies, without doubt, to a wide range of material aspects of life conditions. It is also clear that it applies to certain aspects of behaviour. Marked deviations from the natural sleeping patterns are well known to be a source of maladjustment, and evidence is accumulating that health is likely to be impaired if levels of physical exercise deviate to a marked degree from the natural pattern. It is a reasonable working hypothesis that the principle also applies to many less obvious behavioural and psycho-social aspects of human experience. As will be discussed in chapter 4, the life conditions of hunter-gatherers are usually associated, for example, with the existence of an effective psychological support network, a certain amount of creative behaviour, and situations which promote a sense of personal involvement. It is suggested that factors such as these—aspects of the life conditions of members of the species in the natural habitat—are likely to be conducive to human health and well-being.

In the present context, the relevance of the principle of evodeviation is that it provides a theoretical approach to the identification of the health needs of human beings, suggesting that insights into these needs can be gained from studying the life conditions of recent hunter-gatherers. Although this notion is common sense from the biological point of view, surprisingly little attention has been paid to it by those who are professionally concerned with human health and disease. In fact, the idea has sometimes aroused considerable antagonism, especially among some students of the social sciences who are liable, with cries of 'noble savage', to dismiss it as biological determinism. Let me stress, however, that those who accept the biological point of view do not necessarily advocate a return to the hunter-gatherer lifestyle, nor do they necessarily deny the advantages of civilization. They may well argue, however, that some of the *disadvantages* of civilization could be avoided if more thought were given to the evolutionary background and phylogenetic characteristics of the species.

Some conclusions arising from this approach will be discussed at the end of the next chapter, after a description of the life conditions of hunter-gatherers.

Other approaches

There is little point in dwelling in this chapter on the role of research in medical and biological science in our attempts to learn more about the reactions of human organisms to different kinds of environment and lifestyle, and about the factors which are likely either to promote or interfere with health and well-being. In fact, literature arising from this approach will be one of the main sources of information used throughout this book in our discussions on the interrelationships between life conditions and biopsychic state. It is worth noting here, however, that the results from biomedical investigations are completely consistent with the inferences drawn from the evolutionary concepts discussed above (see, for example, pp. 243 and 268).

The scientific approach has been very useful in improving understanding of the links between health and various material (physical, chemical, biological) aspects of life conditions, but it has not been so successful with respect to less tangible aspects. It is true that epidemiological studies in recent years have provided some evidence of the detrimental effects on health of frequent changes in the circumstances of life and of such stressful experiences as bereavement and social isolation (see chapter 10). In general, however, for insights into the more intangible health needs of people, apart from those derived from evolutionary considerations, we must turn to our own personal experience and to the humanistic literature. Numerous authors have expressed the view that such factors as variety in daily experience, opportunity and incentives for creative behaviour, and an environment which is conducive to a sense of personal involvement, to a sense of belonging, and to a sense of responsibility are desirable for human health and well-being. As in the case of the scientific method, conclusions drawn from this humanistic approach are also consistent with those derived from evolutionary concepts.

Some further general principles relating to human health

It is important to appreciate that there is considerable variability among human beings in both sensitivity and response to potentially adverse environmental or behavioural influences. Part of this variability is due to previous experience and part to genetic factors. It is well accepted, for instance, that nutritional background has a marked influence on the ability of individuals to resist bacterial infection. A good example of the contribution of genetic factors to this variability is provided by the adult form of diabetes mellitus, which, although it is an inheritable condition, is nevertheless also a response to elements of the modern Western diet and lifestyle. This is a clear case of *genetic vulnerability* to life conditions which deviate from those of the evolutionary environment.

Thus, in the case of a potentially detrimental influence there is likely to be wide variation among individuals in susceptibility to the influence, in intensity of reaction to it, and, in some cases, in the kind of reaction. Regrettably, individuals who happen to react to very small doses of a toxic or otherwise harmful agent are too often regarded by the medical profession and by society as mere freaks, to be comforted and treated antidotally for their abnormality. It would be wiser to regard them as particularly sensitive and useful indicators of a potentially harmful environmental change, which may well be producing insidious, subthreshold, and less dramatic effects in many other members of the population. My colleagues and I have previously referred to this notion as the *canary principle*—canaries having been used in coal mines for the detection of noxious gases to which they are more susceptible than human beings. Just as some individuals in human populations play the role of canaries, by being especially sensitive to harmful influences, there are others at the opposite end of the spectrum— the *cockroaches*—who appear remarkably resistant to environmental and behavioural insults. In old age they are able to boast about their indulgences and to pour ridicule on the warnings of the proponents of moderation.

We have already noted that the way individuals *perceive* the environment is in some circumstances an important influence on their biopsychic response. Certain kinds of sound, for example, may be perceived as pleasing by some individuals, but as obnoxious or even as a threat by others.

Perceptions of one's state of health is also important—that is, whether or not one thinks that one is healthy. On the one hand, a hypochondriac's enjoyment of life may be impaired by the belief that he or she is unwell. On the other hand, as pointed out in the last chapter, when the onset of a deleterious change in the environment or in lifestyle is gradual, it is possible for the consequent maladjustment to go unnoticed, so that affected individuals are unaware that they could, in fact, feel much better than they do. A commonplace example is provided by the city man who returns from his summer holiday 'feeling a new man'—indicating that in his ordinary everyday existence he experiences suboptimal health. It can be argued, of course, that this 'boiling frog' effect, like habituation to unaesthetic environments or unpleasant working conditions, is of adaptive value, in that it protects the individual from feelings of discontent in the face of undesirable, but unavoidable, life conditions. On the other hand, it effectively blocks any cultural adaptive response aimed at improving these conditions.

Reference must also be made in these background comments on health and well-being to the concept of *stress*, which was introduced to biomedical science by Selye in the 1930s and 1940s. This subject has received an immense amount of attention in the literature, and an annotated bibliography on the subject produced by Selye in 1976 included around 8000 references [9]. When an individual perceives a threat in the environment, a series of responses, both physiological and psychological, come into play. These represent innate adaptive mechanisms which, under some circumstances, may render the individual better able to cope with the threat. The responses include, among others, the release of adrenalin from the adrenal medulla and of nor-adrenalin from the central nervous system and an increase in the concentration of sugar in the blood. However, in situations in which the cause of the stress persists or in which, for some other reason, a chronic state of stress ensues, undesirable pathological changes may result.

HUMAN BEHAVIOUR

Introduction

Human behaviour—that is, what people actually do—is the true pivotal point of all patterns of culture-nature interplay and as such it is central to our theme; it is human behaviour that keeps civilization going. If people stopped behaving, the whole massive superstructure would collapse. Paradoxically, there appears to be an increasing likelihood that it *will* collapse—*because* of human behaviour.

Serious discussion on human behaviour and its origins is beset with all manner of difficulties. One reason for this is that the subject really is extraordinarily complicated, and many aspects of it are hard to define, analyse, or describe in an incontrovertible and scientifically satisfactory way; and there is perhaps no area of scientific inquiry where the temptation is greater to indulge in extravagant speculation, nor where it is easier to promulgate and cling to fascinating but flimsy hypotheses in the absence of scientific verification. The field is also an unusually sensitive area, and much of the debate is vigorous and emotive [10].

In this book we will keep our approach at a very basic level and focus our attention on human behaviour as it relates especially to the impacts of societal developments on biological systems. Basically, we shall be concerned with two sets of fundamental questions. First: In what ways has human culture affected the behaviour of human organisms and with what implications for human health and well-being? Second: Does the human organism inherit phylogenetic or species behavioural characteristics from its evolutionary past and, if so, what are the implications for individuals or for society?

Cultural influences on behaviour

While there has been much debate about the relative roles of culture and nature as determinants of human behaviour, there can be no doubt that cultural processes affect what people actually do in a wide variety of ways. In each case when this occurs the question naturally arises: What are the consequences of the particular effect of culture on behaviour for the biology of individual humans, for other individuals in the environment, for society as a whole, or for the biosphere and its ecosystems?

Some of the effects of culture on behaviour are relatively *direct*. For example, culture may determine that adult males in a population must wear trousers, or it may decree that food shall be consumed at certain times of the day. In other cases, the effect is *indirect*. Thus the provision by planning authorities of bicycle tracks in a city may have a profound influence on the number of people who travel by bicycle. We should also note that culture may *promote* behaviours, as in the case of dancing, the performance of rituals, or fighting. In other instances, it *prohibits* certain behaviours, such as the eating of pork in some societies or the overt display of emotion in others.

The purpose here is to illustrate, with a few brief examples, the various ways that cultural developments can, through their effects on the immediate environment of individuals, influence individuals' behaviour and, in turn, affect their health and well-being.

The *design of the built environment* may affect behaviour in many ways. For instance, the existence of swimming pools in urban areas may result in many people engaging in regular swimming; and this behaviour can be expected to affect their physical fitness and sense of well-being. The design of buildings in which people in modern society spend most of their time also influences their behaviour patterns in a variety of ways; in particular, there has been much discussion among planners, architects, and social psychologists about the effects of design on social interaction between neighbours and within families and on rates of violent crime [11]. It is also likely that the aesthetic qualities of the designed environment can influence the behaviour of people.

An example of the way that culture-induced changes in the *chemical environment* can influence human behaviour is provided by the effects of exposure of people to lead. In the past this element was used in water pipes, in vessels for holding beverages, and in paint; more recently it has been added, in the form of tetraethyl lead, and as an anti-knocking agent to petrol used in motor vehicles. As a result, lead has found its way into food and drink and into the air which people breathe. The effects on behaviour of mild, subclinical, chronic lead poisoning include lethargy, insomnia, and irritability; and there is evidence which strongly suggests that lead can have a detrimental influence on mental development in children (see Chapter 10). Cultural processes have resulted in the exposure of human beings to many other toxic chemical substances affecting behaviour, especially in specialized occupational settings.

Culturally induced modifications in the *biotic environment* affecting human behaviour include the removal of predators and other threatening species, such as snakes and scorpions, from the everyday environment—a change which has had a marked influence on the way people move around in their surroundings. Another important kind of biological influence is that which results from the introduction into populations of certain infectious diseases as a consequence of human activities. Malaria and schistosomiasis are particularly important examples, both of which greatly modify the behaviour of affected individuals.

The effects of *technological developments* on human behaviour are numerous and obvious. In modern high-energy society important influences in this category include the motor car, the television set, the word processor, and the clock.

It is self-evident that the behaviour patterns of human beings are profoundly influenced by *beliefs, values, and customs*. Influences of this kind include effects on sleeping and eating patterns, intergroup violence, levels and kind of physical exercise, the wearing of clothes, consumer behaviour, sexual behaviour, the learning and practice of skills, and social interaction.

Another general set of influences on behaviour is associated with the introduction of *novel sources of pleasure*. These are linked with the technological influences mentioned above. Thus the lure of the television frequently results in people spending several hours a day in a state of physical immobility as they experience passive entertainment. Modern means of transport have resulted in people seeking and achieving pleasure travelling much larger distances than was previously possible, and hot showers have almost certainly affected the frequency of personal ablutionary behaviour.

The genetic contribution to human behaviour

Fundamentally there are two separate controversies about the genetic contribution to human behaviour, and the distinction between them is important. First there is the question whether we, as members of a species, inherit *phylogenetic behavioural characteristics*. The second area of debate relates to the extent to which *differences in behaviour among humans* can be accounted for by genetic variability. Here we will focus on the first of these questions.

At one end of the spectrum there have been those who have seen human behaviour as largely determined by a series of innate 'instincts' or 'drives'. At the other extreme is the *tabula rasa* hypothesis, which suggests that human beings are born with no predetermined behavioural characteristics or predispositions, and that the behaviour patterns which they develop as they grow up are determined entirely by environmental, and mainly cultural, influences.

Numerous lists have been prepared by the proponents of genetic or biological determinism of behavioural characteristics of humans variously referred to as 'drives', 'instincts', 'tendencies', and 'needs' [12]. There is, as would be anticipated, a great deal of overlap between them. Nearly thirty years ago Allport wrote:

Under the influence of Darwin, personality theorists traded faculties for instincts ... and their derivatives, the sentiments. More than anyone else, McDougall fixed our attention upon the possible existence of uniform motivational units. Freud reinforced the search, though unlike McDougall, he himself offered no clear taxonomic scheme. During this era innumerable instincts were discovered, postulated, invented. In 1924 Bernard reported that more than 14,000 different instincts had been proposed, and that no agreement was yet in sight. Sensing disaster in this direction, psychologists started fishing in fresher waters. The doctrine of drives (a limited form of instinct) continued to hold the behaviourist fort, and still to some extent does so, but most psychologists nowadays seem to agree with Hebb (1949) that to equate motivational structure with simple drives or biological needs is a wholly inadequate procedure [13].

Interest in the whole question of innate 'drives' in the human species received a substantial boost from important developments in the field of animal behaviour, or ethology, and, in particular, from the progress made by the Tinbergen–Lorenz school [14]. Lorenz, who wrote for the popular as well as the academic press, speculated about the significance of his ethological studies in relation to the behaviour of human beings, focusing especially on the theme of aggression. He postulated the inevitable building-up in the human organism of an aggressive urge or force which eventually breaks out in the form of some kind of aggressive behaviour.

In this respect, his views are reminiscent of the Freudian concept of *libido*, although Lorenz was in other respects not a supporter of Freud's ideas [15].

Other workers have applied an ethological approach to human behaviour in quite a different way. They have employed the methods of animal ethology in the direct study of human behaviour, and have been seeking evidence for universality in behavioural patterns [16]. Most of this fascinating work has been concerned with social interactions and especially with non-verbal communication, while the concept of 'personal space' has also received attention. Basically, the research in this area has confirmed the views of Darwin and many others that most facial expressions as well as some ways of using the hands in communication between individuals are innate. Whether *understanding* of the meaning of innate patterns of behaviour is also innate is less certain. However, there is some evidence that, at least in rhesus monkeys, appreciation of the meaning of facial expressions is innate and not learned [17].

In the last ten years or so the culture–nature debate has been refuelled by the development of what has come to be known as *sociobiology*. This 'new synthesis', as its chief exponent, E. O. Wilson, has called it, is defined by him as 'the systematic study of the biological basis of all forms of social behaviour, including sexual and parental behaviour, in all kinds of organisms, including humans' [18]. The approach considers behaviour, and, in the context of the present discussion, human behaviour, in terms of population genetics and evolutionary principles. Thus it assumes that any innate behavioural characteristics that the human species may possess are there because they were of survival value to the genes responsible for them in the evolutionary history of the species. Sociobiology thus has its roots in the writings of many earlier authors including Darwin, Kropotkin, Lorenz, Tiger, and Fox [19].

Sociobiology has been concerned mainly with phylogenetic characteristics of the human species, rather than with genetic differences within the species. Nevertheless, Wilson has stated that 'a key question of human biology is whether there exists a genetic predisposition to enter certain classes and to play certain roles ...' [20].

The proponents of sociobiology have concentrated on such aspects of human behaviour as altruism, sexuality, homosexuality, aggression, and even religion and ethics, and it is not surprising that their speculations have given rise to a much lively debate [21]. There are broadly two kinds of criticism of the approach of sociobiology, although there is clearly some overlap between them. We can refer to these respectively as the scientific debate and the political debate.

The main scientific criticism levelled at the sociobiologists is a predictable one; it is that they put too much emphasis on genes, as opposed to culture, as an influence on human behaviour. Another related criticism is that they have taken entirely unsubstantiated hypotheses about the biological basis of different aspects of behaviour much too seriously. Some of the sociobiological explanations of homosexuality, altruism, warfare, aesthetics, and tribalism are seen to be in this category. There has also been considerable debate about the suggestion by some sociobiologists that the form and direction taken by culture are largely a reflection of the innate behavioural characteristics of humankind and that the phylogenetic characteristics of the species exert an important influence on the selection by society of novel cultural developments or 'culturgens' [22].

The political debate has been initiated by individuals and groups who have linked sociobiology with conservatism and reactionary political attitudes [23]. Some critics feel so strongly on this issue that they suggest that sociobiology, as an area of academic endeavour, should be suppressed [24].

My own view is that the theoretical starting point of sociobiology is sound, and that so long as it keeps its feet on the ground the approach can make a useful contribution to the understanding of human situations and problems. There is no in-built reason why it should contribute to the unjust treatment of minority groups in human societies; indeed, it might very well have the opposite effect. If undesirable consequences are forthcoming it is likely that they will be the result of sociobiologists or their readers allowing their imaginations to run wild and then drawing uncritical and spurious conclusions from the fantasies they create.

Common behavioural tendencies

The approach taken in this volume had a similar starting point to that of modern sociobiology. It is based on the view that human beings do inherit from their evolutionary ancestors some phylogenetic behavioural characteristics. That is to say, it does not accept statements like the following, attributed to Jose Ortega y Gasset: 'Man has no nature. What he has is a history' [25]. However, although our starting point was similar to that of the sociobiologists, the aspects of human behaviour which have attracted the attention of my colleagues and myself are rather different from those on which the sociobiologists have concentrated.

Our approach is based on recognition of the fact that there are certain things that people tend to do in all societies the world over, and presumably always have done, and we strongly suspect that there is a phylogenetic basis for these aspects of behaviour. I shall refer to these things that people tend to do as *common behavioural tendencies*. They range from such physiologically linked behaviours as the tendency to eat when hungry, drink when thirsty, and to copulate when appropriately stimulated, to more psycho-social behaviours such as the tendency to seek the approval of the members of one's in-group and to seek to avoid ridicule, and the tendency to identify with and exhibit loyalty towards in-groups.

It must be stressed that if these common behavioural tendencies are innate, they must have been of selective advantage during the evolution of the human species. That is to say, individuals who carried genes (or combinations of genes) associated with such behavioural tendencies would have been more likely to have contributed genetic material to subsequent generations than individuals who did not carry such genes.

On the basis of personal experience and of information from the anthropological literature and other sources, my colleagues and I have put together a list of what seem to us to be some of the most important common behavioural characteristics of humankind [26]. This list, a version of which is presented in Table 3.1, is not intended to be in any sense absolute or finite—indeed we are forever modifying it in the light of discussion and criticism. We appreciate that many other authors have drawn up their own personal lists of human 'instincts' and 'drives', and we are well aware of the reductionist pitfalls inherent in the shopping-list approach and in attempts to categorize human behaviours. Nevertheless, bearing these hazards in mind, we have found this attempt to recognize and describe the things that people tend to do in common to have been useful.

There is inevitably a good deal of overlap between the items on the list of common behavioural tendencies. This is to be anticipated, because they are not seen as discrete entities and many of them are closely related. Some may well be derived from others. For example, the tendency to show loyalty to members of an in-group may be a secondary behavioural tendency, a manifestation of the tendency to seek the approval of members of an in-group and to avoid their disapproval [27]. It should be pointed out that the list of common behavioural tendencies in Table 3.1 relates particularly to adults; although many of the items also apply to children, there are major differences in emphasis.

The common behavioural tendencies, especially those relating to social interaction, are at a basic or elemental level. In the case, for example, of the approval-seeking set of behavioural tendencies (see Table 3.1), the actual criteria for approval or disapproval are usually largely culturally determined. Thus the specific behavioural manifestation of approval-seeking is likely to be different, for example, in a business executive than in a member of a youth street gang. The underlying behaviour is referred to as *basic behaviour*, while the particular culturally determined manifestation of this basic behaviour is referred to as *specific behaviour* [28].

There is a link between common behavioural tendencies and a sense of enjoyment and of satisfaction. People tend to seek the approval of members of their in-group, and to seek their praise; and such approval and praise, when it is forthcoming, is a source of enjoyment and satisfaction for them. Similarly, people tend to seek novelty, and novelty is also, up to a point, a source of enjoyment. This relationship between common behavioural

Table 3.1. *Common behavioural tendencies.*

Physiological
to eat when hungry, to drink when thirsty
to copulate when appropriately stimulated
to avoid pain, discomfort
to seek comfort
to rest or sleep periodically in response to the urge to do so
to avoid unnecessary exertion; to take the easiest path to an objective
to avoid prolonged periods of very low or very high levels of sensory stimulation, but to seek
   stimulation to an optimum level

Social—approval seeking
to seek attention and companionship within the in-group
to seek approval of members of the in-group, to seek their praise
to avoid ridicule from and disapproval of members of the in-group
to show approval for forms of behaviour considered advantageous to the in-group
to show disapproval for forms of behaviour considered disadvantageous to the in-group
to compete with peers for status, respect, and influence; and to aim to out-perform peers in
   approved activities
(Overall effect: the tendency for individuals to strive to succeed in activities considered
   advantageous to the in-group).

Social—general
to make choices perceived to be of benefit to oneself or to those one loves the most
to support, protect, to show trust in and loyalty to the in-group
to co-operate with members of the in-group
to exhibit care-giving behaviour towards immature members of the in-group, especially one's
   own offspring
to exhibit care-eliciting behaviour towards members of the in-group in times of anxiety
to exhibit care-giving behaviour towards members of the in-group, especially in response to
   care-eliciting behaviour
to establish hierarchies (not necessarily inflexible and lasting)
to accept the beliefs and values of the society into which one is born and in which one lives
to defend perceived (culture-determined) rights (e.g. property, conjugal, territorial)
to seek to acquire what others have
to behave aggressively, on occasion, towards others, especially in response to the frustration
   of desires
to act with suspicion and distrust towards out-groups
to seek sexual relationships
to imitate (especially in the young)
to play games (especially in the young, but including dancing, singing, etc.)
to avoid prolonged solitude and loneliness, but to seek solitude on occasion
to form pair bonds for mating and reproductive purposes

Others
to seek to modify the environment
to exhibit interest in and to explore the unknown (despite a certain suspicion and fear of the
   unknown)
to seek novelty
to seek new information
to seek and accept reasonable short-term challenges (i.e. challenges of a kind which, with
   sufficient effort, stand a reasonable chance of being successfully met or overcome in a short
   time) and to seek to solve problems
to seek explanations
to follow habit

---

tendencies and enjoyment is entirely consistent with evolutionary considerations: animals, including *Homo sapiens*, tend to enjoy doing the kinds of things that are likely to contribute to their survival and successful reproduction in the habitat in which they evolve. In the words of Hamburg:

.. we may find useful guidance in the principle that *individuals seek and find gratifying those situations that have been highly advantageous in survival of the species. ... .* Another side of the evolutionary coin may be stated in the principle that *individuals avoid and find distressing those situations that have been highly disadvantageous in species survival* [29].

There is also an important connection between common behavioural tendencies and the universal health and survival needs of the human species. This connection is particularly obvious in relation to the behavioural tendencies which relate directly to physiological processes; but it probably applies also to the various socially oriented behavioural tendencies which often lead, for instance, to a sense of personal involvement and of belonging, both of which are postulated to be important components of health and well-being (see page 79).

It is clear that the cultural environment can either *reinforce* or *suppress* common behavioural tendencies. Thus, while the tendency to compete with peers for status, respect, and influence seems fairly universal, in some societies this tendency is greatly reinforced by socialization, and in others it appears to be suppressed. Similar effects of culture can be seen with respect, for example, to the tendencies to establish hierarchies, to act with suspicion towards out-groups, and to co-operate with members of in-groups. There is also considerable variability within human populations in the extent to which the various common behavioural tendencies are manifest. For example, an occasional, individual may appear not to seek the approval of members of his or her in-group, nor to seek attention and companionship within the in-group. Such variability is likely to be due to both genetic and cultural influences.

Differences between the sexes with respect to common behavioural tendencies have not been considered in the formulation of the list in Table 3.1. However, there is reason to suspect that some differences do exist [30]. There is evidence, for example, that the balance of male and female sex hormones in the body has an important influence on sexual behaviour patterns [31]. However, it is also likely that these differences are often greatly exaggerated through cultural reinforcement.

As in the case of all other animals, human beings select different categories of stimuli to respond to at different times [32]. Thus, while cooperation with members of an in-group may act as a source of enjoyment and satisfaction to an individual for a certain period of time, after a while he or she may derive more enjoyment from a state of solitude.

In concluding this section on common behavioural tendencies, mention must be made of an important behavioural principle which I call the *principle of the flaming moth*. It is a common observation that at night-time moths often fly directly into exposed flames, thereby incinerating themselves. This behaviour, which is apparently an expression of an innate behavioural characteristic of moths, is clearly not, under the changed or artificial circumstances, of survival value to the moth's genes. And yet the instinct which underlies this behaviour is the product of natural selection, and consequently can be assumed to be of selective advantage in the environment in which moths normally live and in which naked flames are a rare occurrence. In the changed or evodeviant situation, however, the instinct represents a distinct survival disadvantage to individual moths, and to moth populations. The important question which the case of the flaming moth raises is whether the same principle applies to humankind. If humans in civilization have inherited from their hunter-gatherer ancestors certain behavioural characteristics which were of selective advantage in the natural habitat of the species, is it possible that some of these tendencies might, in the changed societal setting, be leading to forms of behaviour (i.e. specific behaviours) which are to the disadvantage of individuals or of populations?

Some specific issues

Human aggression and territoriality

The literature which has grown up in the last twenty years on human aggression reflects a good deal of disagreement and some confusion. One of the factors contributing to this state of affairs has been a tendency to discuss aggressive behaviour at the level of the individual and aggression between large organized groups of human beings, such as nation states, as if they were aspects of one and the same process [33]. In fact, aggression between nations is a form of corporate behaviour, and it has little to do with human aggressiveness on the individual level. Another source of confusion is the use of the term 'aggression' for various kinds of assertive behaviour in, for example, economic and commercial spheres of activity.

In this section we are concerned with aggressive behaviour at the level of the individual or, at most, at the level of small-scale face-to-face group situations. For our purposes we can define aggression as behaviour which is intended to inflict injury on some person or persons. This includes not only physical injury, but also psychological injury resulting, for example, from verbal attack. It is therefore necessary, when speaking of aggressive behaviour between individuals, to qualify the form of injury inflicted or attempted.

Most of the literature on aggression in human beings has been concerned with the motivation of this kind of behaviour. Some authors have taken the view that aggressive behaviour in humankind is an inevitable consequence of our phylogenetic make-up and have tried to explain this characteristic in terms of our evolutionary background [34]. It has been suggested, for example, that it is a consequence of the 'predatory adaptation' achieved by our australopithecine-like evolutionary ancestors and that our species has a 'carnivorous psychology' [35]. This idea has been criticized by other authors who have drawn attention to the fact that intraspecific aggression is not especially a feature of other carnivorous species. Moreover, there is doubt about the extent to which early hominids were carnivorous [36].

There can be no doubt, however, that human beings inherit a *potential* to behave aggressively under certain circumstances; but it is equally clear that they can live peaceably with each other for long periods of time without engaging in physical violence or even, in some cases, in verbal exchange aimed at inflicting personal injury [37]. The situation is succinctly summarized in the words of Campbell:

'... human aggression and violence is an adaptive behaviour (in an evolutionary sense), which arises in response to certain aversive stimuli and ceases upon their removal' [38].

Broadly speaking, culture can influence aggressive behaviour in two ways. On the one hand, it can affect the level and intensity of 'aversive stimuli', as well as the individual's perception of the aversiveness of stimuli. On the other hand, it can influence attitudes towards aggression itself. In some cultural or subcultural situations, aggressive acts are valued as praiseworthy behaviour in their own right, and individuals may engage in aggressive behaviour simply to achieve the approval of their peers and to gain status.

Before leaving the topic of aggression, some reference must be made to the concept of territoriality. Ardrey popularized the idea that *Homo sapiens*, like some other species, is phylogenetically programmed to defend territory (in the geographical sense of that word) [39]. Others argue that territoriality is a learned behaviour and that it is not an innate species characteristic [40]. My own view is that geographical territoriality is not an in-built characteristic of humankind. Nevertheless, there does seem to be a universal tendency among humans to resent intrusion into any kind of *perceived right*; and perceived rights are largely culturally determined. They may relate to land or other forms of property, to conjugal situations, to areas of academic pursuit, or to many other aspects of human experience. If such perceived rights are encroached upon, the affected individual is likely, in one way or another, to defend them vigorously.

Learning behaviour

It is self-evident that the human capacity for culture is dependent upon the extraordinary innate learning ability of the species, especially in the younger years. At a basic level, learning behaviour can vary in a number of ways, each influenced by both biological and cultural factors, the latter in most cases being the most important. For example, there is wide variation in *what* people learn. Biological forces play some role and, as has been pointed out by Tinbergen, 'learning is not random, but is often a highly selective type of interaction with the environment' [41]. In young humans biologically determined sources of pleasure and displeasure and biological propensities, such as the potential for bipedal locomotion, obviously affect what is learned. The extent to which phylogenetic factors play a selective role in the learning of language and other cultural variables is a matter of discussion and debate [42]. However, there can be no doubt that culture is the main determinant of what is learned, and consequently there is very wide variation both between societies and within societies in what people come to know and to believe through learning.

How much people learn is also influenced by both biological and cultural factors. Thus some individuals, apparently for genetic reasons, seem able to assimilate much more information than others. However, the cultural environment is probably the main determinant of quantitative variation in learning.

Altruistic and selfish behaviour

There has been an immense amount of discussion among evolutionary biologists about the possible origins of altruism in evolution. While it is clear that altruistic behaviour is likely to be of benefit to animals at the population level, some difficulty has been experienced in explaining how it could contribute to the Darwinian fitness of individuals. This problem does not exist with respect to altruistic behaviour by an animal directed towards its own offspring, because this behaviour would be likely to favour the survival of its own genes; but this would not apply to altruistic behaviour to individuals other than progeny.

In 1964, Hamilton contributed to the discussion on this topic by drawing attention to the concept of *inclusive fitness*. This notion is similar to the idea that the survival of an individual's own genes is favoured by behaviour which is likely to favour the survival of offspring. Hamilton suggested that an individual will invest energy in activities that are to the benefit of other individuals according to their degree of genetic relatedness. This would result in improved chances of survival of the individual's own genes not only in that individual's offspring, but also in other closely related individuals carrying the same genes—the phenomenon of *kin selection* [43]. Another contribution came from Trivers who put forward the idea of *reciprocal altruism* [44]. He suggested that an altruistic act towards another unrelated individual, while not itself contributing to the survival of the individual's own genes at the time, might in the long run have this effect, by virtue of the fact that a 'reciprocal relationship' would have been established between the two individuals. As a result, the second individual, at a later time, might behave altruistically to the first individual, thereby contributing to the Darwinian fitness of the latter.

The literature on this topic is extensive, and no attempt will be made here to review it; I will simply summarize my own views on the subject. I suggest that, in the evolution of humankind, *co-operation* with other members of a band was of selective advantage to the individual. That is to say, an inherited tendency to co-operate may be seen as the consequence of a 'selfish gene'. Co-operation, however, demands consideration of the needs of other individuals in the co-operating group, for, without this, co-operation would be impossible. Thus altruism becomes an inevitable consequence of the selection pressures favouring co-operation.

Another factor to be taken into account is the fact that there is Darwinian advantage in altruistic behaviour to one's own offspring, not only when they are infants but also when they are grown up. Thus natural selection might be expected to result in individuals displaying altruistic behaviour towards other individuals with whom they are particularly close. For an individual *not* to have a tendency to behave altruistically to other members of the in-group (i.e. not offspring) would therefore require a special mechanism, developed in evolution, for categorizing other close acquaintances in the in-group as 'not offspring', and the existence of such a mechanism in humans would seem to be rather unlikely [45].

Whatever the evolutionary mechanism behind the present situation, human beings do tend to behave altruistically on occasion, especially towards other individuals who are close to them—that is, towards other members of their in-group. There appears to be no universal tendency, however, to behave altruistically towards members of out-groups, although in some situations cultural developments have encouraged such behaviour.

NOTES

1. The *biopsychic state* is equivalent to the *phenotype*. The former term is used because it is more descriptive of the aspects of human experience which are our special concern in this book.
2. The work most quoted in the literature is probably that of Maslow 1954. He postulated five levels of needs (he uses the term in a different sense than that in which it is used in this book), which he ranks in a hierarchical order. He suggested that satisfaction of the more basic needs at the lower end of the hierarchy is usually necessary before the 'higher' needs can be satisfied—and he emphasized the importance for 'full health and desirable development' of the satisfaction of the need for 'self-actualization'. Let us note, however, the following extraordinary statement by Maslow: 'However, I should say that I feel less confident in speaking of self-actualization in women' (see Oppacher 1977, p. 20).

Maslow's five levels of needs are as follows: i. Physiological needs: hunger and thirst. ii. Safety needs: security, order, freedom from pain, discomfort, and threat. iii. Belongingness and love needs: love, sex, affection, friendship, identification. iv. Esteem needs: fame, prestige, recognition, success (self-esteem and respect of others). v. Need of self-actualization—Man's desire for self-fulfillment; 'to become everything that one is capable of becoming'. Later Maslow added to this list the 'desire to know and understand' and 'the aesthetic need'. For a useful short summary of the 'needs' literature see Mikellides 1980.

3. A very different definition of the word *need* is as follows: 'A condition which, if continued, will lead to the death of an individual or the disappearance of the species; a state of physiological disequilibrium and homeostatic imbalance' (see *Chambers Dictionary of Science and Technology* 1974). According to our definition, water is a need (for health and survival); according to the *Chambers* definition, absence of water is a need. Some authors use the term 'need' in the sense in which I use the term 'behavioural tendency' (see later in this chapter). Another different definition of need is 'a state of physiological or psychological want that consciously or subconsciously motivates behaviour towards its satisfaction' (*A Supplement to the Oxford English Dictionary* 1972).

4. The term *contingent needs* used in this sense was introduced by Malinowski 1944, p. 171. His term *fundamental needs* more or less corresponded to our universal needs.

5. Boyden and Millar 1978; Boyden, Millar, Newcombe, and O'Neill 1981.

6. Many authors use the word *want* to convey this meaning. However, others use it to denote the thing which is wanted, which in the present text will be referred to as the *object of want(s)*. Objects of want may be material commodities or they may be intangible factors, such as a trip around the world.

7. The 'official' World Health Organization definition of *health* is as follows: 'Health is a state of complete physical, mental and social well-being and not merely the absence of disease or infirmity' (World Health Organization 1947, p. 13). Another definition, and one which appeals to me, is that of Pericles (died 429 BC): health is 'that state of moral, mental and physical well-being which enables a person to face any crisis in life with the utmost grace and facility' (Furnass 1976, p. 5).

8. Dubos 1965, p. 187.

9. Selye 1976.

10. There has been a considerable number of quite different approaches to the study of human behaviour, each with its own particular conceptual framework. Among these approaches we can recognize, for example, those of psychology, sociology, social psychology, psychoanalysis, psychiatry, clinical psychology, experimental psychology, ethology, and, recently, sociobiology.

11. See Gans 1976; Newman 1972.

12. See reviews by Madsen 1959 and Maddi 1968.

13. Allport 1958, p. 240.

14. See Tinbergen 1966 and Lorenz 1966.

15. Lorenz 1966, 1974.

16. Eibl-Eibesfeldt 1970; Hinde 1974.

17. Eibl-Eibesfeldt 1971, p. 20.
18. Wilson, 1978, p. 2.
19. Darwin 1873; Kropotkin 1904; Lorenz 1966, 1974; Tiger 1969; Tiger and Fox 1972.
20. Wilson 1975, p. 554.
21. See, for example, Barash 1977; Caplan 1978; Blurton Jones and Reynolds 1978; Gregory, Silvers, and Sutch 1978; Midgley 1978; Chagnon and Irons 1979; Harris 1979, chapter 5; Lewontin 1977; Ruse 1979; Albury 1980; Reynolds 1980; Singer 1981.
22. Lumsden and Wilson 1981, p. 12.
23. Albury 1980, p. 525.
24. Barash 1977, p. 279.
25. J. Ortega y. Gasset, quoted by Dubos 1970, p. 207. Although couched in terms different from those used here, the following words of Malinowski (1944, p. 85) are more consistent with those expressed in this book:

   So far we have learned that human nature imposes on all forms of behavior, however complex and highly organized, a certain determinism. This consists of a number of vital sequences, indispensable to the healthy run of the organism and to the community as a whole, which must be incorporated in each traditional system of organized behaviour. These vital sequences constitute crystallizing points for a number of cultural processes ... . As for the impulse, it is clear that in every human society each impulse is remolded by tradition. It appears still in its dynamic form as a drive, but a drive modified, shaped and determined by tradition.

26. This list is similar in many ways to McDougall's list of behavioural *propensities* (see Madsen 1959, p. 64).
27. It is not always obvious which particular behavioural tendency lies at the root of a given behaviour. In the words of Maslow (1954, p. 72): 'An individual going through the whole process of sexual desire, courting behaviour and consummating love making may actually be seeking self-esteem rather than sexual gratification'.
28. The significance of this concept was recognized by Maslow (1954, p. 67) when he wrote:

   There is now sufficient anthropological evidence to indicate that the fundamental or ultimate desires of all human beings do not differ nearly as much as do their conscious everyday desires. The main reason for this is that two different cultures may provide two completely different ways of satisfying a particular desire, let us say, for self-esteem. In one society, one obtains self-esteem by being a good hunter; in another society by being a great medicine man or a bold warrior, or a very unemotional person and so on. It may then be that, if we think of ultimates, the one individual's desire to be a good hunter has the same dynamics and the same fundamental aim as the desire of the other individual to be a good medicine man. We may then assert that it would be more useful for psychologists to combine these two seemingly disparate conscious desires into the same category rather than to put them into different categories on purely behavioural grounds. Apparently ends in themselves are far more universal than the roads taken to achieve those ends, for these roads are determined locally in the specific culture.

29. Hamburg 1964, pp. 312–14.
30. This is a theme which has given rise to a great deal of attention among the sociobiologists. See, for example, Wilson 1975.
31. Hamburg and Brodie 1973.
32. Hinde 1974, p. 24.

33. A number of authors have drawn attention to the fallacy of this approach. See, for example, Hinde 1974 and Rapoport 1975.
34. See, for example, Ardrey 1961, 1966, 1970; Lorenz 1966. See also a strong attack on Lorenz's view by Barnett 1973.
35. Freeman 1964, p. 115.
36. Zihlman and Tanner 1978.
37. Montagu 1973, 1976; Hinde 1974; Reynolds 1980.
38. Campbell 1971, p. 567. See also useful discussions on this topic in Hinde 1974 and Reynolds 1980.
39. Ardrey 1966, 1976.
40. See, for example, Montagu 1968, p. 16.
41. Tinbergen 1972, p. 397.
42. Hinde 1974.
43. Hamilton 1964. See also Maynard Smith 1964 and discussions in Davies and Krebs 1978.
44. Trivers 1971.
45. It has been reported, however, that females of a species of ground squirrel at one year of age distinguish between half-sibs and full sibs (even when reared together in the same burrow). See the article by Lewin 1984.

4

THE PRIMEVAL PHASE OF HUMAN EXISTENCE

INTRODUCTION

The human species evolved under conditions which were very different from those which the great majority of people experience today. For several reasons, knowledge of these conditions is relevant to the study of the biology and ecology of humankind in the modern world. In the first place, it provides a rational starting point from which to begin discussion about the implications for the biosphere and for human health and well-being of the culturally induced changes that have occurred in human society over recent millennia. Moreover, some understanding of this long primeval phase of human existence engenders a certain sense of perspective. In an era of ever-accelerating environmental change and sophisticated technological innovation, it is sobering to recall that the primeval conditions, hazardous as they were, satisfied the survival, health, and reproductive needs of our ancestors for many thousands of generations. It was during the Palaeolithic period, long before the introduction of farming, that the human species spread, apparently from Africa, to all the continents of the world and became established in regions as different ecologically as the tropical forests of the equatorial belt and the frozen landscapes to the north of the Arctic Circle.

Another reason for interest in the conditions of life characteristic of the first ecological phase of human existence relates to the principle of evodeviation. As discussed in earlier chapters, when the conditions of life of an animal deviate significantly from those to which the species is adapted through evolution, some signs of physiological or behavioural maladjustment are likely to occur. It is worthwhile, therefore, to try to identify the basic ways in which the life conditions of human beings living, for example, in modern urban society differ from those which prevailed in the evolutionary habitat of the species; and, in the case of each significant deviation from the evolutionary conditions, it is pertinent to enquire whether the change may be the cause of some form of phylogenetic maladjustment.

Our knowledge of the ecology and lifestyle of people living in the primeval phase of human existence is derived from a number of sources, including palaeontology, archaeology, and the anthropological literature on contemporary or recent hunter-gatherer societies. The last is an especially important source [1]. For a number of reasons, however, we must be cautious in our utilization of information about recent hunter-gatherers. For example, there is, as there certainly was also in the past, wide variability in the life conditions of primeval people from place to place. Another reason for caution lies in the fact that all hunter-gatherer groups in existence in recent times have been affected in one way or another by farming and urban societies. Nevertheless, in spite of these difficulties, there are certain features common to all recent hunter-gatherer societies, including one or two groups which have survived in relatively fertile areas, and it seems most likely that these features were also shared by societies of our ancestors before the domestic transition.

The information given below about recent hunter-gatherers is drawn mainly from the literature on groups living in Africa, Asia, Australia, and the Arctic [2]; we will begin by considering the ecological interrelationships between hunter-gatherer societies and the rest of the biosphere. This will be followed by discussion about the organization of society itself and about the interrelationships between hunter-gatherer bands. Finally, we will consider the life conditions and patterns of health and disease of primeval people.

RELATIONSHIPS BETWEEN PRIMEVAL SOCIETY AND THE BIOSPHERE

Recent hunter-gatherer societies have existed in a wide variety of different habitats, including such extremes as the Kalahari Desert in southern Africa, the tropical forests of the Congo Basin, and the ice-covered plateaus of Greenland and north Canada. This diversity of habitats is clear evidence of the effectiveness of cultural adaptation, in the broad anthropological sense of the term, in the primeval phase of human existence.

We know that humankind lived in a similarly wide range of habitats in prehistoric times, before the domestic transition. It is important from the ecological viewpoint, however, to bear in mind that, although some people in the late Palaeolithic era probably lived in relatively unfavourable fringe habitats, the great majority were presumably enjoying the fertile and climatically hospitable areas of the world—areas that have since been taken over almost entirely by the economies of the second, third, and fourth ecological phases. For this reason, the facts relating to the Eastern Hadza who live in Tanzania near Lake Victoria are especially interesting, because this group has maintained a hunter-gatherer lifestyle in fertile savannah countryside which abounds with animals and edible plants of many kinds— a countryside probably not very different from that in which a high proportion of our ancestors lived in primeval times.

Hunter-gatherers are usually nomadic people, although the length of time that they stay in any one camp varies according to circumstances. They may erect temporary shelters, which are usually of a crude nature, although the Eskimos have relied for a long while on fairly substantial dwellings in the form of igloos and other structures. In prehistory, the situation was probably much the same as among modern hunter-gatherers, and how long people stayed in their camps is likely to have varied from time to time and from place to place; and there is evidence from the south of France that some quite large permanent structures made out of timber were constructed in Palaeolithic time—possibly as long ago as 300 000 years [3]. Natural shelters in the form of caves were also inhabited on a semipermanent basis.

Despite their capacity for culture and their use of tools and weapons, including such innovations as the bow and arrow, hunter-gatherers fit into the ecosystems of which they are a part in much the same way as other omnivorous species, except for one important factor—their use of fire (see below). In the food chain, they play the role of first-order consumers (deriving energy from plant materials), of second-order consumers (deriving energy from the tissue of herbivorous animals), and, to a lesser extent, of third-order consumers (deriving energy from the tissue of meat-eating animals).

Some idea of the likely density of human populations during the primeval phase may be gained from the following estimates: for Palaeolithic France, one person per 5500 Ha; for Australian Aborigines, one person per 1000–20 000 Ha (depending on region); American Indians in prairies of North America, one person per 2000–2500 Ha; Kalahari Bushmen, one person per 1040 Ha [4].

The most conspicuous ecological difference between humans in the primeval setting and other animal species is the fact that humans, as an aspect of their cultural activities, make regular use of fire. In some situations this practice has important ecological impact [5]. In the words of Woodburn:

... the ecological effects of, for example, those hunters and gatherers who systematically use fire to burn the vegetation from very large areas in order to drive game or in order to induce fresh growth of young grass that will attract the game, are very substantial—far more substantial sometimes than the clearance and cultivation of some few acres by the farmer [6].

In using fire, hunter-gatherers are exploiting an additional source of energy, distinct from the *somatic energy* which is acquired in their food, involved in their metabolism, and necessary for muscular work. The use to which this *extrasomatic energy* is put is much influenced by the local environmental conditions: it may be used, for example, to provide warmth, to clear forest, to drive game, and it is universally used for cooking. It has been estimated that the per capita use of extrasomatic energy in the form of fire in hunter-gatherer societies is roughly the same as that flowing through human organisms as somatic energy, that is, about 10^4 KJ per day, bringing the total energy flow through human communities to about 2×10^4 KJ per capita per day [7].

In tropical savannah country in central Africa, for example, the amount of energy, somatic and extrasomatic, used by human beings in the primeval phase is likely to be around 0.003 per cent of the total somatic energy flow through the local ecosystem—that is, 0.003 per cent of the total energy fixed by photosynthesis and used for respiration, growth, and reproduction by all the plants, animals, and microorganisms in the system [8]. The yield per hectare, in terms of human food energy, in hunter-gatherer communities ranges from about 0.6 to 6.0 MJ per year, depending on the ecological conditions.

PRIMEVAL SOCIETY

Hunter-gatherers are social beings and, with very rare exceptions, each individual belongs at any one time to a group or band, the members of which all know each other personally and interact together every day. The size of these groups is variable, and is influenced to a large extent by prevailing ecological conditions [9]. Thus, in the Kalahari Desert, bands of Bushmen may consist of more than seventy individuals at certain times of the year when food is plentiful, but when seasonal conditions make it necessary they may split into smaller groups based on the extended family, and eventually into still smaller groups consisting of nuclear families. In the case of Eastern Hadza, for whom food always seems to be fairly plentiful, the people are spread over the terrain in small groups in the wet season, while in the dry season they gather together around water holes in groups of about one hundred individuals. Most commonly hunter-gatherer groups number between twenty-five and fifty individuals [10].

In most of the hunter-gatherer societies which have been studied and described, membership of bands is not determined by any strict kinship system, although the building block of all groups is the nuclear family [11]. Groups may vary from time to time both in size and composition, and there is a good deal of coming and going between neighbouring groups. Sometimes a band may split into two parts permanently as a result of ecological pressures. Another cause of such splitting is tensions arising from clashes of personality among members of a group.

One of the features of most hunter-gatherer societies, in marked contrast to later agricultural and urban societies, is the lack of a rigid hierarchical structure. For example, the Kalahari Bushmen, the Hadza, the Mbuti Pygmies, the Eskimos, and the Australian Aborigines have no system of chieftainship. Leadership is usually determined by spontaneous mechanisms, based presumably on personality factors as well as on the known skills or experience of individuals in different fields of endeavour. Thus leadership is often a transient phenomenon, depending on the nature of the subject under discussion or the activity of the moment—one individual emerging as leader of the hunt, another as leader of music-making and dancing, and another as leader in respect of spiritual matters. In some groups, especially among the Australian Aborigines, old age is associated with high status and power, although decision-making in most hunter-gatherers appears to be a relatively democratic prccess.

Several authors have referred to the fact that, in the hunter-gatherer societies they have studied, males and females have equal status in the community as a whole, although the relative influence of men and women at any particular moment is likely to depend on the nature of the immediate activities or circumstances [12]. It is noteworthy that among Kalahari people who have taken up farming and have adopted a sedentary way of life, the change in lifestyle has been associated with a 'decrease in women's autonomy and influence relative to that of the men' [13]. However, equality of status between the sexes is not universal among hunter-gatherers. It is said that among Eskimos males were dominant, and there is some difference of opinion about the situation with respect to the Australian Aborigines [14].

An outstanding feature of primeval society is the fact that, with the exception of an age- and sex-based division of labour, there is no occupational specialism of the kind found in urban societies and in some farming communities. All individuals are jacks of all trades, and all the women participate in the gathering of plant food-stuffs (and at times, all the men as well), and all the men participate in hunting and in making weapons [15]. Nevertheless, most individuals are better at some activities than at others, and they are likely to devote more time to these particular pursuits than do individuals who are less proficient at them.

There has been a good deal of speculation about the nature of the forces that keep people together in primeval society. Some authors have suggested that in *Homo sapiens* and in other primates sexuality is the all-important factor. However, this view is not favoured by most anthropologists at the present time. It seems more reasonable to explain the situation in the reverse manner, on the basis of a fundamental attraction between members of the same species; in certain circumstances this natural bond may be greatly enhanced, as in the case of sexual attraction, while in others it may be almost obliterated, as in the case of the mutual antagonisms that sometimes occur between individuals [16].

The strong desire for companionship and, associated with this, the desire to be approved of by members of one's in-group appear to be important factors in the control of social behaviour among hunter-gatherers. On the whole, antisocial activities in the form, for example, of serious physical violence within the band and the grabbing of food are rare, at least under reasonably favourable conditions. Linked with the desire for companionship and approval is a real fear of ridicule by other members of the group. Consequently, the tendency for individuals in hunter-gatherer groups to conform to the norms of the society is very noticeable. The situation contrasts with that of modern urban societies, in which a form of behaviour which is disapproved of by one in-group may well be approved of by another. For hunter-gatherers, there is often only one in-group.

All this is not to say that there do not occur disputes between individuals in primeval societies, nor that tempers do not flare on occasion. Such behaviour has frequently been observed in all recent hunter-gatherer societies, although the most common causes of such disturbances seem to vary to some extent from one culture to another. Culture is the main determinant of the *criteria for approval and disapproval* and of the nature of *perceived rights*, and consequently of the specific causes of arguments. In the case of the Siriono, for example, disputes are mostly about the distribution of meat, while in the Kalahari Desert food is less often the cause of quarrels, which usually centre around illicit extramarital sexual activities.

It is worth noting a behavioural mechanism which operates in hunter-gatherer societies when disputes develop among the members of small groups. If tension mounts between two individuals, perhaps after some harsh words are spoken, it is common for one of them to move away to join temporarily another family group within the band and to sit around its hearth; the clashing personalities thus avoid each other's company until the tension has declined. While this process is often sufficient to resolve the problem, if such personality clashes between two particular individuals persist and are frequent, the separation will eventually become permanent. This *reaction of mutual avoidance* works effectively as a mechanism for maintaining peace within groups not only in hunter-gatherer communities, but also in other forms of human society in which the extended family exists and provides a social framework for its operation [17].

Material culture and ownership

It is self-evident that the material culture of hunter-gatherers is very simple compared with that of all other kinds of human society. It comprises tools and utensils used for preparing food, tools for other purposes, such as sewing or making nets, weapons, knives, axes, ornaments, and clothing. It also includes shelters, ranging from the temporary structures of the Australian Aborigines and the Pygmies to the more permanent residences of the Eskimos and Siriono.

The concept of personal ownership is weakly developed in recent nomadic hunter-gatherer societies, and is usually restricted to little more than objects which individuals can carry around with them. These would usually include a man's weapons and implements (and sometimes sacred objects), and a woman's digging sticks, cooking utensils, and ornaments. Land and its resources are never owned by individuals or families. Ownership becomes more important in the more sedentary hunter-gatherers, such as the Ainu, who live in the southern part of the Kurile Islands (part of Japan), and the Haida, who inhabit the Queen Charlotte Islands [18].

Relationships with neighbours

Many people picture our ancestors of the Palaeolithic period as being constantly at war with their neighbours. This image has received encouragement from some books which have popularized the notion of territoriality in humankind and have put great emphasis on a supposed innate 'aggressive drive' in the species [19]. However, the information collected by anthropologists over the past fifty years throws considerable doubt on this assumption. The following quotations from the literature referring to recent hunter-gatherers serve to illustrate this point. Silberbauer emphasizes the importance, for the Kalahari Bushmen, of the maintenance of 'friendly and harmonious relationships' [20] between bands. Woodburn, referring to the Hadza, has written: 'People, singly and in groups, move freely from region to region. Any individual Hadza may live, hunt, and gather anywhere he or she likes without any sort of restriction and without asking permission from anyone. Neither individuals nor groups hold exclusive rights over natural resources, over land and its ungarnered produce' [21]. Concerning the Mbuti Pygmies, Turnbull has written: '... there is little interaction between bands, the chief areas being marriage, trespass and visiting' [22]. Holmberg, writing about the Siriono, has said that 'The Siriono have a very weakly developed tribal sense ... . When contacts between bands do occur ... relations are peaceful ... the Siriono are not a warlike people' [23].

In sum, the evidence from recent hunter-gatherer societies does not support the notion of primeval society being characterized by continual violent hostilities between neighbouring bands. Nor does this idea make much biological sense, in that constant intraspecific killing is not a feature of other mammalian species in their natural habitats, and on evolutionary grounds, we would not expect it to be. On the other hand, it would be equally unreasonable to assume that violent interaction, sometimes resulting in death, never occurred in hunter-gatherer societies. We know well that humans are capable of killing other humans, and there is no reason to suppose that they did not do so, under certain circumstances, in primeval society. On the whole, however, there is much to be said for the view that organized intergroup violence is much more a feature of communities living in societies of ecological phases two, three, and four than of hunter-gatherer society, and that tribal warfare became important after the domestic transition, when people had property, in the form of animals and stored grain, which could be coveted by others [24].

Although violence is apparently not a constant feature of intergroup relations in recent hunter-gatherer societies, there does seem to be a universal tendency to behave with suspicion towards strangers. In the Kalahari Desert, for example, when a group of Bushmen resident in an area is approached by a band of newcomers, the two groups may remain apart, although in sight of each other, for several days before contact is properly established. The first interactions are of a tentative nature and involve only a few individuals; but once this period of adjustment is over, the two groups may well share the same campsite.

Religion

All recent groups of primeval people practise some form of religion, but in comparison with the situation in early farming, early urban, and modern high-energy societies, their beliefs are usually relatively simple. With the exception of the Australian Aborigines, religious ceremonies and rituals are also simple, or almost non-existent. In hunter-gatherer societies there is no special class of priests or sharmans, although in matters of religion, as in other aspects of life experience, a leader often emerges. For example, when the Mbuti Pygmies set off for a hunt, one of the hunters may construct a small altar on which is placed a token offering to the spirit of the forest.

Whereas belief in spirits seems to be universal among hunter-gatherers, their attitudes towards these spirits vary greatly. The Mbuti Pygmies regard the spirit of the forest as benevolent, and when things go wrong they assume the spirit to be asleep, and they may perform a ceremony to gently awaken it. In contrast, the Siriono are terrified of the spirits which they believe lurk in the forest—so much so that they will not venture beyond the confines of their camps after dark.

Belief in 'ghosts' and in continued life of the spirit after death also appears to be almost universal among primeval people [25].

HUMAN EXPERIENCE

The comments which follow about the life conditions and biopsychic state of people living in primeval society are based partly on knowledge of contemporary or recent hunter-gatherers and partly on palaentological and archaeological evidence.

There is one important aspect of human experience in primeval communities which must be stressed at the outset. Although hunter-gatherers in different regions of the world are exposed to a wide range of different environments, the individual members of any particular society all experience very similar life conditions to one another, except for those differences attributable to age and gender. This is in marked contrast to the situation in early urban and modern industrial societies, which are characterized by major disparities between different occupational and socio-economic groups in such aspects of life experience as diet, size of dwellings, material wealth, behaviour patterns, and learning experience.

Diet and acquisition of food

The energy and chemicals required for growth, basal metabolism, and muscular work in hunter-gatherer communities are derived from food of both plant and animal origin. The relative proportion of meat to vegetables varies according to circumstances, although the major part of the diet, at least by weight, of most hunter-gatherers consists of vegetable matter [26]. The Eskimos are an obvious exception to this generalization, and in some areas their diet consists almost exclusively of mammalian meat—seals and walruses in the winter, caribou in the summer, and occasional fish and birds.

It is noteworthy that the intake of animal fat by hunter-gatherers, again with the exception of the Eskimos, is relatively low as compared with that of most people in modern high-energy societies (see page 263).

A wide range of different kinds of animals are used as a source of food by hunter-gatherers, although under normal conditions most groups do not make use of all the forms of animal protein in their environment. There is a general, but not universal, tendency to avoid the meat of carnivorous mammals, although carnivorous reptiles, fish, and birds are readily eaten. As in our own society, cultural traditions and taboos of various sorts exist with respect to the perceived suitability of foodstuffs. Archaeological studies in Tasmania have shown that fish was a common component of the diet of the coastal inhabitants of that island until about 3800 years ago [27]. After that, fish was apparently not eaten again until the arrival of Europeans. While all hunter-gatherers who have access to honey seem to be extremely partial to it, their taste for the bee larvae and other insects varies from culture to culture. Eskimos are said to have a particular liking for decomposed food, and those of the Bering Straits 'bury fish heads and allow them to decay until the bones become of the same consistency as the flesh. Then they knead the reeking mass into a paste and eat it'; they also enjoy the 'fat maggoty larvae of the caribou fly served raw ... deer droppings, munched like berries ... and marrow more than a year old, swarming with maggots' [28].

Cannibalism in recent hunter-gatherer societies is rare or non-existent. Some authors claim, however, that it has been practised in the past in some groups under conditions of extreme starvation.

In the case of foodstuffs of plant origin, the chief characteristic of the hunter-gatherer diet, in contrast to that of many other societies, is its diversity. It includes a wide variety of roots, tubers, seeds, berries, leaves, fruits, and nuts, its actual constitution varying from one region to another and from season to season. Although plant foodstuffs often represent the major part of the diet, no hunter-gatherer groups are vegetarian.

We may conclude, then, that *Homo sapiens* in its natural habitat is an unspecialized eater, capable of surviving on a diet consisting mainly of meat or on one consisting almost entirely of vegetable matter [29]. However, the typical diet of most hunter-gatherers living in relatively fertile areas of the world consists mainly of a wide range of different foods of plant origin plus some meat, which makes up about one-fifth to one-third of the diet by weight.

These comments apply to all age-groups in hunter-gatherer societies, with the exception, of course, of infants, whose sole source of food in the early months of life is human milk. The length of the period of breast-feeding varies according to circumstances, and weaning usually occurs at some time from about two to four years after birth.

With respect to the acquisition of food, there are a number of practices common to all recent hunter-gatherer societies. One of these is the universal tendency to bring food back to the camp base for eating. This is not to say that no food is consumed where it is collected or killed; but in most situations the bulk of the food acquired on a hunting or gathering expedition is taken back to the camp and shared out with other members of the group [30]. However, it has been reported that among the Hadza the females engaged in gathering activities may consume about half of the berries and roots that they gather before returning to the camp—where they divide what remains between themselves and the others in the camp. Similarly, it is not uncommon for a male Hadza who has killed a small animal to light a fire and to cook and eat it on the spot.

Another factor common to all recent hunter-gatherer societies is the sharp gender-based division of labour seen in food collection. With the exception of the collection of small animals like lizards and grubs, all hunting in most groups is carried out by males, although in some societies females may help as beaters. The gathering of vegetable material is primarily a female occupation, although males may assist on occasion [31].

Whether the division of labour in food acquisition is based on innate, genetically determined differences between men and women or whether it is a universal cultural effect is a matter of debate [32]. If it is innate, and thus the product of natural selection, it may be due to some selective advantage in the division of labour *per se*, or it could be a fortuitous consequence of sexual dimorphism which preceded hunting in human evolution. Like any other phylogenetic behavioural tendency, it may be reinforced or suppressed by cultural pressures. But whatever the mechanism behind the division of labour in hunter-gatherer society, this economic arrangement has been highly successful in terms of the satisfaction of the biological requirements of the species.

The actual techniques of hunting vary according to the nature of the habitat, the kind of edible food species available, and the technological history of the people. For hunting, the Australian Aborigines use spears and non-returning boomerangs, the Hadza and the Kalahari Bushmen use bows and arrows, Eskimos use spears and harpoons, and the Pygmies use bows and arrows as well as nets, which are useful in the forest. The hunting of larger animals is usually a co-operative effort, but individuals often hunt singly for smaller animals [33].

A common misconception, shared by some scientists and non-scientists alike, is the idea that human beings living in Palaeolithic times were always on the verge of starvation [34]. From the biological viewpoint, this is most improbable. It is much more likely that, as in the case of other species in their natural environments, most of the time most of the people had a plentiful supply of food and were well nourished. Certainly, this conclusion is well borne out by the observations of anthropologists on recent hunter-gatherer societies [35]. The average amount of time spent in food-collecting activities in Hadza society, for example, is about two hours per day per individual, and the population shows no signs of malnutrition. The same holds true for the Siriono, the Pygmies, the Eskimos, and the Australian Aborigines [36]. In the Kalahari Desert there are occasional seasons when the Bushmen show signs of malnutrition, especially ascorbic acid deficiency. Even in this relatively harsh habitat, however, the amount of time devoted to collecting food is usually not excessive. An individual may spend almost a whole day on the food quest, but most of the next few days will be spent doing other things, such as making ornaments or weapons, chatting with members of the in-group, or visiting relatives. It has been estimated that the ratio of the amount of somatic energy collected by hunter-gatherers in the form of food to the amount of energy spent in collecting it is from between eight and fifteen [37].

A feature of hunting and food-gathering which deserves mention is the fact that these activities appear to be generally enjoyed by the participants. This is to be expected, since in nature all animals enjoy behaviours which contribute positively to their survival or to their reproductive success. Presumably selection pressures operate against genotypes which do not enjoy, and hence tend to avoid, such activities. This important but somewhat neglected evolutionary principle has wide implications for the study of animal and human behaviour and for the understanding of human health and well-being. Of course, no single behaviour would be enjoyable all the time; excessive hunting, for example, beyond that necessary to satisfy nutritional requirements would have a number of disadvantages, ranging from an overkill of game to interference with other survival activities. In fact, the hunter-gatherer lifestyle involves considerable variety in daily experience—which includes a range of quite different behaviours of survival and reproductive advantage, all of which appear to be enjoyed when practised for limited periods.

Another aspect of food acquisition common to all recent hunter-gatherer societies is the fact that in no case does a group collect more vegetable foodstuff or kill more animals than are required for satisfying its metabolic requirements. Hunting and gathering may indeed be enjoyed, but no virtue is seen in expending unnecessary physical effort, and neither hunting nor gathering takes place unless the urge to do so is reinforced by hunger or by cultural pressures. The innate tendency in humankind towards indolence thus plays an essential role as a part of the set of mechanisms, behavioural and physiological, which have the effect of ensuring that human beings in their natural habitat do not over-exploit their resources, internal or environmental.

With respect to the consumption of food, this is characteristically a group activity among recent hunter-gatherers, and eating appears to have an important social as well as nutritional function. Customs differ from culture to culture with regard to the sharing of food, but in general the successful hunter and his family have first claim to the meat that he brings back, or at least to the best parts of it. There are no regular mealtimes, although daily patterns of feeding tend to develop in response to circumstances.

An essential activity associated with the food quest is the manufacture of weapons for hunting and of implements for cutting the meat and scraping animal skins. It is noteworthy and perhaps rather surprising that, as in the case of hunting and gathering, a fairly sharp gender-based division of labour applies to the manufacture of tools and weapons. The making of stone spear-heads, arrow-heads, axes, and most other tools is the work almost exclusively of males, as well as the manufacture of spears and bows and arrows. Ornaments may be made by men or women.

Sleep, rest, and physical activity

In hunter-gatherer societies people tend to rest or to sleep when they feel like it and when there is nothing better to do. The patterns of sleep and rest thus vary according to circumstances. For example, in the summer-time in the Kalahari Desert, Bushmen are active in the cooler parts of the day— that is, in the early morning and evening—and they rest and doze in the shade during the hot midday hours. In the winter they usually stay around their camp-fires until mid-morning and carry out their various subsistence activities during the middle part of the day.

Features of the sleeping pattern common to all recent hunter-gatherers are that most sleep is taken during the hours of darkness and that short periods of sleep also occur during daylight. The pattern, unlike that typical of contemporary Western society, is thus a polyphasic one.

The nature and level of physical activity varies somewhat according to circumstances. The picture can be summed up by describing the pattern as one which involves a good deal of walking and carrying and moderately frequent bouts of very vigorous muscular work in, for example, hunting and dancing.

Reproduction and sexual behaviour

Under this heading we will briefly consider various aspects of the reproductive process, including sexual behaviour, marriage patterns, fertility rates, attitudes to extramarital sexual relationships, and the rearing of offspring.

We can deal quickly with the subject of sexual behaviour, because remarkably little information is available on this aspect of life experience among primeval people. No investigation on the frequency of sexual intercourse and on the intimacies of sexual relationships have been reported for hunter-gatherers comparable to the numerous reports which have been published on sexual behaviour in modern Western society. We are even more in the dark with respect to the sexual behaviour of our ancestors of the late Palaeolithic period; the only thing we can be certain of is that there was some.

A point which deserves mention is the almost universal tendency for couples to seek privacy for their sexual activities. Among the Pygmies, for example, love-making is not performed openly and usually takes place away from the camp in the forest—although no great attempt is made to conceal it. This pattern is typical of hunter-gatherer societies in general, with the obvious exception of Eskimos, whose environment makes retirement from the camp for this purpose impractical. Nevertheless, in some groups, such as the Siriono, couples often indulge in advanced sexual play in the company of others.

The overriding impression from the literature on marriage patterns in hunter-gatherer peoples is that human beings are flexible in this regard, and this can be illustrated with a few examples. First, however, let us note some constant or universal features common to all primeval peoples. In all hunter-gatherer societies there is evidence of a positive physical attraction between males and females of reproductive age, and this results in sexual intercourse, cohabitation, pregnancy, and the birth of offspring. Another universal factor is the relative permanence of most unions between males and females, once a baby is born. In other words, such couples tend to stay together, and the 'husband-wife' relationship is recognized and respected by the rest of the group. However, while many such 'marriages' are permanent, others are not successful, and in all hunter-gatherer societies permanent separation also sometimes occurs and is usually a simple affair. Permanent homosexual relationships have not been reported. Another common denominator lies in the fact that, while the vast majority of marriages are monogamous, polygamy is permitted and occurs in all groups.

This is about the extent of the universal patterns among hunter-gatherers with respect to marriage; beyond this, there is much variability. In parts of the Kalahari region, for example, a female becomes married when she is between seven and nine years old, usually to a boy who is about seven years older than herself. However, no sexual intercourse is supposed to take place until the girl's breasts begin to develop, and in the meantime the husband may enter into a temporary relationship with a widow or a divorcee. In contrast, among the Hadza, males and females establish permanent relationships when they are about twenty years old, and any man and woman who cohabit are regarded as 'husband and wife', whether or not various socially accepted obligations associated with marriage have been carried out. Among the Siriono, while the majority of marriages are monogamous and permanent, sexual intercourse is permitted with a range of different 'potential spouses'. For a male, this would include his wife's sister, and for a female her husband's brother. Most adult individuals have from eight to ten potential spouses with whom he or she may have sexual relations without fear of censure (usually after the evening meal and before retiring to their hammocks).

The attitudes of hunter-gatherer people to extramarital sexual relationships among married people are also variable. Pre-marital sexual activity seems to be more or less the rule, except in the Kalahari Desert where the age of marriage precludes it in the case of females. Among the Kalahari people an occasional case of adultery may lead to some argument between two or more individuals, but it is considered by the group as a whole to be a private affair. However, if an individual persists in this kind of behaviour he or she eventually becomes generally unpopular and will probably be forced to leave the band. Among the Siriono, too frequent sexual intercourse between more distantly related potential spouses may lead to sexual jealousies and eventually to other troubles. For instance, a wife might be chased away by her husband and so become the object of social ridicule. A more extreme attitude is reported for the Demang in Malaya, who are said to punish adultery with death. Eskimos also regard secret adultery as a major offence, although wife-lending is an established custom.

Modesty, as expressed in terms of the covering of the genitalia, varies somewhat between different hunter-gatherer societies, and in some cases it does not exist. Among hunter-gatherers, and perhaps among all people, the Eskimos are unique in being equally at home in the company of others either fully clothed or totally naked. All clothes may be removed in the igloo. A similar transition from a state of being clothed to one of nudity in the presence of other people also occurs, for example, in Japanese baths and in Finnish sauna, but in both instances this is a more ritualized and less spontaneous behaviour than that of the Eskimos. It is abundantly clear that attitudes towards exposure of the body are not innate characteristics of the human species, and are entirely culturally determined [38].

Turning to the act of giving birth, there is little to say except that it is usually a relatively easy and uncomplicated process, at least in comparison with the situation in modern high-energy society. It is true that the mortality rates both for infants and for mothers are higher among hunter-gatherers than in our society; on the other hand, they are certainly lower than the rates would be in contemporary society were it not for the sophisticated medical facilities of the Western world.

No reliable figures have been published for fertility rates in hunter-gatherer societies. However, the most common picture seems to be for fertile couples to have three or four children, two or three of which can be expected to reach adulthood and to become parents themselves. There is no reason to doubt that, as in the case of other species, there were natural regulatory mechanisms operating in the evolutionary environment of *Homo sapiens* which tended to adjust birth rate in accordance with the availability of foodstuffs. It is likely, in fact, that several different physiological and behavioural mechanisms existed, but there is little justification for the view that the most important of these was infanticide [39]. It is worth noting that among hunter-gatherers females seldom begin producing offspring until they reach their late-teens or early-twenties, although menstruation usually begins at a considerably younger age.

Other aspects of life conditions

Although the overall population density of populations of hunter-gatherers is very low relative to later societies, the population density actually experienced by the individual in the camp may be quite high. It has been reported that members of bands often cluster very close together in their camps and there is a good deal of spontaneous and unselfconscious physical contact. In contrast to the situation in some later societies, however, the hunter-gatherer's experience of high population density is neither constant nor unavoidable.

The individual hunter-gatherer is a member of a very close-knit community, in which there is a free and constant exchange of information on matters of mutual concern and interest. There are few secrets among the members of the band. The individual is much aware of his or her responsibilities to the community, as determined by the social norms of the particular culture. Contact with people belonging to other bands or out-groups is not a feature of daily experience. There is, however, a good deal of contact and interaction with other species—often involving considerable excitement.

Reference was made earlier to the fact that Palaeolithic people devoted considerable effort to creative work with their hands. Similarly, it is evident that among recent hunter-gatherers, most if not all adults are engaged in some form of creative activity on most days—mainly, but not entirely, taking the form of shaping objects by hand. It is reasonable to conclude, therefore, that creative behaviour is an important characteristic of life experience of individuals in the primeval situation. While it is easy to imagine, in terms of evolutionary theory, that a certain capacity for learning creative manual skills might have been a selective advantage in the Darwinian sense, it is less easy to explain why the creative potential of humans appears to be so much greater than would have been necessary to ensure survival and successful reproduction in the hunter-gatherer situation.

In the words of Marshack: '... what is the adaptive value of being able to make music like Mozart?' [40].

The life conditions of children in primeval societies reflect the spontaneous nature and the relatively uncomplicated societal organization of the communities. Babies are kept close to their mothers for the first year or so of life—but after this they may be left at the camp to be minded by relatives when the mother goes gathering. Generally speaking, it seems that for the first few years children are indulged by their parents and other members of the band and they are seldom severely reprimanded or punished for transgression of norms. However, customs differ from one hunter-gatherer society to another with respect to the control of behaviour in children over the age of five or six years. In most of the societies studied, the *laissez-faire* attitude persists, but in some societies older children are severely punished for misdemeanours.

The learning experience of children does not involve any formal programme of teaching. The process appears to be entirely spontaneous, and is based on such universal behavioural characteristics as the tendency to mimic and to seek approval or praise through doing things considered by the group to be good. Much of childhood learning takes the form of listening to and observing and copying slightly older children, who are in turn learning from their older siblings or peers [41]. The playing of games based on mimicry of adult behaviour is universal among hunter-gatherer children [42]. Aggressive behaviour is not uncommon, although it seldom involves injury [43].

A general point must be made about behaviour in hunter-gatherer societies. The descriptions provided by anthropologists of the activities of hunter-gatherers, and the observations in this chapter, are consistent with the concept of *common behavioural tendencies* which was introduced in the last chapter. All the common behavioural tendencies listed in Table 3.1 (p. 00) are characteristic of behaviour in primeval societies. Especially important in relation to social behaviour are the tendencies to seek the approval of the members of the in-group and to try to avoid their disapproval and ridicule. The criteria for approval and disapproval are largely (but not wholly) culturally determined, and some variability therefore exists in the specific behaviours which reflect the underlying common behavioural characteristics. However, because of the fundamental similarities in the ecology and lifestyle of hunter-gatherer groups, the criteria of approval and disapproval are likely to be more similar among them than they are among different cultures and subcultures in modern times.

Physical health

In accord with the Hippocratic postulate, the patterns of health and disease in hunter-gatherer societies are variable, reflecting different ecological conditions. Nevertheless, some generalizations are justified [44]. There are, broadly, three sources of information on this matter. First, there are some data from palaeobiological studies based on the remnants of bones of palaeolithic people. Second, we have the general impressions of anthropologists who have worked with recent hunter-gatherers. Third, some information is available from a few isolated biomedical studies on specific aspects of physiological and health parameters in recent hunter-gatherer societies. However, no extensive and comprehensive health surveys have been carried out.

As will be appreciated, the palaeontological evidence is far from satisfactory, partly because it will reveal only those forms of maladjustment which affect the structure of bones. Moreover, apparent deviation in fossils from normality is often difficult to interpret in pathological terms. This fact, however, has not deterred many authors from some bold speculation about the causes of apparent abnormalities in the bones of some of our prehistoric ancestors.

With respect to the impressions of anthropologists the picture is complicated by the fact that no recent hunter-gatherer population is free of contact with potential sources of disease associated with village and urban society. Nevertheless, there are numerous statements in the literature to the effect that most people in hunter-gatherer communities appear to be in a remarkably good state of health. According to all reports, undernutrition, malnutrition, and obesity are, in normal circumstances, rare or non-existent; and before contact with people from phase three and four societies, hunter-gatherers were free from such infectious 'diseases of civilization' as cholera, typhoid, typhus, measles, smallpox, influenza, and the common cold.

As would be anticipated, there are exceptions to this broad generalization. For example, in the Kalahari Desert, which is a relatively unfavourable habitat compared to that where most hunter-gatherers have lived, in years when the rains come late the condition of the people is said to deteriorate markedly in the middle of the summer. They lose much of their body fat, develop sores easily from minor body wounds, and a fungus infection of the scalp sometimes makes its appearance. Conjunctivitis also becomes much more common, possibly associated with a deficiency of vitamin A in the diet.

With regard to such organic disorders as appendicitis, duodenal ulcers, and cancer, all that can be said from the literature is that reference to the existence of these conditions is virtually absent, and some anthropologists remark upon the fact that they are seldom, if ever, observed.

Hunter-gatherers are, of course, susceptible to the contagious diseases of modern urban society when they come in contact with them. Indeed, their resistance to these conditions is in general much lower than that of populations from which the infections are brought. There are numerous accounts in the literature of the devastating effects of infectious diseases introduced into communities of hunter-gatherers and other isolated peoples by explorers, anthropologists, and other itinerants from the industrial world. In such cases the mortality rate is often very high, affecting all age-groups, and disrupting the whole social fabric.

Another relevant impression of observers of recent hunter-gatherer societies is that the average age of populations is often low as compared with modern high-energy societies. There are two reasons for this. First, the primeval pattern of life is a much more hazardous one than that typical of our own society. The chance of severe injury (acquired in hunting or in other ways) is greater, and the likelihood of a serious wound becoming infected and of gangrene or septicaemia setting in is relatively high. It is also probable, at least at some periods in human history, that large carnivores represented an important cause of death. Another factor is that disablement due to injury or illness is of considerably greater survival disadvantage in primeval society than the same degree of incapacitation would be under the protective conditions of modern civilization. Furthermore, hunter-gatherers have not had the benefit of the artificial antidotal measures, in the form of surgery, antibiotics, and other pharmaceutical products, which are so effectively brought into play to counter physiological disturbances in the high-energy societies.

The specific aspect of the biopsychic state which has received most attention in biomedical investigations on hunter-gatherer populations is blood pressure. A small number of well documented studies in both hunter-gatherer and subsistence farming communities show that, unlike the situation in modern urban society, blood pressure tends to remain constant in adults after the age of about twenty years (see p. 242). Another aspect of health that has been the subject of some investigations is the state of the teeth. The factor which stands out clearly in these studies is that dental caries is only rarely encountered in hunter-gatherers. Palaeontological evidence suggests that this was also the case in primeval society in the Palaeolithic period. Periodontal disease has been reported, but it is not common.

This is about the extent of the information available on health and disease in hunter-gatherer societies. Beyond these facts, some clues to the general state of health of primeval people can be deduced from fundamental biological principles. Our understanding of evolutionary processes and our observations on other species in their natural habitats would lead us to expect that most human beings living in hunter-gatherer societies would be mentally and physically healthy. Nevertheless, because of certain in-built hazards in the primeval life conditions and the lack, as compared with the modern situation, of societal molly-coddling and of the antidotal measures of medicine, the chances of death at any stage in the life cycle are considerably higher than in contemporary Western society. The general picture, under moderately favourable conditions, is one of good health, but of short lives coming to an end after brief illnesses.

Before leaving the subject of health, a few words must be said about hygiene. On the whole, it seems that standards of hygiene among hunter-gatherers vary somewhat from one society to another, and they are certainly different from those of most modern urban communities. As we shall discuss later, there are good reasons why we have had to modify our habits to suit the new conditions under which we live. In most hunter-gatherer societies, solid excrement is deposited more than a hundred metres away from the camp site. In the case of the Siriono, however, fear of the spirits of the forest leads them to defecate at night immediately outside their huts. This behaviour may well account for the fact that intestinal parasites represent an important cause of ill health among these people.

It is self-evident that primeval people do not wash their bodies with soap or detergent. Some of the Bushmen in the Kalahari Desert are careful to wash their hands before eating but, unlike the practice in modern Western society, they do so in their own urine (which, of course, is normally bacteriologically sterile when voided). Urine was also used in the past by the Eskimos of Greenland for washing their hair.

Intangible aspects of the biopsychic state

As discussed in the last chapter, the biopsychic state of an individual includes a range of attributes which are difficult, if not impossible, to measure quantitatively, but which are nevertheless of great importance as influences on behaviour and as aspects of health and well-being. These include values, feelings, and the components of knowledge. The following brief comments on this aspect of human experience in hunter-gatherer society are based on my own interpretation of the impressionistic descriptions provided by anthropologists.

With respect to values, there is a good deal of variability among recent hunter-gatherers reflecting differences in the cultural backgrounds of the various groups. Nevertheless, some commonalities can be recognized. For instance, such traits as loyalty towards members of the in-group and gentle behaviour toward young children seem to be universally regarded as good, and the opposite behaviours are regarded as bad. There does not, however, appear to be any universal concern for the well-being of people beyond the in-group.

Turning to the difficult question of feelings, in particular those which relate to health and well-being, I suggest that the following are especially characteristic of hunter-gatherer society: a sense of personal involvement, a sense of purpose, a sense of belonging (to a place and to an in-group and community), a sense of responsibility, a sense of interest, a sense of excitement, a sense of challenge, a sense of satisfaction, a sense of comradeship and love, a sense of confidence.

It would not be sensible to imagine that all individuals experience all these positive feelings all the time. Indeed, such a state of mind might not be the most desirable in terms of well-being. The principle of the optimum range (see page 37) operates with respect to most, if not all, of these items. If a sense of satisfaction were never relieved by periods of dissatisfaction and occasional frustration, life would indeed be dull. But by and large, 'negative' feelings, such as sense of alienation, sense of anomie, sense of loneliness, sense of boredom, and sense of resentment do not appear to be common features of primeval people under usual circumstances.

Optimum life conditions for human beings

According to the principle of evodeviation, when animals are exposed to life conditions which differ from those to which their species is genetically adapted through evolution, signs of phylogenetic maladaptation are likely to be manifest. The hypothesis was put forward that, in the case of the human species, the principle of evodeviation applies not only to the physical or material aspects of life conditions, but also to less tangible behavioural and psycho-social aspects. It follows, assuming this hypothesis to be correct, that consideration both of the material and of the behavioural and psycho-social aspects of the life conditions of primeval people could provide important clues to the nature of the biologically determined or *universal health needs* of the human species.

With these thoughts in mind, a list has been prepared which is a summary both of the life conditions of hunter-gatherers and, accepting the principle of evodeviation and the hypothesis that it applies to intangible aspects of life experience, of the *optimum life conditions* for members of the human species in general. The list, which is presented in Table 4.1, begins with the more tangible material aspects of life conditions and ends with the more intangible psycho-social and behavioural aspects. With respect to many of the postulated health promoting aspects of life conditions, including the intangible aspects, the principle of the optimum range is applicable; that is to say, too little or too much of a given condition may be detrimental to health.

Table 4.1 *Life conditions conducive to health in* Homo sapiens

---

Clean air (i.e. 'palaeolithic air'—not contaminated with hydrocarbons, sulphur oxides, lead, etc.)

Environmental temperatures within the range of those experienced in the 'natural habitat'*

Exposure to visible light (duration and intensity) within the range of that experienced in the natural habitat

Noise levels within the range of those experienced in the natural human habitat

Diet:
    Calorie intake neither less nor more than metabolic requirements Social norms which allow the individual to eat when hungry, but which do not encourage overconsumption of calories in response to ritual, habit, or, for example, boredom.
    Foodstuffs providing the full range of nutritional requirements for the human organism. In the primeval situation this is usually provided by a diverse range of different foodstuffs of plant origin and some lean meat (cooked).
    A diet which is balanced in the sense that it does not contain an excess of any particular kind of chemical constituent or class of foodstuff.
    Foodstuffs with a physical consistency of that of natural foods and containing fibre.
    Foodstuffs devoid of potentially noxious contaminants or additives.

Clean water—free of contamination with chemicals or pathogenic microorganisms

Minimal contact with microbial or metazoal parasites and pathogens

An effective emotional support network providing a framework for spontaneous care-eliciting, care-receiving and care-giving behaviour.

Frequent interaction on a daily basis with members of the extended family and in-group on matters of mutual interest and concern.

Opportunities and incentives for small-group interaction on projects of mutual interest and concern

A social environment which confers responsibilities and obligations on the individual towards the in-group

Opportunities for the individual to move spontaneously and freely from one small group to another, and to and from a state of solitude.

Levels of sensory stimulation which are neither much less nor much greater than those of the natural habitat.

A pattern of physical work which involves some short periods of vigorous muscular work and longer periods of medium muscular work, but also frequent periods of rest

A polyphasic sleeping pattern, and the opportunity to rest or sleep in response to the urge to do so

Opportunities and incentives for the learning and practice of manual skills and for creative behaviour in general

Opportunities and incentives for active involvement in recreational activities

An environment which has high interest value and in which changes of interest to the individual are continually occurring (and at a rate which can easily be handled by the human psyche)

Opportunities for considerable spontaneity in behaviour

Considerable variety in daily experience

Short goal-achievement cycles

Aspirations of a kind likely to be fulfilled

An environment and lifestyle which are conducive to a reasonable degree of:
    a sense of personal involvement
    a sense of purpose
    a sense of belonging
    a sense of responsibility
    a sense of interest
    a sense of excitement
    a sense of challenge
    a sense of satisfaction
    a sense of comradeship and love
    a sense of enjoyment
    a sense of confidence
    a sense of security

---

* The term 'natural' habitat is used to mean habitat with the characteristics of those inhabited by human beings in ecological phase one societies.

NOTES

1. For discussion on the use of data about modern hunter-gatherers for interpreting remains of prehistoric societies, see Peterson 1971.
2. Except where otherwise indicated, the information provided in this chapter is derived from the following sources: Holmberg 1950; Thomas 1959; Weyer 1932, 1959; Turnbull 1961, 1965; Tobias 1964; Silberbauer 1965; Berndt and Berndt 1964; Woodburn 1968a, 1968b, 1980; Lee 1969; Hayden 1981; and various papers in *Man the Hunter* (Lee and DeVore 1968) and in *Woman the Gatherer* (Dahlberg 1981). Except for the Siriono, a group living in the forest in the central part of South America, data relating to American Indians are not used.
3. de Lumley 1969; see also Klein 1973.
4. Clark and Haswell 1970; Leach 1976; Peterson 1978.
5. Jones 1969.
6. Woodburn 1980, p. 100.
7. Newcombe 1976.
8. This is calculated from the figure for photosynthesis in tropical savannah of 4800 $Kcal/m^2/year$ and in grassland of 2000 $Kcal/m^2/year$ (see Golley 1972), and assumes a population density of one per 500 Ha (which is twice the population density in the Kalahari) and a daily energy use by humans of 2000 Kcal. The proportion of energy flowing through the human population would thus be 0.003 per cent in tropical savannah and 0.007 per cent in grassland.
9. There has been a good deal of theoretical discussion and debate in recent years on the relationships between ecological conditions and societal organization in hunter-gatherers. See, for example, the review article by Barnard 1983.
10. See Barnes 1970. For a discussion on *why* people stay together in groups, see the discussion in Hayden 1981, pp. 360-70. The fact that an individual who lives as a member of a group—rather than in isolation—is less likely to be attacked by predators is another important factor not taken into account in this discussion. See also Tiger 1969; Tiger and Fox 1972.
11. Reynolds 1962.
12. There has been considerable literature recently about the role of women in hunter-gatherer societies; see, for example, *Woman the gatherer* edited by Dahlberg (1981). The main theme of this book is the importance of women as gatherers, transporters, and processers of food.
13. Draper 1975, p. 109.
14. See, for example, Dahlberg (1981, pp. 16, 17) who refers to wife-beating among Australian Aborigines. On the other hand, Rohrlich-Leavitt, Sykes, and Weatherford (1975) have claimed that status and power in Australian Aboriginal women is higher than earlier studies have suggested.
15. In most groups, women do not engage in hunting large animals, except sometimes in the role of beaters. In the Agta in the Philippines, however, women do participate in the hunt. See Dahlberg (1981) for a review of the literature and a discussion.
16. Writing about the Pygmies, Turnbull (1965, p. 29), refers to 'that elusive principle that binds together the apparently lawless and un-kin-conscious forest society'. Whatever the nature of this elusive principle, it is certainly very powerful, and all authors agree that hunter-gatherers are very fearful of loneliness and long

periods of solitude. Thus it seems that companionship, or rather the desire for companionship, is the most powerful force acting towards the unity and coherence of the groups and subgroups in hunter-gatherer society. The importance of approval and a sense of belonging to the !Kung Bushmen has been well described by Marshall (1976, p. 351), who has written: 'Their desire to avoid both hostility and rejection leads them to conform to a high degree to the unspoken social laws. I think that most !Kung cannot bear the sense of rejection that even mild disapproval makes them feel'. Nevertheless, a good deal of argument occurs among !Kung people and occasionally serious quarrels flare up.

17. Special mention should be made of the Ik, a hunter-gatherer group in East Africa written about by Turnbull (1972). These people had been forced to leave their traditional habitat as a consequence of the establishment of a National Park and were experiencing a great deal of difficulty rehabilitating themselves in their new home in the highlands. According to Turnbull, their personal interactions were mainly antagonistic and individuals showed little concern for the well-being of other members of their in-group. This was seen as a reaction to the social disturbance and to the difficult ecological situation in which they found themselves (but see also comments on the Ik by Wilson, 1975, pp. 549–50).

18. Woodburn 1980, p. 796.

19. See chapter 3. See also Wilson 1978.

20. Silberbauer 1965, p. 52.

21. Woodburn 1968*b*, p. 105.

22. Turnbull, 1986, personal communication.

23. Holmberg 1950, p. 51.

24. For a useful discussion on this point, see Reynolds 1966.

25. Simmons 1945.

26. Lee has surveyed data on 58 societies which depend entirely on hunting, gathering, and fishing for subsistence and concludes that 29 depended mainly on gathering, 18 depended mainly on fishing, and 11 depended mainly on hunting. With only one exception, in all groups at least 20 per cent (by weight) of the diet was meat. In general, the colder the climate, the higher the proportion of meat in the diet (Clarke and Hindley 1975, pp. 30–1).

27. Jones 1978.

28. Weyer 1932, p. 72. It is not surprising, in the light of these eating habits, that botulism has been a relatively common cause of death among Eskimos (Cockburn 1971, pp. 47, 48).

29. See the excellent review by Hayden 1981.

30. For a discussion on food preparation and food-sharing in hunter-gatherers, see Hayden 1981.

31. Hunn 1981; Dahlberg 1981.

32. See, for example, Wilson 1978, pp. 138–9 and Dahlberg 1981 (introductory chapter).

33. See the discussion by Woodburn 1980, p. 801.

34. See the discussion in Harris 1978, pp. 10–11.

35. The fact that the search for food does not take up a great deal of the typical hunter-gatherer's time is emphasized by Sahlins 1972.

36. These statements refer only to the actual collection of food; the time spent preparing the food for consumption is not taken into account (see the discussion

by Hayden 1981). However, even when food preparation is considered, it is clear that most hunter-gatherers have plenty of spare time.

37. Carruthers, Clark, De Smet, Freckleton, McClelland, Pike and Roe 1973; Leach 1976.
38. Henriques 1959, pp. 36, 46, 76.
39. For discussions on possible cultural means of controlling hostility in hunter-gatherers, see Divale 1972 and Bates and Lee 1979.
40. See Lewin 1979.
41. For an especially good description of the life experience of children in a hunter-gatherer society, see Silberbauer 1965. See also an excellent account of learning among the Siriono by Holmberg 1950.
42. Doke 1937.
43. Eibl-Eibesfeldt 1975.
44. For discussions on health and disease in hunter-gatherer groups, see (apart from references given in note 2) Dunn 1978; Powles 1973.

5

THE EARLY FARMING PHASE

BACKGROUND

The transition from an economy based on gathering and hunting to one based on farming involved some profound changes both in the ecological relationships between human populations and the biosphere and in the actual life experience of individual human beings. This chapter is concerned with these changes and with the principal features of the patterns of culture–nature interplay in the early farming phase of human history.

The early farming, or second ecological phase, of human existence is taken to include all farming societies from the time of the introduction of agriculture and horticulture around twelve thousand years ago up until the introduction of the energy-intensive methods of farming associated with the fourth, high-energy ecological phase. It includes the cultivation of cereals, legumes, fruits, and root crops for food production as well as the farming of sheep, goats, cattle, swine, chickens, ducks, and other animals, whether kept in enclosures or ranging in flocks or herds over the countryside. Since the time of the first cities, some farming communities have existed alongside early urban, or phase three, societies. While the urban populations were dependent upon the farming societies and influenced them in important ways, the early farming economy nevertheless retained its essential ecological and biosocial characteristics throughout the whole period.

Although the range of farming methods developed over this period in the many different ecological and climatic settings has been very wide, these methods were no more than variations on a single basic theme. This theme consisted, in essence, of human beings putting physical effort into cultivating edible plants and breeding and tending edible animals—with the result that a higher proportion of the chemical energy fixed by photosynthesis in a given area flowed through the human population [1]. As a consequence of this increase in food supply a larger number of people could be supported in this area than had been possible under primeval conditions.

It is not known when people first began to experiment with cultivating plants and domesticating animals for food production; but we do know that, by around ten thousand years ago, mixed farming was a well established aspect of the economy of some populations in south-western and south-eastern Asia. This deliberate, conscious, and purposeful manipulation of natural processes by human beings represented the beginning of a new era in the ecological history of the human species, an era characterized by a progressive intensification of culture–nature interplay [2]. Since this new way of life involved the domestication of plants, animals, and human beings themselves, the transition from the hunter-gatherer economy to the early farming economy is referred to here as the *domestic transition*.

Where and when?

There has been much speculation about whether or not the domestication of plants and animals for food production began in one location and spread to others, or whether it arose independently in different parts of the world. Today, a minority view holds that farming in all its forms originated from a single spot in south-west Asia, and that even the cultivation of crops for food production in America was the result of early contacts between Amerindians and people from Europe and Asia. Certainly, the earliest definite evidence of farming based on the cultivation of cereals, notably wheat and barley, and the domestication of goats, sheep, and cattle, comes from relatively high country extending from Greece in the west to an area just south of the Caspian Sea in the east, and from the uplands which flank the valleys of the Tigris and Euphrates Rivers. Most contemporary authors, however, believe that farming originated independently in a number of locations [3].

It is important to appreciate that the change-over from an economy based on hunting and gathering to one in which populations depended mainly on farming for their sustenance was by no means a sudden happening. It was a 'revolution' only on the biological time-scale and in terms of its ultimate importance to humankind and to the rest of the biosphere; but on the time-scale of human experience it was a very gradual process indeed. It is now apparent that cereal grains had been an important component of the diet of some human communities for a long while before farming developed as the mainstay of the economy of populations in the Near East. Some recent excavations have shown that people living in southern Egypt as long ago as seventeen and eighteen thousand years were collecting and grinding grain from barley. The tools of these people included not only grinding stones, mortars, and pestles, but also what appear to be blades of sickles, presumably used for reaping the barley and perhaps other grasses. The possibility has even been suggested that the inhabitants of this area at the time were already cultivating barley, although there is so far no direct evidence for this [4].

About 11 000 BC there lived in Palestine an interesting group of people who are referred to as the Natufians. They still depended on hunting and gathering for their food supply, although many of them lived in houses arranged together in small hamlets. Like the people in southern Egypt four or five thousand years earlier, they used sickles for cutting the stems of wild cereals, which apparently contributed substantially to their food intake, and again the possibility cannot be ruled out that they were deliberately cultivating these cereals. The Natufians were artistically inclined, and they liked to decorate themselves with pendants and necklaces and on occasion they wore fine shell-bead head-dresses [5].

Gradual changes were occurring in the climate in the Near East around this time. Although in Europe there had been a pause in the retreat of the ice sheets, south-western Asia was already entering the post-glacial phase. Following a cold and dry period, there had been a gradual rise in temperature and an increase in rainfall, and the forests of the Mediterranean coastal areas were expanding eastwards. Where the forest encroached upon the steppes to the east and south there was an intermediate zone of open woodland. Excavations at a place called Abu-Hureyra in this zone show that hunter-gatherers were living at the site from around 10 000–8800 BC, and it is not impossible that they practised some agriculture [6]. The site was reoccupied about 7500 BC, and it soon became one of the largest, if not the largest, settlement in the Levant at the time. It had a population of several thousand, but it covered an area of no more than twelve hectares. The people cultivated both einkorn and emmer wheat, barley, and lentils. The animal bones found at the site are mainly those of gazelles, goats, and sheep—the same species that had been hunted by the inhabitants of the settlement two thousand years earlier. It is believed, however, that these animals were now being herded rather than hunted. As time progressed, the proportion of gazelles slaughtered appears to have fallen drastically, and soon after 6000 BC cattle and pigs had been added to domesticated animals kept at the site. Nevertheless, the people still derived a significant proportion of their diet from hunting and gathering until soon after 6000 BC when they came to rely almost exclusively on farming.

The houses at Abu-Hureyra were rectangular, consisted of several rooms, and were made of mud-brick. They were built close together and were separated only by narrow lanes or courtyards. By around 7000–6000 BC trading relationships had become established, not only within the local region but also with other regions. For example, the inhabitants regularly received the volcanic glass, obsidian, as well as jadeite, serpentine, agate, and malachite from Turkey. They also received turquoise from the Sinai, soap-stone from the Lagos mountains, and cowrie shells from the Mediterranean or the Red Sea.

It is clear also from excavations at other sites, including Jericho in Palestine, Jarmo in Iraq, and Çatal Hüyük in southern Turkey [7], that significant changes were occurring throughout this region around 8000–7000 BC. The excavations at Çatal Hüyük have proven to be exceptionally interesting. Çatal Hüyük is a mound in Anatolia which covers about twelve hectares. It is situated on the Konya plain just north of the Taurus Mountains and about 300 Km south of modern Ankara. Excavations carried out at the site in the early-1960s show that between 7000 and 6000

BC there existed a thriving community here which, at its peak, must have consisted of around 5000 people [8]. The economy of this settlement was primarily based on the cultivation of barley, wheat, peas, and lentils, and the breeding of sheep, goats, and dogs and, later in the period, cattle. Pigs do not seem to have been domesticated. However, the people still hunted wild cattle, red deer, wild asses, sheep, pigs, and leopards.

Twelve building levels have been uncovered, spanning a period of about one thousand years. There is no sign of any invasion or sudden change of culture during this period, and accidental fires may have been responsible for the necessity for rebuilding. The mud-brick houses were rectangular and usually had a store room attached which contained a bin for holding grain, and in each case the entrance was a hole in the roof. The buildings were very close together, and little sign has so far been found of lanes and passages, although courtyards, which were apparently used as a depository for rubbish, are plentiful.

It is evident that the inhabitants of Çatal Hüyük, or at least some of them, had plenty of time for activities other than farming and hunting and house-building. Excavation of burials have revealed that the women and children wore necklaces, armlets, bracelets, and anklets of beads and pendants made from a wide variety of different kinds of stones and shells and, toward the end of the period of occupation, from copper and lead. It seems that cosmetics were widely used, consisting of red ochre, blue azurite, green malachite, and possibly galena; and mirrors were made of highly polished obsidian. There is also evidence of a high level of technical competence and sophistication in the manufacture of textiles. At least three different types of weaving were practised, apparently using wool from sheep.

The complexity of the material culture of these people can be brought home by the following quotation from an article by Mellaart:

It is singular that with all these products of human workmanship we have found so few traces of the workmen ... . We have thousands of finely worked obsidian tools but only two small boxes of chips, thousands of bone tools but no piles of waste or splinters. Somewhere in the mound there must be the workshops of the weavers and basketmakers; the mat-makers; the carpenters and joiners; the men who made the polished stone tools (axes and adzes, polishers and grinders, chisels, maceheads and palettes); the bead makers who drilled in stone beads holes that no modern steel needle can penetrate and who carved pendants and used stone inlays; the makers of shell beads from dentalium, cowrie and fossil oyster; the flint and obsidian knappers who produced the pressure-flaked daggers, spearheads, lance heads, arrowheads, knives, sickle blades, scrapers and borers; the merchants of skin, leather and fur; the workers in bone who made the awls, punches, knives, scrapers, ladles, spoons, bowls, scoops, spatulas, bodkins, belt hooks, antler toggles, pins and cosmetic sticks; the carvers of wooden bowls and boxes; the mirror makers; the bowmakers; the men who hammered native copper into sheets and worked it into beads, pendants, rings and other trinkets; the builders; the merchants and traders who obtained all the raw material; and finally the artists—the carvers of statuettes, the modellers and the painters. ...

The unusual wealth of the city of Çatal Hüyük, as manifested by this great variety of sophisticated workmanship, is a phenomenon as yet without parallel in the Neolithic period. .... . One cannot possibly be wrong in suggesting that it was a well-organized trade that produced the city's wealth. Moreover, it appears likely that the trade in obsidian was at the heart of this extensive commerce. This black volcanic glass, which first appeared in the preceding Mesolithic period, became the most widespread trading commodity during the Neolithic period in the Near East. It has been found in the 'proto-Neolithic' and prepottery Neolithic periods at Jericho; it occurs as far south as Beidha near Petra; it reached Cyprus in the sixth millennium. The origin of this obsidian, which was the best material of the time for cutting tools, was also certainly central Anatolia, and it is extremely likely that the city of Çatal Hüyük controlled this source and organized the trade' [9].

At this point, a comment on perspectives is not out of place. I referred above to the fact that significant changes were taking place in the Near East between 8000 and 7000 BC. This is a short period in the history of humankind as a whole, and an even shorter period in the history of life on earth. But in terms of human experience, the period spans some forty generations, and the degree of change from one generation to the next may well have been almost imperceptible.

Pastoral nomads

It is not known when pastoral nomadism, in which small groups of people moved across the landscape with their flocks or herds of domesticated animals, first came into existence. At one time it was thought that this lifestyle had evolved directly from hunter-gatherer society. However, most authorities now take the view that it came later than crop cultivation, and that from the beginning it was practised mainly on the margins of agricultural areas [10].

The beginnings of farming in other parts of the world

It is now clear that people in south-east Asia have been cultivating plants and keeping animals for food production for a very long while. However, the climate in this region was very different from that in south-west Asia, and so were the local vegetation and fauna. Accordingly, the first farmers in these parts were not cultivating wheat, barley, and oats, but cucumber, peppers, bottle gourds, and various kinds of beans. When this practice began is not known, but one author believes that people living in Thailand around 10 000 BC already had an advanced knowledge of horticultural techniques [11]. By 3000-4000 BC, rice was well established as a cereal, and pigs and cattle were domesticated. In all probability, the transition to farming was as gradual a process in this part of the world as it seems to have been in south-western Asia.

On the American continent the cultivation of plants for food may go back about 11 000 years, but certainly by 5000 years ago the cultivation, in Mexico, of maize, squash, chili pepper, and avocado was well established. Farming in Peru, where the potato eventually became the staple food, appears to have become established during the second millennium BC [12]. The people in America did not keep and breed animals as a source of food.

Why?

The facts discussed above provide a general picture of what was happening as human societies gradually shifted from hunting and gathering to farming; but they do not tell us *why* people made this change in their lifestyle and economy. Explanations put forward for the introduction of farming range from the view that it was a natural and inevitable consequence of human intelligence and inventiveness to the suggestion that people only turned to farming when forced to do so—when, for one reason or another, the amount of food that could be gathered by hunting and gathering was insufficient to support them [13]. The various arguments for the different hypotheses will not be discussed here; instead I will present the explanation which, in my view, best fits both the known facts and our understanding of human nature.

The post-glacial conditions in south-west Asia and south-east Europe were generally favourable for the hunter-gatherers of the region, so that some of them could afford to lead a relatively settled existence, especially in lightly wooded areas with a good water supply. Several significant consequences of this development can be envisaged: for instance, here and there wild grain accidentally spilled on the ground around settlements will have sprouted and produced more grain. Furthermore, in the relatively sedentary conditions people had the opportunity to experiment with sowing grains, not necessarily because of any urgent need for additional food, but rather perhaps simply for amusement. Patches of earth were set aside for these experiments, and when they were successful the grain was harvested and eaten.

We can imagine, too, that now and again orphaned lambs, kids, or calves may have been brought back alive to the settlement, and the sedentary conditions of life would have allowed sympathetic individuals to care for these young animals, which would then have grown up to be relatively tame and manageable [14]. The initial motivation may well have had nothing to do with food production, but when the animals grew up, their food potential would have been appreciated; and we can imagine differences of opinion within the community about whether or not they should be killed and eaten.

As time went on and the techniques of cultivation of grain improved, and as the number of animals kept in and around the settlements increased, these domesticated sources of food came to constitute an increasing proportion of the diet of the inhabitants. Meanwhile, because the new lifestyle was somewhat less hazardous and transient, illness was less of a survival disadvantage, the population was growing; and this larger population could now be supported by the additional food provided by farming. Thus it transpired that the cultivated plants and the domesticated animals came to be *needed*, and farming became a serious business. As populations became increasingly dependent on food grown at home, more deliberate effort was aimed at improving the techniques of food production.

If this scenario is a moderately accurate one, it provides an early example of a biohistorical principle of great importance—*the principle of technoaddiction*. In human history it has frequently been the case that, when new techniques have been introduced into a society, they have not been really necessary for the satisfaction of the survival and health needs of the population. Sometimes they have been introduced simply for curiosity, and sometimes because, in one way or another, they have benefited a particular individual or group within the society. With the passing of time, however, societies reorganize themselves around the new techniques and their populations gradually become more and more dependent on them for the satisfaction of basic needs. Eventually a state of complete dependence is reached. Clearly, already by 7000 BC, the population of Çatal Hüyük, and that of Abu Hureyra, had become entirely dependent on farming for their survival, despite the fact that a small part of the food supply still came from hunting. The dependence of the populations of high-energy societies on machines driven by fossil fuels is a more recent example.

The situation in Europe

At the same time that the Natufians and the people of Abu-Hureyra were settling down in small hamlets, the inhabitants of northern and western Europe were leading a relatively mobile existence. These people, the Magdelenians as they are now called, were specialized hunters of reindeer, the herds of which they followed over the open grassland which covered the plains. In some areas in Germany up to 99 per cent of the animal bones which these people left behind were from the reindeer. The Magdelenians not only consumed the meat of these animals, but they used other parts of their bodies as a source of raw materials for clothing, tents, ropes, and various tools [15]. Then, relatively suddenly, around 8300 BC, the ice sheets in Scandinavia made their final retreat. In the course of a few generations the temperature increased and forests invaded the open grazing lands, and the deer hunters of the northern European plains were forced to adapt to a new ecological situation. They did so by reverting to an economy based on mixed hunting and gathering and by developing new techniques for extracting food from rivers and the sea. This cultural response to ecological change resulted in people living a somewhat more sedentary existence close to or within the forests, especially along river valleys and the sea-shore.

The people in north-western Europe, including Britain, who adopted this *mesolithic* economy are known as the Maglemosians, and their counterparts in the east as the Kunda. Their cultures have been described as extraordinarily vigorous, and they introduced an impressive series of technical innovations appropriate to the changed economy. Like the Natufians, they were an artistic people, decorating objects of bone, antler and amber with engravings of animals and humans, and with various geometrical designs reminiscent of the art of their Magdalenian ancestors. They also made small amber carvings of the important animals in their environment, such as bears and elks.

When Çatal Hüyük and Abu Hureyra had become fully fledged neolithic settlements with economies based on agriculture and the farming of domestic animals, the people in the northern parts of Europe were still mesolithic hunter-gatherers—the Maglemosians to the west and the Kunda to the east.

Farming spread into Europe from Anatolia, Greece, and south-west Asia by two means. On the one hand, groups of people who practised agriculture migrated northwards and westwards, especially along the valleys of the great rivers, exploiting the vast beds of loess that had been laid down in the area from Serbia and Poland in the east to Belgium in the west by the strong winds of the glacial epoch. These people pushed along the Danube into the area which today incorporates central Hungary, lower Austria, and Bohemia, and then on westward as far as the Rhineland and the Maas. By the fifth millennium BC they had reached southern Germany and Holland. They grew einkorn, emmer, barley, peas, beans, and flax, and they kept goats, sheep, cattle, and pigs. Meanwhile, other farming people were also moving up through Spain, Switzerland, France, and eventually to Britain. By 2500 BC they had reached the southern edges of the great coniferous forests of Scandinavia and Russia.

The other means by which farming spread was by acculturation, as distinct from colonization. The Maglemosian people eventually took it up, adapting it to their local situation and also to their own cultural background. Needless to say, these two processes were not likely to have been entirely separate; the Danubians moving north and west will have met, and probably mixed with and interbred with, local people.

About the same time as farming was being introduced into western Europe, it was also creeping down the valleys of the Tigris and Euphrates Rivers. Here, in the so-called 'Fertile Crescent', it was later to provide the economic base for the early cities in Sumer, and later for the Akkadian and Babylonian Empires.

Two general trends are discernible in farming in northern and western Europe from the time of its introduction during the fourth millennium BC through to the mid-19th century AD, both associated with increases in the human population. The first was the gradual spread of agriculture to cover more and more of the land of the continent; the second was the development of increasingly intensive forms of farming [16]. Thus, when the Danubians first entered Europe they practised *shifting agriculture*, and they restricted their farming activities to areas with well-drained loess soil which was relatively easy to cultivate with the hoe. In so doing, they exhibited an obvious but very important common behavioural tendency of humankind, the tendency to take the easiest path. They cleared a patch of forest and then sowed their crops of wheat and barley, and sometimes beans, peas, lentils, and flax. They also kept some cattle and sheep. When, after a few years, yields declined due to the exhaustion of fertility in the soil, they abandoned the plot and cleared another patch of forest. Eventually, when all easily accessible land around the village had been used, they moved away altogether and set up another village somewhere else and started the whole process over again [17]. In those early days there was plenty of land and few people around, and this form of adaptive response was a feasible proposition. After many years when the soil had recovered its fertility, the original village site might be reoccupied.

However, as the population of Europe increased, people found it more difficult, when the time came to abandon a village, to find unoccupied land with favourable soil. In this situation, two options existed. They could move further afield to places with less favourable soil, or they could adopt the second approach—that is, to try to modify their techniques of farming in such a way as to maintain or even increase the yield of foodstuffs at the original site. In fact, both these responses occurred. From the time of its introduction into Europe during the fourth millennium BC until the mid-19th century, farming spread outwards from the first areas settled by the Danubians, where the soil was easy to cultivate, to less favourable areas with heavy clay soil—a movement which was particularly active during the last few centuries of the first millennium AD, but which continued for many centuries afterwards. The movement was facilitated by a technological development—the introduction into Europe of the simple wooden plough (or ard), and more especially, somewhat later, the heavy wheeled metal plough [18].

The development of more intensive techniques of farming involved a change from shifting agriculture to settled agriculture with fixed fields. In essence, this took the form of a considerable shortening of the period of fallow made possible through the effect of the manure of farm animals in restoring soil fertility. The spread of the ard for lifting the soil contributed indirectly to this change, because it was drawn by oxen, which in turn contributed manure to the system. The end result was a general improvement in the methods of cultivating soil as well as improved techniques for maintaining and improving soil fertility—leading to bigger yields per unit area and allowing people to stay in one place, and at higher population densities than had previously been possible.

Already by the time of the Roman Empire the two-field system, in which the fields alternated between spring or winter grain and fallow, was practised in some regions, although there was still much shifting agriculture. By the 6th century AD, the two-field system existed over most of the continent of Europe. There is evidence that by the 8th century AD the three-field system had come into use [19]. This involved using one field for a winter-sown cereal crop, one for a spring crop, while the third lay fallow; this system resulted in four harvests in two years from an area which would have yielded only three harvests under the two-field system.

The early part of the second millennium AD saw some important ecological developments in Europe. From the 11th to 14th centuries AD the population of England doubled and similar trends occurred elsewhere on the continent. This period has been referred to as the 'great age of land clearance', when farming spread at an accelerated rate into the forest areas which, up until the end of the first millennium, had still dominated the European landscape. Although there were no particularly important novel technological developments at this time, generally improved farming practices resulted in an improved yield per acre. It is said that until around this time, the farmer seldom achieved a yield greater than two grains of cereal for every grain sown; but by the 14th century the yield had probably increased to three or four grains per grain sown [20].

By the beginning of the 14th century AD the farming of cereals and livestock in northern, western, and central Europe had taken on the broad ecological features which it retained until the early influences of the high-energy phase of human society began to be felt around the middle of the 19th century.

The range of plant species cultivated for food production by early farming societies in Europe is extensive. Eventually it came to include a number of varieties each of wheat, oats, barley, rye, and maize as well as different kinds of beans, lentils, peas, turnips, carrots, squash, beets, cabbages, lettuces, cauliflowers, potatoes, marrows, cucumbers, parsnips, spinach, peppers, onions, leeks, apples, pears, and, in southern parts, oranges and lemons [21].

The range of animal species reared for food production is more restricted, the most important being cattle, sheep, goats, pigs, chickens, ducks, and geese. Two main methods have been used, the enclosure method, in which the animals are kept in pens or within fixed fields, and pastoralism, in which the animals move in herds or flocks over large tracts of land, followed and tended by farmers who, in some instances, are truly nomadic, taking their families with them.

INTERRELATIONSHIPS BETWEEN HUMAN SOCIETY AND THE BIOSPHERE

This section is concerned with the changes that farming brought about in the ecological interrelationships between human society and the other components of the ecosystems of which human societies are a part. The discussion will be arranged under the following subheadings: general comments; energetics; redistribution of species; genetic changes; things that go wrong.

General comments

The disturbing influence of farming on ecosystems can usefully be discussed in terms of *ecological equilibrium*. For the purpose of this commentary, we can say that an ecosystem is in a state of dynamic equilibrium when the interrelationships of the components of the system are such that the overall rates of bioproduction through photosynthesis, on the one hand, and of decay, on the other, are more or less equal and more or less constant from one year to another. In such a system the amount of energy fixed in photosynthesis in one year is about the same as that used up in the metabolic processes of the living components, including the micro-organisms, of the system in one year. Such equilibrium in natural ecosystems is the product of the evolution both of the system and of the living organisms within it [22].

In nature the equilibrium of an ecosystem may be disturbed by a variety of factors, such as unusual climatic events and eruptions of the earth's crust. When this happens there is often a period in which the system can be described as seeking a new equilibrium. Initially, if there has been great destruction of life forms, there will be a period when the rate of decay greatly exceeds the rate of bioproduction. But this is likely to be followed by a period in which, as some plant species re-establish themselves, the rate of fixation of energy in photosynthesis exceeds the rate of decay, so that there occurs a general build-up of biomass. A system with these characteristics is referred to as an *immature ecosystem*. Eventually, the system again reaches a state of equilibrium, in which the rate of bioproduction and the rate of decay are about the same—that is to say, it becomes a *mature ecosystem* [23]. Usually the mature ecosystem is more diverse, in terms of the number of species it contains, than is the immature ecosystem.

There is much variability among mature ecosystems with respect to rates of bioproduction. This variability is mainly the result of differences in the water and nutrient content of the soil and in environmental temperature. In fact the rate of bioproduction in a disturbed ecosystem, once it has achieved a new equilibrium (i.e. after the immature phase has passed) may, as a result of long-lasting changes in the system, be considerably less than it was in the original mature ecosystem before the disturbance. This may occur, for example, after massive loss of top-soil through erosion.

Here we are concerned mainly with disturbances in ecosystems resulting from the processes of human culture, and in particular from farming. The clearing of forest and the introduction of cultivated food plants obviously constitute a major disturbance in the system. The natural forest is usually a mature ecosystem in which the amount of biomass produced in a year is relatively small in relation to the total biomass in the system, and in which the amount of energy from sunlight fixed and converted into chemical form is about the same from year to year as that released as heat through the metabolism of plants, animals, and micro-organisms (i.e. the rate of bioproduction is about equal to the rate of decay). In farming, however, the aim is to maximize production so that the energy fixed by the food plants is considerably in excess of that used up in their respiration and decay; the farm thus has the characteristics of an immature ecosystem. The intention is to produce a big increase in biomass during the growing season and up to the time of harvest. It is only as a result of continuous human effort that this situation can be maintained. If left to its own devices, the system would gradually mature and the rate of production would progressively decline, and the cultivated edible plants would be largely replaced by wild, mainly inedible forms.

The level of productivity of a farming system and the sustainability of this level depend on a multitude of factors. Broadly, these are of two kinds. First, the basic characteristics of the original ecosystem, including such variables as the climate and the chemical and physical properties and depth of the soil, clearly influence the response of the system to different farming practices and the ultimate yields of the system. Some ecosystems are much more 'fragile' or 'brittle' than others. Second, the particular methods of farming which are applied affect the productivity of the system and the ability of the system to maintain its initial rate of bioproduction. As has been learned at cost so many times, agricultural practices which have proved successful and sustainable in one region may be disastrous in another. To be successful in the long run, this cultural interference with ecosystems must be sensitive to, and take account of, the inherent characteristics and limitations of these ecosystems. If it does not do so, the initial disturbance and the chain reactions which this sets in motion may push the systems beyond the limits of their resilience, leading to progressive and sometimes irreversible degradation. This principle has been recognized, at least by some farmers, since the earliest days of agriculture. As Xenophon wrote in the fourth century BC: 'The earth willingly teaches righteousness to those who can learn; for the better she is served, the more good things she gives in return' [24].

In cases in which the productivity of the food-producing system is sustained over long periods of time, a new equilibrium has been established, so that the level of production of the desired species is maintained at a more or less constant level for a long period of time. Examples of this outcome are provided by the farming systems of western Europe, where fairly constant levels of production per unit land area were maintained over many centuries. In the words of Clark:

Were it possible to visit prehistoric Europe the chances are that one would find stable forms of economy prevailing over wide territories, the result of a more or less perfect adjustment between social appetites, technical capacity and organization and the resources of the several regions. Such a state of equilibrium and stability has ... often been observed among modern primitive peoples by anthropologists working in the field [25].

Another example is provided by the mixed subsistence farming in the Peruvian highlands, where a constant human population apparently achieved a state of ecological balance with the natural environment that lasted for several thousand years [26]. A similar equilibrium state was achieved in many horticultural societies, such as that of Papua New Guinea [27].

We will consider later, under the heading *Things that go wrong*, some of the basic reasons why farming activities have sometimes been unsuccessful, as well as the various adaptive responses of human societies when this happens.

Energetics and societal metabolism

An animal must acquire, in its food-seeking activities, more units of energy than it spends in these activities. This is partly because it needs energy for basal metabolism and for activities other than food-seeking and partly because, in some instances, there are dependent individuals to be fed which do not participate in the business of food acquisition. With respect to human beings, we noted in chapter 4 that the energy ratio for hunting and gathering—that is, the energy content of the food collected over the energy spent in acquiring it—has been estimated to be between 8 and 17. In farming systems, however, the picture may be complicated by the input into the process of energy used by draught animals or machines. Thus the complete energy budget of a farm in which draught animals are used must include the somatic energy which flows through these animals. Nevertheless, since the energy used by these animals is, like the somatic energy used by their human masters, derived from the system itself, the principle stated above still applies; that is, the total energy input must be less than the total energy yield. As we shall discuss later, this principle does not apply to farming in modern high-energy society in which energy from outside the system—in the form, for example, of fossil fuel—is used to drive machines to do physical work of a kind that had previously been done by human beings or animals. In such systems the total energy input into farming may exceed the energy yield.

A number of estimates have been made of the energy ratios of early or subsistence farming regimes. Inputs in these estimates include not only the somatic energy used in human labour and that used by draught animals, but also the extrasomatic energy used in the manufacture of farming implements made of metal. Some reported ratios of energy yield over energy input in crop production in early farming systems are as follows: mixed farming communities in the Andes, about 11; horticulture in Papua New Guinea, about 20; subsistence farming in India, about 15; peasant farming in China (1935–7), 24–41 [28].

Another important statistic from the demographic point of view, because it relates to the number of people that can be supported in a given area, is the food-energy yield per unit land area per year. Estimates of this figure must take account of the land which is lying fallow, but which is nevertheless part of the farming system. In the case of some shifting agricultural systems, the fallow land may be as much as 95 per cent of the area of the system. Some figures for energy yield (in MJ/Ha/y) are as follows: shifting rice (Iban), 850; horticulture (Papua New Guinea), 1390; maize (Mexico), 29 400; wheat (India), 11 200; peasant farming (China), 281 000 [29]. Leach, from whose book these figures are derived, points out that the example of peasant farming in China is very exceptional, a system unique in the history of farming, with extraordinarily intensive manuring, inter-cropping, and double-cropping.

We noted earlier that the energy yield in hunting and gathering ranged from about 0.6 to 6.0 MJ/Ha per year. Thus it can be calculated, assuming the above figure for energy yield in Chinese peasant farming to be correct, that this agricultural system is theoretically capable of supporting nearly 50 000 times as many people per unit area as is a hunter-gatherer economy in a particularly fertile area. Wheat production in India should be able to support about 1800 times as many people as the hunter-gatherer system.

In terms of energetics, then, the fundamental ecological change which occurred with the domestic transition can be described as follows. Cultural changes, in particular the development of farming technology, resulted in local concentrations of food-producing plants and animals. Human beings used part of their somatic energy to till soil, to sow seeds or plant seedlings, to keep out weeds and unwanted animals, to maintain and harvest food plants, or to tend domesticated animals. This new arrangement resulted in an increase in the amount of somatic energy available for human consumption per unit area, so that higher population densities were possible than had been the case under primeval conditions. The change did not in general mean, however, that the efforts of each individual farmer yielded much more, or any more, food than those of the individual forager. It is true that some early farming societies produced a food surplus which permitted non-farming populations in cities to exist. However, in most cases the surplus required was not very great, since only a very small proportion of the total population in the region lived in cities [30].

There are a few other characteristics of the technometabolism of early farming society worth noting. First, the level of extrasomatic energy was, in general, not very different from that of primeval society. Fire was still used for heating and for cooking and, in some situations, as an aid in clearing land; in the later part of the early farming phase it was also used for the manufacture of metal implements, weapons, and ornaments and for baking pottery. Nevertheless the per capita use of energy in the form of fire in early farming societies, while varying considerably from place to place and from time to time, was probably not on the whole very different from that in hunter-gatherer communities. With the introduction of metallurgy, there was a qualitative change in the use of non-renewable resources; but the per capita consumption of metals was very low compared with the situation in modern high-energy societies.

An ecologically interesting development in some regions was the use of extrasomatic energy in the form of wind power and water power for driving machines for various subsistence activities—mainly for grinding grain. The water-mill, for instance, although a Mediterranean invention, came to be widely used in northern and western parts of Europe towards the end of the first millennium AD. The *Domesday Book* lists 5624 water-mills in England alone; and by the late 12th century windmills were also common in north-western Europe. These additional uses of extrasomatic energy had little effect, however, on the overall energy budgets of human society.

The per capita use of total energy (i.e. somatic and extrasomatic) in most early farming societies was in the order of 20 MJ per day [31]. Consequently, unlike the situation in modern high-energy societies, the overall rate of energy use by humankind over the years or the centuries ran more or less parallel to the rate of growth of the human population.

Redistribution of species

One of the most obvious effects of early farming on the biosphere was the series of changes it brought about in the distribution of many different species of plants and animals, not only locally, but eventually also globally. On the local level, the cultivation of food crops involved the clearing away of unwanted species of plants from the selected area and their substitution with desired food plants. In the case especially of cereal crops, the aim was to produce as pure a stand as possible of the desired plant—*a monoculture*—and part of the work performed by farmers was directed at keeping out other unwanted plants. In other words, a specialized ecosystem was created and maintained in which the *diversity ratio*, that is, the ratio between the number of species and number of individual organisms, was low [32]. As we shall discuss later, the existence of monocultures of food plants and the dependence of societies on such monocultures for sustenance has given rise to various kinds of ecological problems for human communities.

The domestication of animals led to local concentrations of sheep, goats, cattle, and pigs and the expulsion from the area, as far as possible, of unwanted animal species. This expulsion was easier in the case of large animals than small ones. Rats, for example, were able to exploit the new situation and to share with people the fruits of human labour in the fields. Wherever farming people went, so did the rats, and the rat population increased enormously as a consequence of the domestic transition. This association was to have far-reaching demographic, societal, and experiential consequences for humankind, especially through its role in the great epidemics of bubonic plague.

The cases of introduction of species of plants and animals into new habitats can be divided into those which were deliberate and those which were accidental. In Europe the early deliberate introductions included barley, the two forms of wheat, lentils, sheep, goats, and cattle, the wild forms of which existed in south-west Asia. In the Mediterranean area plants like cucumbers and squash were introduced from south-west Asia, whence also came domestic fowls, ducks, and geese [33]. The accidental introductions in Europe included numerous weeds which, like rats, exploited the new environmental situation, growing prolifically in the newly cultivated soil; plantains, for example, were common wherever early farmers grew crops. Among the weeds which accompanied farming communities as they spread northward in Europe and which appeared among the wheat and barley were two of particular importance, namely wild rye and wild oats. In colder climates these two plants actually grew better than wheat and barley, and people eventually turned to cultivating them for grain production in preference to wheat and barley. The weeds became the crops and the crops became the weeds [34].

After the exploits of Columbus and other sea-going adventurers, Europe received a new influx of deliberately introduced species from the Americas, the most important of which were maize and the potato. Both these crops produced bigger yields, in terms of their calorific content, than did the grain crops of Europe at that time. Potatoes became an extremely important food source in the northern parts of the continent, as did maize in the Mediterranean area, and both contributed significantly to the nutritional support of the growing population of Europe from the 17th to 19th centuries. In turn, the cultivated varieties of the cereal complex which had originated in south-west Asia were taken to America. By the middle of the 19th century the various plants and animals, which had been domesticated in different corners of the earth between two and ten thousand years before, were to be found all over the world.

Just as new plants and animals were introduced, so were local native ones displaced. In Europe, the most striking example of this displacement was that of the dominant group of organisms—the trees of the great forests which enveloped so much of the continent when the Danubian farmers first moved in around 5000 BC. As discussed, these people cleared small patches of the deciduous forests for their farming, and as the fertility of the soil eventually declined, they moved elsewhere and the forest regenerated. But as the intensity of farming increased, less regeneration was possible, and some fertile areas remained more or less permanently cleared; and in the less fertile areas to the north and west of Europe where there was a high rainfall, deforestation and pastoralism led early on to the permanent clearing of areas which frequently became open moorland covered with heath and other low shrubs [35]. Even in Roman times, however, dense forests still covered most of central and western Europe. As mentioned above, it was not until the early part of the second millennium that 'the great age of land clearance' began—associated with a significant increase in population.

It was not only the need for land for producing food that brought about the disappearance of forests. As time went on, the demand for timber as fuel and for various other purposes became increasingly important as a motive for felling trees. The pace of industrial activities was quickening. Timber was used for props in the mining of tin, copper, and iron, and charcoal was used for smelting metals in the metal industries, and wood ash was needed for glass works and soap works. By the 16th century there were complaints in many places about the shortage of timber. In 1548 a commission was set up to inquire into the disappearance of wood in the iron-making area of the Weald in south-eastern England and many Acts of parliament were introduced in attempts to control the destruction. But these cultural adaptive responses of society to the scarcity of timber were unsuccessful, and the problem was eventually solved, at least partially, by the introduction of a new fuel in the form of coal. Coke was first used around 1709 for smelting iron-ore, and this practice became common by the middle of the 18th century. This is an example of cultural adaptation to scarcity through *substitution*, a response which has been common in human history and which has often proved effective in the short run. As we shall discuss in a later chapter, substitution is not always desirable, and in certain instances it is not possible. In some cases, the only long-term solution to increasing scarcity lies in phasing out the particular societal activity which depends on the commodity in question.

In the Mediterranean region the clearance of forests came about earlier than it did in the north. The process of forest clearance was already well under way at the time of Homer in the 9th century BC, who likened the noise of a battle to 'the din of woodcutters in the glades of a mountain' [36]. Early in the 4th century BC, Plato, referring to the disappearance of forests in Attica, wrote: 'What now remains compared with what then existed is like the skeleton of a sick man, all the fat and soft earth having been wasted away, and only the bare framework of the land being left' [36].

The earlier disappearance of trees in the Mediterranean area came about partly because the evergreen forests of the region were easier to clear than the deciduous forests further north, but more particularly because the ecological conditions were less favourable to their rapid regeneration. Among the components of these ecological conditions, along with the nature of the climate and the soil, was the ubiquitous goat, which roamed the cleared areas in large numbers, effectively destroying any seedlings that might otherwise have grown into trees to replace those which had been removed by human beings [37]. So it was that much of the evergreen forest in the Mediterranean region was transformed early on into the seminatural brushwood which is known as *garigue* or *maquis*. Some authors believe that deforestation in Italy and Greece was an important factor contributing to the decline of Roman and Greek civilization [38].

Genetic changes

As distinct from the broad changes on the level of ecosystems, involving modifications in the numbers, distribution, and relative proportion of different organisms, farming and other cultural developments gave rise to some significant changes in the genetic constitution of populations of certain plants and animals. In fact, as a consequence of societal activities, a whole range of new genetic forms was brought into existence—forms which would never have seen the light of day were it not for human culture; and this occurred well before the development of the sophisticated methods of genetic selection, cross-breeding, and genetic engineering of the modern era. Most of these novel, culture-induced genotypes would disappear quickly from the face of the earth if humankind were suddenly to vacate the scene. Thus many of the new genetic forms are not only the product of cultural intervention, they are also dependent on human culture for their survival.

The selection by human beings of certain genetic variants of plants or animals for cultivation or breeding may be deliberate or incidental. A good example of incidental selection is provided by the selection of cereals which took place at the very beginning of the early farming phase. In populations of the wild forms of many cereals, the great majority of the individual plants shed their seeds immediately upon maturation. However, such populations usually contain a small number of 'pathological' mutants which retain the seeds on the stalk for some time after maturation [39]. When people collected grain from wild wheat, for example, they must have taken home with them a disproportionately large number of grains of these particular mutants, simply because these grains remained accessible on their stalks for a longer time. When grains collected in this way from wild plants were used for sowing crops, the new population would, in turn, have contained a disproportionately large number of the 'pathological' mutants. From archaeological evidence it is clear that, from a very early date, cultivated forms of such cereals as wheat and barley differed from wild forms in that the seed was released only after threshing. Cultivated maize in middle America differed from wild forms in a similar way [40].

Later, as a result of trial and error and with an increasing intuitive understanding of the nature of heredity, deliberate selection of desired types of animals and plants came to be widely practised—long before Mendelian theory was developed or accepted. In eastern Australia, for instance, deliberate selection and cross-breeding of forms of wheat which were suitable to local conditions and which were resistant to rust was well under way and effective at the end of the 19th century and was carried out in the absence of any scientific information on the nature of the genetic process. Deliberate cross-breeding and selection of disease-resistant strains of potatoes was described in Britain around the middle of the 19th century.

In the case of animals, selective breeding has given rise to a fantastic array of varieties of cattle, sheep, goats, pigs, and poultry. The effects of human culture in this area are particularly obvious in the case of dogs, in which the criteria for selection have not been, or at least have seldom been, related to food value. It is true that many breeds, such as sheep-dogs and cattle-dogs, are the result of selection for behaviour of practical or subsistence value; but others, like the pekinese, the chihuahua, and the toy poodle simply reflect the frivolous whims of dog-fanciers.

Things that go wrong

Many of the culturally induced changes that farming brought about in the ecological relationships between human populations and the natural environment were of a kind that most people would regard as being of benefit to humankind. Most importantly, the new economy made it possible for many more people to exist on earth at one time. Other advantages included the protection that the village lifestyle provided against predators. There has, however, also been a negative side to the picture. While some food-producing systems have proven sustainable for many hundreds or thousands of years, others have become irreversibly degraded, some of them to such a degree that they are now useless for farming purposes [41]; and when this has occurred, it has almost always been due to the fact that the methods of farming have not been considerate of the ecological sensitivities of the local regions and have, in one way or another, overburdened the ecosystems.

The main factors responsible for declining productivity in agricultural systems are: depletion of nutrients in the soil; the accumulation of chemicals in the soil; soil erosion; diseases of cultivated plants.

Depletion of nutrients in the soil

The cultivation of crops for food production involves a relatively rapid rate of incorporation of nutrients from the soil into the substance of the growing plant. As discussed above, the agricultural system is, in this respect, akin to an immature ecosystem in nature. It differs from a natural immature ecosystem in that the crop is eventually harvested, and the nutrients it contains are taken out of the system instead of being returned to the soil, thus breaking the natural biogeochemical cycles [42]. Consequently, the rapid rate of incorporation of nutrients into plant tissue, together with their removal from the system at harvest, results in a relatively rapid depletion of nutrients in the soil, so that productivity eventually declines sufficiently to render farming no longer worthwhile.

Declining productivity due to this cause must have been universal in the experience of farmers at the beginning of the early farming phase; and cultural adaptation to this adverse change in the natural environment took a very obvious form: people simply abandoned the site and tried again somewhere else. This is an example, in its simplest form, of a common class of cultural adaptation, which can be referred to as adaptation through *giving up and moving on*. From the point of view of the group involved, this kind of adaptation is perfectly satisfactory, so long as there is somewhere left to go to.

The early discovery that if the exhausted plots were left alone, so that wild plants could regenerate and grow again, they eventually recovered their capacity to support satisfactory growth of cultivated crops, was in part due to the fact that the deep roots of some of the wild plants transported nutrients up from the depths, ultimately to be scattered on the ground as litter. Moreover, through the practice of burning, much of the nutrient content of the wild plants was made available in the soil for the crops [43]. Thus a system of rotation was worked out and the original give-up-and-move-on adaptive response was modified, through improved understanding of natural processes, to a cyclical pattern which was sustainable over long periods of time [44]. However, this method requires that at any given time most of the land available to a community for cultivation—perhaps over 90 per cent—is lying fallow and, as mentioned earlier in this chapter this pattern eventually became unworkable in Europe and other areas simply because of the increasing human population. It was in this situation that people learned that when manure from farm animals was spread over the land, the soil recovered its productivity in a much shorter time.

The accumulation of chemicals in the soil

Productivity in agricultural systems can be lowered by cultural practices which result in increasing concentrations in the soil of certain substances which interfere with plant growth. The most important environmental change of this kind in the early farming phase was the salinization of soils in areas where irrigation was practised. It involved the excessive accumulation of sodium chloride which was brought to the system in very dilute solution in river water and which was left behind in the soil as the water evaporated.

An important instance of culturally induced salinization with far-reaching societal consequences is the case of Mesopotamia. We have discussed how farming spread down the valleys of the Tigris and Euphrates Rivers and how techniques of irrigation were developed in this area. The early cities of Mesopotamia which were flourishing as independent city states in the latter part of the 4th millennium BC and which were later united under the kingdoms of Sumer and Akkad, were dependent on this irrigation farming system. The main cultivated crop in the early days of this Mesopotamian civilization was wheat. However, archaeological evidence indicates that around 3000-2500 BC the wheat was replaced by barley, and it seems likely that this was a cultural adaptive response to increasing concentrations of sodium chloride in the soil, for barley will grow well in considerably higher concentrations of salt than are tolerated by wheat. This change from wheat to barley is another example of cultural adaptation to an unsatisfactory development through *substitution*. It is an antidotal rather than a corrective response, and it suffers from the potential disadvantage of antidotal responses in general in that, because it involves no attempt to correct the fundamental change which is responsible for the difficulties experienced, the underlying causal factor is permitted to persist or, indeed, to worsen, so that eventually more extreme cultural adaptive processes must be devised and implemented. In the Mesopotamian situation this is precisely what happened. Salt continued to accumulate and it eventually reached such concentrations that even the substitute, barley, would not grow. This was the end of the road as far as substitutional cultural adaptation was concerned—the only course left was adaptation through the give-up-and-move-on response. The kingdoms of Sumer and Akkad became depopulated, and power in the region shifted further north to Babylon where salinization was less of a problem [45].

Soil erosion

Soil erosion has been defined as the progressive removal of soil or rock particles from the parent mass by a fluid agent [46]. The 'fluid agents' in this context are water, glacial ice, and air. The process of soil erosion is a natural one, in the sense that it occurs in the absence of human culture. Soil itself is merely a transient phase in the sequence of events which begins with the breakdown of bedrock and ends with its eventual arrival, in solution or in suspension, in the oceans. The soil particles removed from the parent mass by the fluid agent may be temporarily deposited elsewhere on the land, before reaching their ultimate oceanic destination. The loess soils of Europe which were so favoured by the early Danubian farmers provide an example; they are the product of both glacial action and the power of the wind, which deposited them in the areas where they now exist.

Soil, for all its transience, is an essential prerequisite for terrestrial life. In turn, biological processes affect both the rate of formation of soil and the rate of its movement towards the sea. On the one hand, the roots of plants, especially of trees, may help to break up the bedrock. On the other, a cover of vegetation slows down the rate of removal of soil particles by water or wind. In mature natural ecosystems, the rate of soil formation is usually at least as great as the rate of loss of soil through erosion. The trouble occurs when human activities, such as some kinds of farming practice, result in a significant acceleration of the natural process of erosion. During the period between the removal of the natural cover of vegetation for the cultivation of crops and its replacement with a cover of crop plants, the exposed soil is especially vulnerable to the eroding effects of water and wind; and water erosion is most likely when crops are cultivated on hillsides. When the erosion occurs more or less evenly over a sloping surface (sheet erosion), the change results in a general lowering of productive potential. When deep gullies are formed, the terrain becomes unworkable.

Severe soil erosion is more likely to occur in some regions of the world than in others. In the relatively stable ecosystems of central and north-western Europe, soil erosion resulting from human activities has been much less of a problem than in the Mediterranean region where rainfall is seasonal and often very heavy when it occurs. Even in regions such as this, serious soil erosion has not been an inevitable consequence of farming on hillsides. Farmers in Greece in the Bronze Age and early Iron Age were successfully growing olive and fruit trees, vines, and vegetables in hilly country which had been cleared of primeval forest. They terraced the land and through their deliberate efforts they succeeded in maintaining productivity and ecological equilibrium. It was only when the area became depopulated as a result of Roman slave-hunts that erosion set in and the productivity of the land rapidly declined. Eventually the area degenerated into useless wasteland, and a new ecological equilibrium became established involving much lower rates of fixation of solar energy and a smaller biomass.

Similarly, areas which are now desolate in southern Italy and Sicily were apparently farmed successfully until around 1300 AD. At about this time, Spanish methods of sheep breeding were introduced and this resulted in overgrazing and, eventually, in disastrous erosion [47]. These examples illustrate the point made earlier in this chapter that the sustainability of a farming system depends both on the characteristics of the original ecosystem and on the particular methods of farming. Prudent agricultural practice can be successful even in fragile ecosystems.

Thus, in the early farming or second ecological phase of human existence, soil degradation due to accelerated erosion brought on by farming practices which abused the natural environment was a common, and often ultimately catastrophic, occurrence in certain regions of the world. Let us note, however, that erosion as a consequence of mismanagement or misuse of land did not cease with the advent of the modern technological phase—witness the serious culturally induced erosion which is taking place in parts of Australia today [48]. The difference is that in the early farming phase this undesirable change in the environment was the result of the activities of people who were largely unaware of the relevant ecological principles. It was a case of *ignorant abuse of nature*. No such excuse exists today, and the erosion which is taking place in modern high-energy societies is an example of *informed abuse of nature*—with human beings ignoring, but not ignorant of, the sensitivities of the biosphere and the fundamental principles of ecology.

Diseases of crops

One of the hazards faced by human populations which are dependent on agriculture for their sustenance is the possibility of crop failure due to disease. The likelihood of difficulties from this cause is influenced by two important aspects of the interrelationships between the biosphere and human society. First, there has been a tendency in early farming and early urban societies for populations to become increasingly dependent on a single food source—the staple diet. Communities in south-east Asia depended on rice, those of Central America and Mexico depended on maize, rural communities of western Europe on rye, and the peasants of Ireland on potatoes. In such situations, if disease strikes the staple crop and significantly reduces the yield at harvest, the population which is dependent on the monodiet will be in serious trouble. Societies of hunter-gatherers or of farmers who grow a significant amount of a range of different crops are relatively immune to this problem, since most diseases of plants, like those of animals, are relatively host-specific and attack only one kind of plant.

The other relevant ecological factor which influences the likelihood of serious outbreaks of disease in food crops is the fact that farming, by its very nature, involves growing large numbers and large concentrations of the same plant in a given area. The existence of these monocultures greatly increases the likelihood of infective agents or parasites settling on a susceptible plant; and it also results, when an infestation occurs, in the release of very high concentrations of infective agents into the environment. Thus monocultures represent an especially favourable situation for the spread of diseases of plants caused by pathogenic microbes or parasitic animals.

From time to time, disease has attacked and devastated crops of all the main plants grown for food production—including wheat, oats, barley, rye, maize, rice, and potatoes [49]. The disease-producing organisms include fungi, bacteria, viruses, and various parasitic animals such as aphids, mites, and nematodes. Sometimes the parasitic animals act as vectors of the microbial pathogens. Here I will comment briefly on only one plant disease, potato blight, selected because, in the context of the Irish potato famine, it well illustrates some important principles and concepts of culture–nature interplay.

Potato blight and the Irish potato famine

It has been written of the Irish potato famine that:

... it would be impossible to exaggerate the horrors of these days, or to compare them with anything which has occurred in Europe since the Black Death of 1348 ... . Considered as a whole, the annals of the Irish famine constitute one of the most tragic memorials of which we have record, in man's long chronicle of suffering [50].

It is likely that the potato was first brought to Ireland near the end of the 16th century, not long after its first introduction into Europe [51]. Within fifty years it had become the universal and staple component of the diet of the peasant population of the greater part of Ireland, and it has remained very important in that land ever since. There were several reasons for the ready acceptance of the potato in Ireland, some ecological, others cultural and societal. The ecological reasons include the fact that the rich, deep, and friable soil and the good rainfall were ideal for growing potatoes. Moreover, the moist atmosphere and the prevailing westerly winds discouraged infestation with aphids, which transmit various virus diseases of potato plants.

The societal reasons were complex and were associated with the cultural history of the Irish people, their political relationship with the English, their incredibly impoverished life conditions, and the system of land tenure. The peasants lived off very small plots of land, and most or all of the cereal crops they grew were sold in order to pay the rent. For their own sustenance, the potato proved ideal. It grew well, and a relatively small input of tubers and labour in the spring resulted in a substantial harvest in the autumn. Moreover, the potato, especially when stored underground, was relatively safe, as compared with grain, from the attentions of the marauding bands of soldiers which were so much a feature of the Irish scene in those days. In this way the potato became as important in the diet of many of the peasants over the greater part of Ireland as rice was important for the inhabitants of southern China. About a third of the Irish population depended solely on the potato for survival, and in County Mayo, the proportion was nine-tenths.

In 1845 Ireland was one of the most densely populated regions of Europe. The population had doubled in fifty years, and at this time stood at around eight million; and the population density of potato plants in Ireland must surely also have been the highest in the world. The spring of that year was mild and pleasant, and there was nothing abnormal about the summer. Suddenly, however, in early autumn, the temperature dropped to well below normal and it began to rain incessantly. When the peasants went to dig up the potatoes, they found that many of them were rotten, and most of those that appeared normal soon rotted away in storage. Fortunately, a few potatoes had been dug up before the rains came, and here and there a sound potato, or a partially sound one, could be found and ground up to make potato flour. Although there was a considerable quantity of grain in parts of Ireland, it belonged to landlords and was intended for export; and in any case the peasants had no money to buy it. So the people in their desperation were forced to search the natural environment for nuts, berries, the roots of dandelions and of ferns, as well as other wild foods. Amazingly, the population as a whole survived the winter and, with some help from American maize, which had been bought by the British Government, they also survived the spring and summer of 1846.

Early in 1846 potatoes were planted again and the plants grew well until June, when the rains came again. Within a month all the potatoes were blighted and the whole situation rapidly became much worse. It has been estimated that at least one million people, and probably considerably more, died during the ensuing famine. Many succumbed directly from starvation or from scurvy, while others, their resistance lowered by malnutrition, died from cholera, dysentery, or typhus. These microbial diseases were exacerbated by the societal disorganization which resulted from the movement of people away from their homes and their crowding together in work-houses.

The adaptive responses to the famine took a variety of different forms. First, there were the efforts of individuals and families who were directly affected by the food shortage to try to find alternative sources of sustenance in the natural environment. This approach—a reversion to a hunter-gatherer lifestyle—may have saved a few lives, but in general it is unlikely to have been very effective, since the population density in Ireland as a whole was about 150 persons per square kilometre. We noted earlier that even in favourable environments the hunter-gatherer economy seldom supports more than one person per ten square kilometres.

Some landlords were sympathetic and compassionate, and participated in the adaptive response by dropping their demand for rent from the peasants, and even sharing what little surplus food they had with their tenants. There is one report of a woman in County Tipperary buying up all the bread and meat in the town and distributing it to the people who lived on her estates. But many others were selfish and callous in the extreme, even to the point of evicting starving tenants who could not pay their rent from their homes.

In the long run, the most important adaptive response of people who were directly affected was emigration. It was an example of adaptation through giving up and moving on. Between the years 1845 and 1855, about 2 million Irish peasants emigrated to North America.

The other aspect of the societal adaptive response to the famine was that which took place in England, and it consisted of actions both of the Government and of voluntary bodies. The scene was characterized, as so often in similar situations in more recent human history, by a vigorous controversy between *the concerned*, who believed that urgent action was necessary, and *the unconcerned*, who argued with vehemence against government aid for the victims of the famine. The arguments of the latter group included the now familiar warnings about the risk of people becoming conditioned to government hand-outs and hence unwilling to make any effort themselves, and of the economic dangers of interfering with the market forces of supply and demand.

Nevertheless, already in 1845, the Government in London, under the Prime Minister, Sir Robert Peel, accepted in principle its responsibility to take action of some sort. By November 1845, Peel had ordered $100 000 worth of maize to be shipped to Ireland from America. An Irish Relief Commission had been established in Dublin to co-operate with landlords and other local residents in the organization of relief work on roads and docks, aimed at providing employment and income so that people could buy the cheap food, including the maize, which the Commission was distributing. The Government spent about three-quarters of a million pounds on various grants and loans at this time. The anti-relief lobby was, however, still powerful. In fact the man appointed by the Government to head the Relief Commission, Charles Trevelyan, was clearly sympathetic to the *laissez-faire* school of thought. He managed on various pretexts to slow down the distribution of American maize and, in June 1846, he even rejected a cargo of maize that was on its way to Ireland. Soon after this episode, when it was already apparent that the new potato harvest was diseased, he wrote as follows:

The only way to prevent people from becoming habitually dependent on government is to bring operations to a close. The uncertainty about the new crop only makes it more necessary ... . Whatever may be done hereafter, these things should be stopped now, or you run the risk of paralysing all private enterprise ... [52].

In fact the Government, which was now under Sir John Russell, also took the view that matters should be left to market forces. Eventually, however, the situation became so bad that, reluctantly and belatedly, the Government felt obliged to take further action. More maize was ordered from America, more relief work was organized, and, following the Soup Kitchen Act, more than three million people were eventually receiving free soup and other basic foods. Another measure was the new Poor Law Relief Act of 1847 which allowed work-houses to take people in and to feed them in return for work. However, the so-called Gregory Clause, which stated that individuals who were in possession of one English quarter acre or more were not eligible for such help, meant that starving peasants, in order to survive, were forced to give up their plots of land and their homes.

After the situation had become very much worse late in 1846, the Government spent a further $8 million on relief. The total response of the Government was unprecedented in English history. It was possibly the first time that a national government in Britain had accepted the responsibility to do something to allay suffering among a section of the population under its rule, and it was certainly the first time that a government had deliberately planned and financed relief measures on a large scale.

The Irish famine elicited responses not only from the Government in England but also from concerned individuals and groups in the community. Among various voluntary organizations that tried to help, the Society of Friends was perhaps the most noteworthy. The Society organized public collections for funds and set up soup kitchens in Ireland. Taking the view that potatoes might have failed for all time, it introduced carrots, turnips, and parsnips to the peasants' plots: 40 000 lb of turnip seeds were distributed in 1847. The Society also drew attention to the unsatisfactoriness of the land tenure system, and predicted that the social situation would remain unstable so long as this system remained in force.

The potato famine of Ireland thus elicited what may well be the first example in human history of a cultural adaptive response which, despite its inadequacies, was truly transpopulational.

All these adaptive responses took place in the absence of any understanding of the nature of the potato disease. Many interpreted it as divine punishment for the sins of the Irish people, and concluded that the only useful response would be prayer, penitence, and personal reform. Although a few individuals in both England and France who possessed microscopes had observed the fungus on the potato, it was not until 1861 that experiments showed conclusively that this fungus, under certain conditions of temperature and humidity, actually caused the disease. This finding in itself was not at first of any help. In 1890, however, this knowledge led to the testing of a remedy which had been used in France for controlling a fungus disease of grapes. This was a mixture of copper sulphate and lime, known as Bordeaux Mixture, and this was found to be effective in killing the germinating spores and hyphae of the potato fungus. The other scientific, cultural, adaptive response came later and involved many decades of selective cross-breeding which has given rise to countless new varieties of potato plant, many of which are markedly resistant to blight.

The societal and political consequences of the potato famine in Ireland have been numerous and far-reaching, and it would be impossible to try to trace or document them in this book. In America the Irish invasion, which began in earnest with the potato famine, has had immense political and societal impact. Among its effects was the rapid growth of Roman Catholicism in the USA—to become a major religious force in the country. Indeed, at an early date this prospect worried militant protestant groups which, for religious reasons, actively advocated the deportation of the Irish immigrants. Many of the Irish who arrived in America nurtured a deep hatred of the English, whom they blamed for all the misfortunes which they and their ancestors had experienced in Ireland, including the consequences of the famine itself. It has been suggested that this embitterment has on many occasions influenced the political relationships between Britain and the USA. Moreover, while the chronic human tragedy which bedevils Ireland and Britain today and the relationships between the Catholics and the Protestants in Ulster is certainly not due to the potato famine, one wonders whether, if the famine had not occurred or if the response of the British Government had been more effective, a different scenario might have developed.

Thus, in the Irish potato famine we see biological and cultural factors combining to create a human tragedy of immense proportions and of lasting impact. A more enlightened, far-sighted, and vigorous response on the part of the Government in Britain could have been more effective in alleviating the situation and could have reduced the suffering and mortality, but it could not have prevented the disaster. *Post hoc* (or after-the-event) adaptation could not have saved the situation.

It should be mentioned that the failure of potato crops due to blight in 1845 and 1846 was not restricted to Ireland. It also occurred in England and throughout a great deal of northern and western Europe. But people in these parts were not so dependent on the potato as were the Irish. Indeed, in England there had been considerable resistance to the new food source, which was regarded with suspicion and reputed to be poisonous and to have alarming aphrodisiac qualities. In 1916, during the First World War, potato blight struck particularly savagely in Germany, and 'struck a blow ... as decisive as any campaign planned by the Allied generals' [53].

HUMAN SOCIETY

Let us now briefly consider the implications of farming for human society itself. There has clearly been enormous variation among early farming communities with respect to hierarchical structure, systems of exchange and of land tenure, relationships with neighbours, and many other aspects of societal conditions. Here we are concerned mainly with those characteristics which are universal, or at least common, among ecological phase two societies and which are the result of, or particularly associated with, the farming relationship with the biosphere.

The economic mode of the early farming phase of human society existed for at least five thousand years before the cities of Mesopotamia came into being around 5500 years ago. After that time, however, and until the technoindustrial transition, two very different kinds of society, the early farming and the early urban societies, existed side by side. Indeed, the early urban societies depended on, and necessarily interacted with, early farming societies, which were consequently affected in various ways. Ecologically, the most obvious and universal aspect of this influence was the fact that a portion of the food produced in the farming systems was diverted to the non-farming city dwellers; but there were certainly other influences, and these varied in degree and in kind from region to region. Opinions differ with respect to the significance of these influences. At one extreme, there is the view of Jacobs that farming itself was the product of city-like agglomerations of human beings which had formed in response to trade in obsidian and other commodities between hunter-gatherer groups. She suggests, in essence, that since cities are and have always been the source of new ideas they must have existed before the notion of farming dawned on people [54]. This argument is not very convincing: one might as well claim that cities were in existence before the use of fire by humankind or before the invention of the bow and arrow. Nevertheless, the effects of urban societies on rural ones over the centuries have certainly been numerous and diverse. They range from such direct influences as rural depopulation resulting from slave-drives to the subtle economic effects of the introduction of a cash economy. No attempt will be made here to analyse and describe these influences, although reference will be made to some of them as we discuss various aspects of human society and human experience in the early farming situation.

There has been considerable debate among social scientists about the nature of the relationship between agricultural development and population growth. Broadly, we can recognize two schools of thought. The first may be referred to as the Malthusian view, which assumes that population size has been determined by the intensity of agricultural activity; that is, by the productivity of the farming system:

The second view is associated with the name of Boserup, who suggests that the relationship is the reverse of the Malthusian explanation; that is, the intensity of the farming practice is determined by the size, and hence nutritional needs, of the population:

population growth → farming
and size   practice

Boserup argues that there is considerable elasticity with respect to the carrying capacity of land but that, because of the 'law of least effort', people do not practice a more intensive form of agriculture than is necessary to provide sufficient food for the existing population [55]. Comparing the long forest fallow of shifting agriculture with much more intensive systems of cultivating, she points out that, among people who know both systems, the forest fallow method is usually referred to as the 'easy system'—it yields a good crop with little input of labour. While this relationship between intensity of farming practice, as reflected in yield per unit land area, and hard work for the individual farmer is probably in general true for pre-industrial farming, there appear to be some exceptions [56]. Extrapolating from the calculations of Leach, the traditional Chinese peasant in the 1930s would seem to have had to work in the fields only about one hour per day on average in order to support himself and his family (see page 00).

In fact, it seems most likely that both the Malthusian and the Boserup models are correct; that is to say, in some situations food supply has strongly influenced population size, and in other situations population pressure has resulted in increases in food production. Nevertheless, I suspect that the Boserup model has more often been the operative one, and that in general the intensification of farming has been a response to, rather than a cause of, increase in population. This hypothesis is at least consistent with the view that one of the most important of the common behavioural tendencies of humankind is the tendency to take the easiest path.

Increased production can thus be seen as a cultural adaptive response to population overload. As in the case of cultural adaptation in general, the distinction is important between short- and long-term adaptation. For example, the introduction of a large number of sheep into an agricultural system might well, in the short run, provide extra sustenance for a growing human population. However, overgrazing may then lead to degradation of the ecosystem so that, in the long run, the system supports fewer people than was the case before the adaptive response.

Whatever the nature of the causal relationship between farming and population size, the end result is clear: large populations of human beings entirely depend on farming, at varying degrees of intensity, for their survival.

Societal organization

It can be argued that farming itself does not necessarily demand any important change in societal organization, as compared with the hunter-gatherer society, beyond some arrangements which determine who is responsible for working a particular portion of land and who has the right to keep and use the food grown on it [57]. However, the concomitant of farming—comparatively large groups of people living together in a single society—may be seen as a factor requiring some additional organization and rules of behaviour beyond those necessary in small unstratified hunter-gatherer communities. In fact, the range of patterns of societal organization and also of land tenure of early farming communities has been wide. Nevertheless, there is one underlying common denominator and basic determinant of societal organization, and it is one which is also characteristic of primeval societies. This factor is kinship. The family network, whether the society is patrilineal or matrilineal, is the basis of the economic system, and economic cooperation is mostly between members of families, and most social interaction is based either on family connections or on shared residential area.

With respect to social stratification—in terms of the distribution of power and material wealth—there is much variation among early farming societies. On the one hand, tribal societies have existed, such as some of those of early neolithic Europe and those of the highlands of Papua New Guinea in recent times which are essentially egalitarian. It is true that in Papua New Guinea communities have their 'big men', but these hold little real authority [58]. On the other hand there have been many early farming societies, such as those of feudal Europe and of the chiefdoms of Africa, which are highly stratified. Recent archaeological evidence from burial sites in southern England suggests that, contrary to what was earlier thought to be the case, the distribution of power and wealth within communities in the region was very uneven as early as 3000 BC [59].

Clearly, when urban societies came into being they sometimes imposed their power on nearby rural communities, and sometimes on communities which were not nearby, as in the case of the Roman Empire. Control of farming societies by a centralized authority was also a feature of situations where populations depended for their sustenance on farming based on large-scale irrigation programmes, as in Egypt and China [60].

There is usually little occupational specialism in early farming societies beyond that which is reflected in gender-based and age-based division of labour. However, in some farming communities a few specialists exist. This was so even in very early times; for instance, the impressive array of artefacts used at Çatal Hüyük in the 7th millennium BC must surely have been the work of specialist craftsmen: but Çatal Hüyük is not typical of early farming societies, and perhaps should be seen rather as an embryo city.

While sex-based division of labour in early farming communities is very common, the actual pattern of division of labour between the sexes is variable. In many societies, such as those of the Ainu in northern Japan, the Kikuyu in eastern Africa, and the tribes of the Papua New Guinea highlands, the women cultivate and harvest the crops. In these societies, the tasks of the men are variable, but may include hunting, or protecting women and children from marauders and, as in Papua New Guinea, clearing land and making fences. In some groups, however, like the Trobriand Islanders, it is the men who tend the crops. In fixed field farming in Europe, men have traditionally worked with the plough, although women have participated in almost all other aspects of the process. Describing the situation in 16th century England, Laslett has written:

The boys and the men would do the ploughing, hedging, carting and the heavy, skilled work of the harvest. The women and the girls would keep the house, prepare the meals, make the butter and the cheese, the bread and the beer, and would also look after the cattle and take the fruit to market. At harvest-time, from June to October, every hand was occupied and every back was bent [61].

Warfare

With respect to war and peace, there is no common pattern among early farming societies. The view has been put forward that it was essentially the farming economy which gave rise to warfare in the first place: it was not until people came to possess commodities which were coveted by others that organized violence between communities became a feature of human affairs [62]. While there may well be something to this hypothesis, warfare is certainly not an inevitable consequence of agriculture; moreover, there seems to be little evidence for the old notion that hostilities eventually developed between agriculturalists, pastoralists, and hunter-gatherers [63]. There have been plenty of early farming societies which have lived at peace with their neighbours for long periods. From archaeological evidence it seems that the farmers of Çatal Hüyük, and those of the valleys of the Tigris and Euphrates Rivers, and the early neolithic people of central Europe belonged to this category [64].

On the other hand, it is clear that violent hostilites were, from very early times, an important aspect of life in some farming communities. The walls of Jericho, first built around 8000 BC, are often cited as evidence of this fact. And towards the end of the neolithic phase in Europe the relative peace was shattered by the aggressive 'battle-axe' people, intent on warfare and on political domination; and recent evidence indicates that around 3000 BC early farming people in the south of England built and attacked fortified settlements [65].

Some prehistorians and others have sought to find explanations, in terms, for example, of local ecological conditions, for the fact that in some regions villagers are constantly at war while in others they are not. My own view is that the chief determinant is cultural. In the words of Reynolds:

Events in Vietnam (e.g. at My Lai), and the activities of the Nazis, as well as the systematic purges and ravages in bygone times such as those of the Viking Norsemen under King Harald, seem to indicate that human cultures or sub-cultures can become obsessed with violence which becomes normatively sanctioned and spares not a thought for the sex and age of the victims, only for the social category they fall into [66].

While we can recognize among the common behavioural tendencies of human beings a tendency to behave protectively and with consideration towards members of the in-group, and especially towards the children, there is no such universal tendency to behave in this way toward strangers. Nor, however, do we see evidence of a universal tendency to behave aggressively toward them, although there does appear to be a tendency to regard them with suspicion. In other words, anything can happen, and cultural norms, including religious beliefs, are a major determinant of whether or not early farming societies are involved in violent intergroup conflict; and no doubt in situations in which there is scarcity of essential resources, this factor also plays a part.

Needless to say, physical combat in early farming societies, when it did occur, was on a face-to-face basis.

Exchange and trade

There is much variability among early farming societies with respect to patterns and degree of trade with communities beyond their borders. The only common factor that is discernible is the beginning of a progressive trend towards increasing dependence on other societies for essential resources, or at least for maintaining a pattern of life to which people have become accustomed and for avoiding societal collapse. This phenomenon, which may be referred to as a 'lengthening of chains of dependency', has become vastly more important in the modern world.

In general, however, early farming societies have been to a large extent self-sufficient. In the early days especially, trade was largely in non-essentials [67]. Life could have gone on without it. In some regions, however, there was trade in a range of staple commodities, such as stone for querns and flint for axes. Nevertheless, the chains of dependency in early farming societies have usually been short, simple, and weak.

While the view has been put forward that rural communities have always been influenced by and dependent on urban societies [68], there are many obvious cases, such as the villages of Papua New Guinea, where this cannot possibly have been so. In other words, we must surely conclude that such dependence is not necessary either for the establishment of a farming economy nor for its maintenance over long periods of time. This is not to say that when and where urban societies exist, they do not influence the ecology, economics, and lifestyle of rural populations.

One of the factors which has affected early farming societies in important ways the world over and which was the consequence of human activities in cities was the introduction of the cash or monetary economy. Important effects of this influence were already felt in farming communities in Europe around the 13th century AD, when landlords began to demand rent instead of services from the tenants, and they have been felt in Papua New Guinea within the last fifty years [69].

Towards the end of the early farming phase in Europe, increased possibilities for trade provided by improved transportation and the cash economy had an important influence on the patterns of activity of farming communities and led to considerable regional specialization. Farmers in Holland, for example, increasingly specialized in the production of vegetables and other relatively perishable foodstuffs, while they imported their grain from other areas which were less densely populated and which were generally more suited to grain production [70].

Religion, beliefs, and values

It would appear from all the evidence emanating from archaeologists and historians that every early farming society that has ever existed has had a religion, in the sense that the members of the society believe in, or profess to believe in, a god or a number of gods. In all cases, there are rituals and ceremonies associated with the worship of a deity or deities and designed to express thankfulness, to ask for forgiveness, or to seek divine help in connection with important aspects of life experience, such as food production, health and disease, and warfare.

Beyond this generalization, no common denominators can be discerned. In some cases there is a single god, in others there are many. Some gods are human-like, others are animal-like; some are female, others are male; and some are benevolent while others are vengeful and terrible. Nevertheless, despite this variability, some beliefs have proved more attractive and more lasting than others. For instance, in the Near East and in Europe, for thousands of years before the advent of cities, the most fundamental and widespread belief was that which regarded the 'spirit of the earth' as the mother of all living things, and the people worshipped 'the great female principle of nature'. Usually the goddess was associated with a male consort and sometimes with a child. The Natufians who first settled Jericho in pre-pottery times made many hundreds of little effigies of the mother goddess out of clay, bone, and stone. She was worshipped in the forty or so shrines and sancturaries which have so far been unearthed at Çatal Hüyük, where excavations suggest that about one in every three rooms was used for religious purposes; and she formed the basis of the religious beliefs and practices of countless generations of neolithic farmers from Egypt to Ireland and from the western coasts of Europe to Siberia [71].

Clearly, religion is one of the aspects of the culture of early farming people which was influenced by urban civilizations, when these came into existence. As a consequence, the mother goddess declined in importance in rural communities in the second part of the early farming phase.

The material environment and machines

In early farming societies resources from the natural environment, other than those used to provide the chemical compounds necessary for the metabolism and growth of the human body, were used chiefly for two purposes: to make buildings and to make implements and weapons. In addition, smaller amounts were used to provide clothing and to make ornaments.

The built environment consisted basically of no more than five kinds of structure and in many societies of only one kind. All early farming societies built *dwellings* for people to live in—usually one family per building. In some communities, special buildings were set aside for a particular age- or gender-group, as in the case of the men's huts in Papua New Guinea and the young men's huts in Samoa. The other four kinds of structure, any of which might be found in early farming societies, are: *shelters for domestic animals* (occasionally part of the dwellings for people); *buildings for storage* (for grain or other produce); *religious shrines or temples*; and *fences, walls, or ramparts* (to protect villagers from attack by other people or by animals, or simply to contain domestic animals).

Implements in early farming societies include containers of various kinds and a range of tools used for the production or preparation of food, for making clothes, and for various other household purposes. These were made of wood and stone and, later in the period, of metal.

Various machines, that is, assemblies 'of interconnected parts arranged to transmit or modify force in order to perform useful work' [72], were developed and used in early farming societies. They were used mainly in connection with the production of food, in which respect the plough was of major importance in Europe [73]. This particular product of human culture, which had so far-reaching an effect on the interrelationships between human populations and the biosphere, had additional ecological significance in that it depended on work performed by draught animals. That is to say, the machine was driven by non-human, although still somatic, energy. The heavy plough in use in Europe in the Middle Ages required a team of six or eight oxen to drag it through the soil.

Reference has already been made to two other machines which existed in some early farming societies and which used non-human, and in this case extrasomatic, energy—the water-mill and the windmill. While these machines based on the use of non-human energy were important technological developments in terms of certain aspects of the relationship between human communities and the biosphere, they had only minimal effect on the per capita use of energy in human populations as a whole.

It is noteworthy that these uses of non-human energy did not give rise to undesirable side-effects resulting from pollution due to by-products of the breakdown of the energy source.

HUMAN EXPERIENCE

We will now consider briefly the impacts of the transition from the first to the second ecological phase of human history on the life experience of individual human beings. For this purpose we will take the principle of evodeviation as a starting point; that is, we shall aim to identify the essential ways in which the life conditions of the early farming people differed from those of their hunter-gatherer ancestors, and we shall consider their reactions to these changes. Let us note at the outset, however, that the degree and extent of evodeviation in life conditions in the typical early farming society was slight as compared with the situation in the early urban and high-energy ecological phases.

A few general characteristics of life conditions and health patterns in early farming societies are worth noting. First, the new conditions meant that a given degree of ill health was less of a threat to survival and reproduction than had been the case in the primeval setting. The village situation provided greater opportunities for recovery in acute disease and it allowed people suffering from chronic illness to live longer, protected from some of the hazards inherent in the hunter-gatherer lifestyle.

Another important feature of most early farming societies is the fact that all people of the same age-group and sex in a community share very similar life conditions, and hence similar patterns of health and disease. In this respect early farming society is like primeval society, but unlike early urban and high-energy societies.

With regard to geographical mobility, the great majority of early farming people spent all their lives within a few kilometres of their place of birth. Important exceptions to this generalization were the great migrations of farming populations, including those of the Indo-Europeans from southern Russia to the south, east, and west; of the Huns and Mongols from central Asia into Europe; and of Europeans from Europe to the east coast, and then to the west coast, of America.

Another aspect of the new lifestyle that deserves mention is the fact that it represents the first step in the process of separation and 'divorce' of human beings from nature. It was only a small step in this regard, because early farming people were necessarily very aware of, and tuned into, the life-giving processes of the biosphere and were appreciative of their dependence on 'mother earth'. On the other hand, to a degree they must have felt less a part of the natural environment than their primeval ancestors.

Material aspects of life conditions

Perhaps the most obvious differences between the life experience of primeval hunter-gatherers and early farming people are those resulting from the increase in population density. The populations of early farming communities numbered hundreds or thousands, as compared with, say, twenty to fifty in the case of the typical hunter-gatherer community. Nevertheless, despite the increase in population density and in social complexity, almost all interactions and transactions are, as in primeval society, on a face-to-face basis, and encounters with strangers are rare.

The increase in population density created a new situation with respect to the interrelationships between human organisms and their various parasites and pathogens. The nature of this change will be discussed in the following chapter, in which we will consider the various biological repercussions of the even higher population densities of the early cities. In brief, we can say that the new conditions increased the likelihood of an individual becoming infected with pathogenic microbes and animal parasites.

From the very early days of farming two particular infectious diseases were of special importance. They were eventually to become, and are still to this day, major causes of death in humankind. They are schistosomiasis and malaria. The former is due to the animal parasite *Bilharzia haematobia* which lives for part of its life cycle in a snail. The disease occurs everywhere in the world where large-scale irrigation is practised, and there is good evidence that it had become common in Mesopotamia, Egypt, and the Indus valley by several thousand years BC [74].

It has been suggested that it was the introduction of farming in tropical and subtropical regions which allowed malaria to become an important cause of ill health and of death in human communities. This change in the economy brought human beings into close contact with populations of mosquitoes breeding in stagnant pools of water and this fact, combined with the relatively high density and immobility of the population, was particularly favourable for the maintenance of the life cycle of the malarial parasite [75].

The close physical association with domestic animals, where this occurred, increased the chances of individuals becoming infected with micro-organisms or animal parasites which are normally carried by these animals and which are potentially pathogenic for people. Examples of diseases due to such organisms are brucellosis, bovine tuberculosis, and infection with the dog hookworm.

Turning to diet, it is reasonable to suppose that most early farming people throughout history have, for most of the time, been adequately nourished. They are unlikely to have been deprived of calories; and, since they have mostly eaten a range of different plant foods supplemented by a little meat, they are unlikely to have suffered from malnutrition due to deficiencies of specific nutrients. However, in some areas, and at some times more than others, agricultural communities have become excessively dependent on a single foodstuff. Such 'monodiets' are unsatisfactory for two basic reasons. In the first place, if a population is almost solely dependent on a single foodstuff, as were the Irish peasants on the potato, and if for some reason the crop of that foodstuff fails, serious famine is inevitable. Second, a single source of food is unlikely to contain suitably balanced proportions of all the essential nutrients required by the human organism, and thus signs of malnutrition due to specific nutritional deficiencies are liable to occur. This problem has been of greater importance in early urban societies and will be discussed more fully in a later chapter. Here we may simply note that this kind of malnutrition has occurred from time to time in early farming societies, pellagra and beri-beri in rural populations in Asia being notable examples.

Behaviour patterns and psycho-social aspects of life conditions

In terms of the discussion on behaviour in chapter 3, the *basic behaviours* of people in early farming societies have been the same as in hunter-gatherer societies. That is to say, people in general followed the common behavioural tendencies of humankind, although in some instances cultural pressures may have reinforced certain of these tendencies, and in others they may have suppressed them. With regard to *specific behaviours*, however, there are some obvious differences, because cultural developments had produced changes in the criteria for approval, disapproval, and success and in specific subsistence behaviours. Thus the farming economy and settled village life meant that people spent at least some of their time doing things, like hoeing, ploughing, weeding, weaving, and building or repairing permanent dwellings, which had not been part of the experience of hunter-gatherers.

In the literature it is common to find two conflicting and opposite assumptions relating to behaviour in early farmers. On the one hand, some authors take it for granted that farming gave people much more leisure time than hunter-gatherers, which could be used for all kinds of enjoyable cultural activities. Others assume that the farmers' life was necessarily very hard, and we frequently find references to their constant toil in the fields. In fact the evidence shows that there is considerable variation in the amount of time that early farmers have had to spend working in their fields. For shifting agriculture in the tropics, average figures over a full year range from 10 to 30 hours a week. This is about the same as the amount of time most people in primeval societies spent hunting and gathering; the difference is that the farming activities may be somewhat less enjoyable. The figure for fixed field agriculture in Europe in the Middle Ages is estimated to have been around 30 to 35 hours each week [76]. There has been some debate on whether or not increasing intensity of farming results in an increase in hours of work in the fields on a per capita basis [77]. It may be noted, however, that the lowest of all estimates—15 to 30 minutes per day—is for the very intensive agricultural system of traditional Chinese peasant society producing rice and beans (see page 00).

The transition from the primeval to the early farming phase of human existence does not appear to have been associated with major changes in the intangible or psycho-social aspects of life conditions at the basic level. For example, it is likely that most people were frequently involved in co-operative small-group interaction, that they practised learned manual skills and engaged in some creative activity, and that they experienced life conditions conducive to a sense of personal involvement, sense of belonging, and sense of responsibility. Two significant differences from the primeval situation are, however, discernible. First, we see the beginning of an important change with respect to variety in daily experience. While there has certainly been a great deal of variability among early farming communities in this regard, in general it does seem that in some situations the farming lifestyle was probably more monotonous than that of hunting and gathering. Second, we noted in chapter 4 that the hunter-gatherer lifestyle was characterized by relatively short goal-achievement cycles. Individuals set out each day to seek and find food in the environment, sometimes successfully and sometimes not; but if they failed, the likelihood was that they would succeed the following day. In contrast, the farmers' goals with respect to food acquisition are often achieved only after a delay of many weeks or months.

NOTES

1. Broadly, with respect to the cultivation of plants, two main kinds of techniques have been used. These are referred to, respectively, as *seed agriculture* and *vegeculture*. The former, which involves growing plants from seeds, is used for most crops where the seed is itself the main food component, as in the case of wheat, barley, rye, oats, maize, and rice; and it is also used for some vegetable crops. It is the dominant method for the cultivation of food plants in the modern high-energy societies. Vegeculture is the vegetative propagation of plants—a method which is used for potatoes and, especially in tropical areas, for a wide variety of root and tuber crops such as taro, cassava, yams, sweet potatoes, and arrow-root. It is still the main method of cultivation used in contemporary early farming societies in parts of South-East Asia, West Africa, and South America. *Tree crops* have also been important in some parts of the world. They include olives, dates, and citrus fruits in the Mediterranean region, and bananas and coconuts in tropical and subtropical regions. See Grigg 1974 (this book provides a very useful account of farming practices the world over, past and present).

2. The cultivation of animals and plants for food production is not restricted to humankind. Various species of ants, for example, cultivate fungi and tend aphids for this purpose. The difference between farming by ants and farming by humans lies in the fact that in the former case it is an innate, phylogenetically determined process, whereas in the latter it is not innate, but a learned, culturally determined phenomenon.

3. For a collection of different viewpoints, see Solheim 1972; Heiser 1973; Reed 1977*b*; Carter 1977.

4. Wendorf *et al.* 1979.

5. Hawkes 1963, pp. 99, 165.

6. Moore 1975, 1979; Reed 1977*a*, pp. 543–67.

7. Hawkes 1963, pp. 222–7 (Jericho); Hawkes 1963, pp. 223–5, 287–8 and Zohary 1969 (Jarmo); Moore 1975, 1979 (Abu-Hureyra); Mellaart 1964 (Çatal Hüyük).

8. Mellaart 1975, p. 16.

9. Mellaart 1964, p. 99.

10. Sauer 1956, p. 58, 1969, p. 84; Grigg 1974, p. 115.

11. Solheim 1972; Heiser 1973.

12. See MacNeish 1964, 1965; Heiser 1973; Bushnell 1976.

13. For useful discussions on this theme, see Ucko and Dimbleby 1969; Reed 1977*c*.

14. Reed 1977*a*, p. 563 has suggested that young girls played an essential role in this process.

15. Clark 1961; Schild 1976; Hawkes 1963, pp. 97–100.

16. There are three main *patterns of land-use* in early farming societies for the cultivation of plants for food production. First, there is swidden or *shifting agriculture*, in which an area is cleared of natural vegetation and crops are planted, grown, and harvested until such time as the growth rate of the plants declines due to depletion of nutrients in the soil. The area is then abandoned and the process is repeated at a new site. Eventually the soil in the original area may regain its fertility, so that the cultivators are able to return, clear it again, and plant their crops. Subsistence farming of this kind is still widely practised in some tropical and semitropical parts of the world today. It has been estimated that, as late as 1957, 200 million people depended for their food supply on this method, and that the total area (i.e. both cropland and fallow) used was more than twice the area given up to other kinds of agriculture. (For the history of shifting agriculture, see Grigg 1974). Second, there is *settled agriculture with fixed fields which relies on natural rainfall*. It displaced swidden agriculture in Europe, where it has been the main form of farming practised for at least two thousand years. The third method is *settled agriculture with fixed fields which relies on irrigation* for the cultivation of cereal crops. It was irrigation and not rain-fed agriculture which provided the food for the populations of the earliest cities in Mesopotamia and in the valleys of the Nile and of the Indus and Yellow Rivers. Today about 20 million hectares of the world's croplands are irrigated (see Skogerboe 1974, pp. 141–3 and Brown and Finsterbusch 1972, pp. 44–7).

17. Clark 1952; Narr 1956; Russell 1968.

18. Clark 1952, pp. 100–5; Darby 1956, p. 190.

19. The three-fold system may have been in use as early as the 4th century BC, in the Hellenistic Age in Greece (see Heichelheim 1956).

20. Grigg 1980, p. 74. Apparently, rather higher yields were obtained in Roman times in Italy, where one grain sown produced up to four grains at harvest (see Derry and Williams 1960, p. 66).

21. Except for swedes, mangel-wurzels, tomatoes, and coffee, no new species have been added to the list of commonly cultivated plants in 2000 years.

22. When applied to a *population* the term *equilibrium* refers to a state in which the relationship between the population and its environment and the resources therein is such that the size of the population tends to fluctuate around a fairly constant mean over long periods of time. This implies that the pattern and rate of resource use and of waste production by the population is one that can be sustained indefinitely in that environment. See Kormondy 1969, p. 84 and Kormondy 1974.

23. Another, less common, factor has been change in climate due to human activities. Recent work has suggested that soil exposure resulting from the reduction of plant cover may have a marked effect on rainfall when large tracts of land are involved (Charney Stone, and Quirk 1975, pp. 434-5; Dimbleby 1976, p. 206).

24. Cited in Hughes and Thirgood, 1982, p. 206.

25. Clark 1952, p. 7.

26. Brooke-Thomas 1979.

27. Clarke 1971, 1973.

28. The figures for the Andes are from Brooke-Thomas, 1979. Other figures are from Leach, 1976 and Newcombe, 1976 (the figure of 24 for the peasants of Yunnan is taken from Newcombe and the figure of 41 from Leach).

29. Leach 1976.

30. It has been estimated that about 1.6 per cent of the population of Europe around the year 1600 lived in cities of 100 000 or more inhabitants (Davis 1965); and that about 5 per cent of the population of central Europe lived in townships with 5000 or more inhabitants early in the 16th century (Pounds 1979).

31. Cook (1971) suggests a figure six-times higher than this for 'advanced' pre-industrial societies.

32. Harris 1969.

33. For good descriptions of the dispersal of food species, see Sauer 1969, pp. 28-39 and chapter 3; Brown and Finsterbusch 1972, pp. 20-1; Grigg 1974.

34. Hawkes 1969; Grigg 1974, p. 16.

35. Dimbleby 1976.

36. Darby 1956, pp. 184-5.

37. Clark 1952, pp. 14, 92.

38. Hughes and Thirgood 1982.

39. Reed 1977*b*, pp. 879-953. For a full description of the relationships between the different wild and cultivated cereals, see Zohary 1969.

40. Baker 1970, pp. 73-6.

41. Dimbleby 1972, 1976.

42. This applies only in situations in which human excreta are not returned to the soil and in which part of the crops are exported from the local ecosystem— that is, it applies to the majority of situations.

43. Burning may, however, have the adverse effect of destroying the humus in the upper part of the top-soil.
44. Rotational cropping is not necessary in the case of wet rice cultivation, in which case nutrients are constantly supplied by the water source.
45. Jacobsen and Adams 1958; Helbaek 1960.
46. Strahler 1956.
47. Heichelheim 1956, p. 171.
48. Woods 1984.
49. For a useful and entertaining account, see Carefoot and Sprott 1969.
50. Salaman 1949, p. 300.
51. Except where stated otherwise, the sources of the information given here on the Irish potato famine are Salaman 1949 and Carefoot and Sprott 1969.
52. Kee 1980, p. 86.
53. Carefoot and Sprott 1969, p. 83.
54. Jacobs 1969.
55. Boserup 1965. (For useful discussions of Boserup's idea, see Spooner 1972.)
56. See Clarke 1973.
57. It has been suggested that food production may have existed for a long time without any important change in societal organization, or even without the existence of permanent dwellings. See Wendorf *et al.* 1979.
58. Clarke 1971, p. 32.
59. Dixon 1979.
60. Wittfogel 1956.
61. Laslett 1971, p. 13.
62. Reynolds 1966.
63. Roper 1975.
64. Bocquet 1979; Hawkes 1963.
65. Clark 1952, pp. 97–8; Dixon 1979.
66. Reynolds 1980, p. 8.
67. Redfield 1953, p. 7; Mellaart 1964, 1975.
68. Jacobs 1969.
69. Grigg 1980, pp. 68–9; Christie 1980*a*, 1980*b*.
70. Skogerboe 1974.
71. Bews 1935, p. 200; Hawkes 1963, pp. 341, 348; Mellaart 1964, 1975; Hughes 1975.
72. *Collins Dictionary of the English Language* 1979.
73. Clark 1952, pp.100–5; Grigg 1980, pp. 36–8.
74. Brown and Finsterbusch 1972, p. 55; Brothwell and Sandison 1967.
75. Livingstone 1962.
76. Waldron and Ricklefs 1973.
77. See, for example, Clarke 1971.

6

THE EARLY URBAN PHASE

INTRODUCTION

It is usually assumed that the first cities came into existence in Mesopotamia in the valleys of the Tigris and Euphrates Rivers somewhere around 3500 BC, in the area known as Sumer [1]. Clearly, however, this development did not take place overnight. We have already noted how the settlement of Çatal Hüyük in Anatolia, which flourished around 6000 BC, had some of the features of a city. In Sumer, the population is believed to have increased about thirteen-fold between 6000 and 4000 BC, and during this time people were coming to live together in larger and larger groups [2]. Eventually, about fifteen cities developed in the area, the best known being Eridu, Uruch, Lagash, and Ur [3].

Early civilization took a somewhat different form in Egypt where, between 4000 and 3000 BC, many large buildings and palaces were constructed and the population became increasingly specialized occupationally. The specialists, however, were living mainly on the estates of wealthy princes and were essentially part of the royal household. True cities did not come into existence in Egypt until considerably later. The Harappan civilization, which began rather suddenly around 2500 BC in the valley of the Indus River, did produce true cities, as did the civilization which developed in the valley of the Yellow River between 2000 and 1500 BC. In the new world, the Olmec civilization was in existence by 1200 BC, in the swampy coastal rain forest near Vera Cruz in Mexico; and it is possible that some elaborate temple structures in the highlands of Peru date back to 2000 BC [4].

The question of *why* cities came into existence is one that has occupied the minds of scholars for many years, and a number of different explanations have been put forward [5]. Whatever the answer to this question, we can be sure of one thing, and it is that the first cities were not the result of a preconceived or master-minded plan. They just happened, and the people who were involved simply adjusted as best they could to the new situation.

Before embarking on a discussion on the ecological and biosocial characteristics of early urban societies, two cautionary comments must be made. Both have to do with perspectives. There is a tendency for us in the 20th century to conjure up in our minds an image of early Sumer as an age of great intellectual excitement and inventiveness, of scholars at their desks working out systems of writing and of mathematics, of public servants formulating new laws to protect the interests of individuals, and of artists and craftsmen enthusiastically improving their techniques. Indeed, the literature abounds with comment on the extraordinary creativity of the gifted people of Sumer, and there has been much learned debate about how and why they achieved so much in so short a time. On the other hand, it should be borne in mind that around two thousand years passed between the time when permanent temples were first constructed in the area and communities settled about them to the time when government, trade, crafts, and writing can be said to have attained 'mature status'— that is, by around 2500 BC [6]. This was about eighty generations, roughly the same period of time as has elapsed between the birth of Jesus Christ and the present day. The degree of change, therefore, from one generation to the next may often have been almost imperceptible. Certainly, it was very slow in comparison with the rate of technological and societal change that has occurred in the past one hundred years.

The second perspective relates to the number of people who, until recent times, have lived in cities. While it is proper to put emphasis on the ecological significance of urbanization, it is also relevant that, while cities have been in existence for 5000 years, only a very small minority of the human beings that have existed during this period have lived in them.

THE RELATIONSHIPS BETWEEN HUMAN SOCIETY AND THE BIOSPHERE

From the ecological standpoint, two fundamental developments lie at the root of all the others that set cities apart from earlier human societies, and both of them have had far-reaching biological and cultural repercussions. First, there was the increase in population density itself, the various consequences of which will be discussed in the next section. Second, most of the people living in cities, except perhaps after the very early days, were not farmers [7]. That is to say, for the first time in human experience, a significant proportion of the population went through life without participating in the intimate interactions with the natural environment which are associated with the food quest. For sustenance, they relied on the activities of others. This was not only a novel development in the history of the human species; it is unique among the vertebrates [8].

It is clear that the impacts of early urban societies on the biosphere were out of all proportion to the number of people living in them. For instance, their use of timber for buildings, for industry, and for shipbuilding contributed importantly to deforestation. City populations also played an important role, indirectly, in the redistribution of species of animals, plants, and micro-organisms around the world. It was events in Rome, for example, which ultimately resulted in the introduction of rabbits and of poplars, elms, and sweet chestnut trees into Britain; and it was happenings in the cities of Portugal, Spain, and Britain which ultimately led to the introduction of horses, wheat, and smallpox into the Americas and to the introduction of potatoes, maize, and possibly syphilis into Europe.

SOCIETY

Until the 17th and 18th centuries AD, the populations of cities in Europe and the Middle East did not, with a few exceptions, exceed about one hundred thousand. Most of the early cities in Sumer are thought to have had populations of around ten to twenty thousand, possibly reaching fifty thousand in the case of Uruk [9]. Athens in the days of Pericles and Socrates probably had a population of about one hundred thousand. Exceptions were Rome and, later, Constantinople. Estimates for the population of Rome at about the time of Jesus Christ range from three hundred and fifty thousand to one million [10]. Three or four hundred years later Constantinople is said to have had a population of around one million [11]. During the Middle Ages and at the time of the Renaissance, most of the larger cities of Europe had populations of between twenty-five and sixty thousand, although Venice and Milan may have exceeded one hundred thousand. During the 17th century, however, a series of factors resulted in large increases in the populations of some cities, and, by the end of the 18th century, Paris had a population of over 670 000, Naples of over 430 000 and London of over 800 000 [12].

During the early part of the Middle Ages, there was a great multiplication of cities in Europe, associated with a considerable increase in the total population of the region. It has been reported that 2500 townships were formed in what is now Germany between the 12th and 14th centuries.

It has been estimated that the population density in the Sumerian city of Ur was between 300 and 500 persons per hectare [13]. This reminds us of the situation in recent times in Hong Kong. In 1971, the population density in squatter areas in Hong Kong was about 830 persons per hectare [14]. The density of populations in the cities of Europe in medieval and Renaissance times is thought to have varied from around fifty persons per hectare to nearly three hundred per hectare [15]. This primary change in the human condition had important repercussions of various kinds—biological, social, and economic.

The other outstandingly important development accompanying urbanization was the shift from relative homogeneity to heterogeneity in human populations. This change was made up of three basic interrelated elements: occupational specialism; political stratification (social stratification in terms of power); and wealth stratification (stratification in terms of material possessions). We may note in passing that material wealth in itself was a new factor in human experience.

From the biohistorical standpoint, one of the most significant consequences of population heterogeneity is the fact that, for the first time in the history of humankind, big differences existed in the life conditions of different groups of people living within the same society. This meant, in turn, that the different groups of people experienced different patterns of health and disease. Moreover, unlike situations in earlier times, new societal developments could be good for one group of people in a community, but bad for another group.

Division of labour, or occupational specialism, has been described as the hallmark of civilization. Already in the cities of Sumer five thousand years ago there were leather-workers, cabinet-makers, potters, metal-workers, basket-makers, shopkeepers, brewers, stone-cutters, weavers, gardeners, artists, music-makers, soldiers, scribes, priests, and other officials in the administration; and the pattern was much the same throughout the whole of the early urban phase of history. In general, the specialist occupations in the early cities were held by males. Adult females were usually in charge of the household dimension of the economy.

The fact that a high proportion of the population in cities was no longer involved in food production meant that societal machinery was necessary to redistribute food and other essential commodities among the people. This machinery necessarily involved the collection and storage of foodstuffs and a system for allocating them among the populace. In the early cities in Mesopotamia, this task was apparently a responsibility of the state—the responsibility, that is, of the group of priests and officials which constituted the administration. One of the outcomes of this arrangement was consolidation of the power of the administration and particularly of the king, who was considered to represent the local god—the giver of all good things. Later in history this redistributive role was taken over by the free market—but by that time kings and their administrations had other means of achieving and maintaining power.

An important distinction between social stratification in most early urban situations and those of primeval society was not only that the hierarchies in the former were more complex and more extreme, but also that they were less transient and were seldom decided spontaneously according to the natural propensities of individuals for leadership in different circumstances. Unlike in the primeval situation, they were determined mainly by birth and they did not vary from hour to hour or from day to day. Certainly, the hierarchy could be destroyed or changed by invasion or revolution, but otherwise it was relatively permanent. An individual could be born dominant, to become a king or queen or member of the nobility, or born middle class or lower class; and for the first time in human history, he or she could be born a slave, completely subservient to the will of others.

The other important aspect of occupational specialism was the fact that some occupations were clearly less pleasant than others. It is not surprising that individuals in the more desirable positions strove to retain them and, reflecting the common behavioural tendency of humans to support and protect their offspring, they did their best to ensure that their children were similarly privileged. The following quotation from the *Teaching of Khety, Son of Duauf*, possibly dating back to the Eleventh Dynasty in Egypt (around 2000 BC), illustrates these points well. A father is taking his son to school and advises him to work hard, emphasizing the advantages of the scribe's profession. He says:

I have never seen the smith as an ambassador, but I have seen the smith at his work at the mouth of his furnace, his fingers like the crocodile's, and he stank more than fishes' eggs ... . The stonemason finds his work in every kind of hard stone. When he has finished his labours his arms are worn out, and he sleeps all doubled up until sunrise. His knees and his spine are broken ... . The barber shaves from morning till night; he never sits down except to meals. He hurries from house to house looking for business. He wears out his arms to fill his stomach, like bees eating their own honey ... . The farmer wears the same clothes for all times. His voice is as raucous as a crow's. His fingers are always busy, his arms are dried up by the wind. He takes his rest—when he does get any rest—in the mud. If he's in good health he shares good health with the beasts; if he is ill his bed is the bare earth in the middle of his beasts. Scarcely does he get home at night when off he has to start again ... . Apply your heart to learning. In truth there is nothing that can compare with it. If you have profited by a single day at school it is a gain for eternity [16].

Often the life conditions of specialist groups deviated significantly from those of the evolutionary environment, and consequently signs of phylogenetic maladjustment were common. The following short list of some well known occupational diseases illustrates this point: baker's asthma, blacksmith's deafness, brassfounder's ague, butcher's tubercle, chain-maker's cataract, chimney-sweep's cancer, coal-miner's phthisis, cotton-twister's cramp, dustman's shoulder, foundry fever, hatter's shakes, lighterman's bottom, miner's nystagmus, nun's bursitis, painter's colic, washerwoman's itch, weaver's bottom, and writer's cramp [17].

Use and exchange of resources

With respect to patterns of resource use and of exchange of resources and commodities, the early urban phase involved some very important developments. This topic is a complex one, and is the theme of economic history. Here we can only briefly touch on some of the more outstanding aspects as seen from the biohistorical standpoint.

One of the most significant changes which accompanied urbanization was a substantial increase in the importance of non-essential or luxury commodities in the economy. This development had a major influence on the growth of trade between communities which were often considerable distances apart. For example, while the earliest cities in Sumer were, with their farming hinterland, self-sufficient with respect to food and simple housing materials (e.g. mud, reeds, and date palm logs), the region was totally lacking in good timber and in mineral resources. With the development of new technologies, especially metallurgy, with the contribution of monumental religious buildings and palaces, and with the growing demand of luxury goods, arrangements had to be made to bring the appropriate raw materials from elsewhere [18].

As demand increased, this trade required more organization. Initially the agents responsible for the exchange were public servants or officials of the city states, and not free entrepreneurs. According to some authors, by the time of Hammurabi in the 18th century BC, a merchant class had come into existence and private enterprise began to play a significant part in the economy [19]. This situation then persisted throughout the early urban period.

Trade between cities and between regions continued to be a feature of the early urban period, and it was mostly in non-essential or luxury items, at least in the first part of the period. Later, possibly as a consequence of increasing scarcity resulting from population pressure, some regions in Europe began to specialize in certain primary products, each putting its main effort into those activities in which it had the greatest comparative advantage over other regions, while buying from other regions the commodities in which it was at a disadvantage. This trend was also discernible on a very local level. In England in the 14th and 15th centuries, for example, as the monetary economy became increasingly important, landlords discovered that they could increase their profits by concentrating on special crops, and as a result the tradition of manorial self-subsistence began to crumble away [20]. These developments represent the beginning of a process of immense ecological significance—a process by which human populations have become increasingly dependent, even for the satisfaction of their universal survival needs, on other human populations in far-away places. This lengthening of chains of ecological dependency has become much more extreme in the modern industrial phase of human existence.

Within the early cities, the new introduction of all manner of material goods—textiles, utensils of many kinds, ornaments, earthenware, metalware, jewellery, furniture, and so on—resulted in changes in the meaning and importance of ownership and of material wealth. Possessions became, in a sense, an extension of the individual's personality [21]. Some of them had practical functions, but more importantly the quantity and quality of possessions became symbols of status, and the idea became entrenched in culture that having large numbers of possessions is a good thing. This notion has, of course, become an immensely important influence on human behaviour throughout history, although it has been questioned again and again by religious leaders and philosophers.

Another profoundly important development in the early Sumerian cities was the extension of the concept of ownership so that it applied not only to land, animals, and material objects, but also, for the first time in human history, to other members of the human species. In the Mesopotamian cities slaves were quite plentiful. Most of them were probably collected during raids into the hills flanking the Tigris and Euphrates valleys, but others were captured in the frequent scuffles between the cities. The concept of ownership also eventually applied to the various members of a married man's family. In Ur, for example, a man could avoid bankruptcy—that is, he could avoid being sold into slavery—by selling his wife or children in order to pay off his debts [22]. Cultural evolution thus gave rise to an entirely new perception of the nature of the relationship between spouses. However, although both slavery and the notion of a wife as a part of a man's chattels persisted in some societies throughout the early urban phase of human history, neither was universal.

A significant economic development associated mainly with early urbanization was the introduction and increasing importance of a cash economy. The desire to exchange things is as old as humankind , but it is not older. I can think of no other species which engages in exchange of material objects, although giving and taking is, of course, common among animals. When such exchange took place in relatively simple economic situations, such as in primeval societies, it was usually by the process of barter. However, in early urban societies (and in some early farming societies) an alternative system was introduced involving a third element— *money*. There have been countless definitions of money, but for our purposes the following quotation from Galbraith is sufficient: '... money is nothing more or less than what he or she always thought it was—what is commonly offered or received for the purchase of or sale of goods, services or other things' [23].

To perform its societal function, money need have no intrinsic value itself. This is very obvious when we consider the paper money, cheques, and computerized money which are in use in our own society. Money is simply a symbol which can be given in exchange for material goods or for services and which can be used by the recipient for the same purpose.

In the early cities, precious metals were used as a form of currency. They had the advantage of being durable, but they had also the disadvantage that the necessary weighing and dividing up was a tedious process. So in time the idea arose of minting metal coins of a definite weight. The first coins were probably struck some time during the 7th century BC, either in the Greek coastal cities of Ionia or by the kings of Lydia. The use of coin money then spread south eastwards towards Persia, and westwards towards the Aegean Islands and mainland Greece. Later it was taken up by the Semitic peoples, the Celts, the Romans, and the peoples of India. Certainly by the 6th century BC the new money economy was becoming well established in Greece, and bankers and commercial go-betweens were becoming important components of society. By the 3rd century BC, a new type of money had appeared—the bill of exchange, payable to any bearer.

The development was to become a major influence in economic affairs two millennia later.

While money, as a convenient medium of exchange, was used consistently from many centuries BC, most transactions, especially in rural areas, were for a long time still based on barter. In Britain, for example, it was not until around 1300 AD that the monetary economy began to spread into and become important in rural areas—a development which contributed significantly to societal and ecological changes. Money came to be increasingly used, not only for material goods but also for services and labour. Landlords were finding that the employment of wage labourers was more profitable to them than that of serfs, and serfdom itself began to give way to tenancy—a change which was accelerated late in the 14th century after the Black Death had reduced labour supply [24].

Warfare

An outstanding feature of the early urban phase of human existence was warfare. In fact more has been written and taught about this aspect of human history during this period than about any other. Here we will consider only a few of the most essential points from the biohistorical viewpoint.

It was with urbanization that warfare became institutionalized and accepted world-wide as a normal aspect of human affairs. As discussed earlier, violence between groups is unlikely to have been a constant feature of the primeval phase of human history, and some authors consider that it was very rare. Similarly, there is plenty of evidence that many early farming communities have lived for long periods at peace with their neighbours, although this has not always been the case.

In the early days of Sumerian civilization there was no standing army, although the king may have had a small bodyguard. The male citizens of the city were organized into a fighting force when the occasion demanded. At first this occurred when wandering bands of barbarians invaded the area. However, as the region came to be more sharply divided into city states, each owing allegiance to a different patron god, and as water and land became increasingly scarce, warfare developed between them. By the beginning of the third millennium BC, military formations with a 'definitely professional look' were in evidence [25]. Foot soldiers were uniformly equipped with bronze or copper helmets, big rectangular shields, and long spears and were depicted as being drawn up in a six-deep phalanx; and other troops, with helmets but no body armour, carried long lances and were arranged in double file. It is not clear whether these were full-time soldiers, but we do know that Sargon, King of Akkad around 2385 BC, the first real empire builder in the region, found it necessary to have a standing army of 5400 men. Hammurabi of Babylon also had a large force of professional soldiers.

So it was that deliberately organized lethal combat came to be regarded as normal, and indeed as a good thing—so long as it was in the name of one's own god—and warriors came to be regarded as heroes in society. The pattern was thus set for the rest of the early urban period, and aspects of it persist to the present day.

Let us note, however, that well before the time of Jesus Christ there were some individuals who, although presumably regarding warfare in the name of a god as inevitable for the time being, looked forward to a day when it would no longer be necessary—an attitude reflected in the following words from the Book of Micah in the Old Testament: 'They shall beat their swords into plow shares and their spears into pruning hooks. Nation shall not lift sword against nation, nor study the art of war any more' [26].

The point which needs emphasizing is that the frequent occurrence of organized warfare throughout the 5000 or so years since the formation of the first cities cannot be taken as an indication of an innate aggressive tendency in humans which must inevitably lead to one war after another *ad infinitum*. Certainly, human beings, and especially males, are quite capable of engaging in lethal combat, sometimes with great enthusiasm; but it is equally true that most people most of the time do not like and do not want to be personally involved in war. Early evidence for this assertion comes from ancient Egypt. When the Pharaohs of the Middle Kingdom, around 2000 BC, decided to embark upon foreign expeditions, they had great difficulty in raising an army from the Egyptian peasantry, the members of which are said to have detested army service, and the Pharaohs were forced to recruit foreign mercenaries for this purpose [27].

One characteristic of warfare in the early urban phase is that combat was carried out almost exclusively on a face-to-face basis. Those that were killed, men, women, or children, literally died at the hands of other human beings. This was still largely true even at the end of the period, in spite of the introduction of firearms.

In-groups and out-groups

Occupational specialism, social stratification, and the large number of people living together in cities resulted in a fundamental change of great importance in the ecology of human beings—a change in the in-group and out-group structure of society. Before discussing this change, some brief comments are necessary on terminology. The expression *in-group* is used here to denote a group of people with which an individual identifies, and the members of which behave supportively and show loyalty towards each other. The relationships in in-groups are usually face to face. The term *primary group* is used exclusively for in-groups for which the relationships are face to face, such as the extended family, the nuclear family, some small occupational groups, and the street gangs of modern Western society.

Reference group has a broader meaning, and the relationships in it are not necessarily face to face [28]. In the words of Nisbet:

Any type of social aggregate or group may be a reference group: small or large, open or closed, personal or territorial, *Gemeinschaft* or *Gesellschaft*. A reference group may be organized around kinship, religion, social status, ethnicity, economic interests, or political values. It may be near at hand, the scene of an individual's incessant participation in its activities, or it may be distant and dispersed. A reference group can be deeply influential in an individual's life even when he is not a member of it. What is essential in the concept of a reference group is that it is the group or social aggregate to which an individual 'refers', consciously or unconsciously, in the shaping of his attitudes on a given subject or in the formation of his conduct. It is the social aggregate toward which he orients his aspirations, judgements, tastes, and even his profoundest moral or social values [29].

The extent to which a reference group is also an in-group depends on circumstances, and is influenced in particular by the degree to which the reference group is perceived by its members to be under threat. As Adam Ferguson pointed out in 1767—the stronger the threat from outsiders, the closer are the internal bonds within a collective [30]. Symbols, such as flags, national anthems, coats of arms, tartans, uniforms, and political emblems sometimes serve to increase or strengthen the bonds within reference groups.

Out-groups are all those groups of people which are perceived by an individual to be external to the in-group or in-groups to which he or she belongs.

In the evolutionary or primeval setting, the in-group and out-group situation was relatively simple. Individuals usually belonged to a single in-group, the band, which was composed largely of members of the extended family. Out-groups were other bands, which were for most of the time out of sight, and probably usually out of mind. Perhaps populations of other species of animal in the local environment were regarded in much the same way as were out-groups of human beings in later societies.

We noted in chapter 4 that, although hunter-gatherers often live at peace with neighbouring bands, they do not exhibit any particular concern for their well-being. This fact, as well as observations on other societies, past and present, leads to the conclusion that, whereas we can recognize in the human species a common behavioural tendency to show loyalty towards and to be supportive of the in-group and its members, there is no corresponding innate or built-in concern for the well-being of out-groups. Indeed, there seems to be a common tendency to regard out-groups with suspicion; and together the lack of concern and the suspicion may sometimes lead to hostility, overt aggression, and violence.

These behavioural characteristics of the human species are especially relevant to the fundamental changes which occurred in the in-group–out-group situation with the formation of cities. The community of which a person was a part could no longer be counted in tens, but rather in thousands, or even in tens of thousands; and there was an enormous increase in the number of daily contacts that the individual experienced with other members of the species. These contacts were not only with members of the family and with workmates, but also with casual acquaintances and with complete strangers. Because of occupational specialism and stratification, these other people came to be classified in the mind of the individual as belonging to distinct categories or classes. The tendency of occupational groups to band together as in-groups, with common interests to be closely guarded, was reflected in the early formation of craft guilds with the purpose, among others, of protecting the secrets of the trade. In Ur, the 'chief of the weavers' and the 'chief of the joiners' were the representatives of their guilds to the government and were supposed to look after the interests of their professional colleagues in the city state [31].

Thus one of the important biosocial effects of urban specialization and stratification on the individual was that the world became full of different out-groups to which he or she did not belong. For the first time, the distinction between 'us' and 'them' became an extraordinarily important feature of the life conditions of the average person. Some of these out-groups wielded power over the individual and were to be feared; others were to be despised. Because of the human tendency to treat out-groups with suspicion, the lack of innate concern for the well-being of members of out-groups, and the lack of an effective innate mechanism to prevent overt hostility between out-groups, the new in-group–out-group structure of early urban society created a potentially explosive situation. It was one in which the average citizen could be severely threatened by various forms of abuse, including physical attack, and it was one that could be extremely disruptive of societal harmony in general. Moreover, the situation opened the door to the possibility that the more powerful groups in society might take advantage in various ways of weaker groups; for example, by extortion of their goods and chattels.

Difficulties of this kind were, in fact, experienced early on in the history of cities. An historian living around 2350 BC in the city of Lagash in Sumeria described the situation which had existed before certain important reforms were introduced in his lifetime. According to Kramer, his description 'provides a grim and ominous picture of man's cruelty toward man on all levels—social, economic, political, and psychological' [32].

In the absence in this new situation of an effective natural mechanism for controlling social behaviour it was necessary for society to compensate by introducing special measures to protect the members of the community from each other. This cultural adaptive response took the form, in part, of the introduction of regulations and the institutionalization of punishments for disobeying them. To a greater or lesser degree, these regulations or laws were regarded as instructions from a deity. The Code of Hammurabi from about 1750 BC, and that of Moses from around the same time, are particularly well known, although they are both essentially revisions of earlier codes. In fact, evidence suggests that the promulgation of laws by the rulers of the Sumerian city states was common already by 2400 BC [33].

The following component of the Hittite Code is of some interest in the ecological context:

> If a man puts filth into a pot or tank, formerly they paid six shekels of silver: he who put in filth paid three shekels of silver [to the owner?] and into the palace they used to take three shekels of silver. But now the king has remitted the share of the palace; the one who puts in the filth pays three shekels of silver only [34].

Thus we see, with urbanization, the introduction of a new way of controlling social behaviour, quite different from that which had operated for tens of thousands of generations of humankind beforehand. Basically, it was necessary because of the fact already discussed that human beings do not appear to have any innate tendency to care very much about the well-being of members of out-groups. However, the new written law enforced by the authority of the state did not mean the disappearance of the older 'natural' mechanisms of social control which depended upon certain of the common behavioural tendencies of humankind. The customs and values of the common people still remained an important factor, partly because they influenced the law itself and partly because they still formed the basis of the criteria for disapproval and censure within in-groups, such as the family and the occupational group; and fear of such disapproval acted as an effective check on social behaviour at this level. This applies not only to attitudes towards various specific forms of behaviour within in-groups, but also to the values held by in-groups with respect to the breaking of the law itself. Thus in some urban families today, but not in all, law breaking is considered a bad thing in itself and consequently elicits in-group disapproval.

Writing and learning

One of the unique characteristics of *Homo sapiens* is the ability to invent and use symbols. In the evolutionary setting, the symbols in question consisted of oral sounds, and they were used in everyday communication. Eventually, proper languages were developed and these have presumably been in existence in some form or another for several hundred thousand years.

The cultural application of this capacity to communicate with symbols seen in the development of writing occurred relatively late in human history, when the biological characteristics of *Homo sapiens sapiens* were well established. In other words, unlike the capacity of speech, writing has not influenced the phylogenetic characteristics of humankind. When it did arise, it was essentially the product of economic necessity. Humans had, of course, been drawing and painting for tens of thousands of years before writing developed; but at a certain stage, after the domestic transition, it became useful to apply this sort of technique to the recording of information—a practice which some authors suspect may even go back eleven thousand years [35]. It was, however, in the very early cities that writing became really important, initially for keeping economic records (tallies of cattle, sheep, jars of butter, measures of grain, etc.) which were clearly seen as necessary for the business of the state. As time went on, however, and as the art of writing became more sophisticated, it was used not only for business documents, but also for royal inscriptions, legal codes, religious texts, and the recording of legends and political events. And during the Mesopotamian period, writing evolved from the initial use of simple pictographs, through hieroglyphs, eventually to phonetic signs. At one stage during the course of this evolution, writing became extremely complicated. Around 3500 BC, for example, two thousand signs were in use. Three or four hundred years later, the number had dwindled to about eight hundred, and by 2900 BC another two hundred signs in these texts had disappeared [36].

Writing thus became an extremely important extra component of culture, and has been used for recording facts, fallacies, and fantasies, and for communicating these not only across national boundaries, but also across generations. And it has made possible, especially since the invention of the printing press, the accumulation and dissemination of a massive amount of information.

It is worth mentioning another consequence of the introduction of writing in Sumer. As writing became increasingly complicated, it was necessary for special steps to be taken to train people in this art. It was in this way that the first schools came into existence, probably about five thousand years ago. Today one of the best known schools from Sumerian times is one discovered at Ur and now known as 'No 1 Broad Street'. It is dated at around 1780 BC, and probably accommodated about two dozen boys. Nearly two thousand tablets have been found there, including several hundred 'school exercise' tablets. Referring to this school, Leonard Woolley has written:

Such, then, was the material setting of the educational system, a setting of the simplest sort; the scene in the Broad Street school, with the boys seated or squatting in rows in the courtyard and guest room, waiting for the headmaster to come downstairs from his private quarters while the ushers or pupil-teachers supervised the preparation of the morning lesson ... [37].

Schooling was another new experience for humankind. In primeval times, learning had been largely spontaneous and through imitation, and children probably acquired most of their knowledge and skills from watching and copying slightly older siblings and peers. In the early farming period a similar situation presumably prevailed, and there was certainly no formal instruction. Even in the early cities, where a whole range of different arts and crafts was practised, occupations were largely hereditary and members of the younger generation learned their professions mainly by watching and imitating other members of the family. But schools, as places where young people were required to devote themselves to tasks which seldom inspired in them the least degree of enthusiasm, were something new. Because the lessons were often dull and meaningless to the students, strict discipline, associated with an air of formality, was necessary to ensure any degree of success. Fear replaced fun as a motive for learning, and the school teacher emerged as another new kind of specialist. Schooling was, from the very beginning, a major deviation from the primeval conditions, and it has continued to be so to the present day. To quote Woolley again [37]:

Discipline was strict. Boys might be 'kept in' over long periods; probably already they were given impositions, though it is only in the Neo-Babylonian time that we find actual examples of pupils having to write 'fifty lines' or 'a hundred lines' by way of punishment; but for the most part correction was by the stick, and the stick was used freely, by masters and by pupil-teachers alike.

Let us conclude this section on learning by referring to a somewhat satirical essay (written in cuneiform script) of a student at the No 1 Broad Street school:

'What did you do at school?' I reckoned up [or 'recited'] my tablet, ate my lunch, fashioned my [new] tablet, wrote and finished it; then they assigned me my oral work, and in the afternoon they assigned me my written work. When the school was dismissed, I went home, entered the house, and found my father sitting there. I told my father of my written work, then recited my tablet to him, and my father was delighted.

The next day the boy was less fortunate:

When I awoke early in the morning I faced my mother and said to her: 'Give me my lunch; I want to go to school.' My mother gave me two rolls and I set out. In the school the 'man on duty' said to me: 'Why are you late?' Afraid, and with my heart pounding, I entered before my teacher and bowed.

The story is taken up by Woolley as follows:

But the teacher was correcting the student's tablet of the day before and was not pleased with it, so gave him a caning. Then the overseer 'in charge of the school regulations' flogged him because 'you stared about in the street', and again because he was 'not properly dressed', and other members of the staff caned him for such misdemeanours as talking, standing up out of turn, and walking outside the gate; finally the headmaster told him, 'Your handwriting is unsatisfactory', and gave him a further beating. The luckless youth appeals to his father to mollify the powers above in the orthodox way, so the father invites the headmaster to his home, praises him for all that he has done to educate his son, gives him food and wine, dresses him in a new garment and puts a ring on his finger; the schoolboy waits upon him and in the mean-while 'unfolds to his father all that he has learnt of the art of tablet-writing', and the gratified teacher reacts with enthusiasm: 'Of your brothers may you be their leader, of your friends may you be their chief, may you rank the highest of the schoolboys. You have carried out well the school's activities, you have become a man of learning.' The schoolboy, claiming now the proud title of 'Sumerian', becomes in his turn a pupil-teacher ... .

Religion, values, and attitudes towards nature

The values and attitudes of a society are important ecologically because they influence human activities and consequently, in turn, affect biological systems. This applies particularly to values and attitudes relating to nature and to humankind's place in nature [38]. They are important not only because they may act, for example, as forces tending to promote activities which may adversely affect ecosystems or human health, but also because they may be influential in either blocking, or in permitting, potentially damaging activities of self-interested groupings or organizations in society. However, as J. D. Hughes writes:

Attitudes alone do not determine the way a human community will interact with the natural environment. People whose religion teaches them to treat the world as a sacred place may still manage to make their surroundings a scene of deforestation and erosion, because good intentions toward nature are not enough if they are not informed by accurate knowledge about nature and its workings [39].

The following comments relate to the early urban phase of human existence in the Western world up until the 17th century, when there occurred a new flurry of philosophical activity relevant to the interrelationships between humankind and nature. Broadly, we can recognize two levels—the emotional and the intellectual. These two aspects are, as always, closely connected, and they found expression in religious practices, in the writings of philosophers, scientists, and poets, and, no doubt, in the behaviour patterns of ordinary individuals.

On the emotional level, feelings ranged from enjoyment of and respect for the processes of nature to, at the other extreme, hostility and disrespect. On the intellectual level, the main concern was the question of how humans fitted into the natural world—and the extent to which they should be seen as subservient to or masters of the natural processes.

According to Hughes (1975, p. 35), the first urban societies in Mesopotamia were 'the first societies to abandon a religious attitude of oneness with nature and to adopt one of separation'. In Mesopotamian mythology nature was depicted as being in a state of chaos, and the task of human beings and their gods was to overcome this chaos and to establish order.

Somewhat later, Judaism made its distinctive contribution. The earth and all the living animals and plants therein, as well as the rest of the universe, were seen to be created by a single god. Again in the words of Hughes:

His creative power could be seen in the things that he had made, and his actions of providence and miracle could be seen in the world of nature. But God was above and beyond his creation, ruling it from on high. ... . Mankind's place on earth, ... is subordinate to God's ultimate dominion. God created human beings, male and female, in his own image, and gave them dominion over 'the fish of the sea and over the birds of the air and over every living thing that moves upon the earth' but only as God's deputies. Human beings are not the lords of creation, free to do with the earth as they please, but God's viceroys, responsible to God for their actions. ... . The grant of dominion was not a license to kill, exploit heedlessly, and pollute, and was not understood as such by the ancient Jews, although later Western thought did indeed take it in that distorted sense. ... . Compared with the nature cults of the surrounding peoples, Judaic thought regarded the natural environment with greater respect and care [40].

The early Greeks were exceptionally nature-conscious. In the religious sphere, all their gods and goddesses lived in the wilderness and all altars and places of worship were originally in the open in groves of trees. Greek religion saw humankind as being a part of, rather than separate from, the rest of nature.

As is well known, intellectuals in ancient Greece were very interested both in the processes of nature in their own right and in the relationships between humankind and nature. There emerged the idea of a design in nature, reflected, for example, in the order in which all the different kinds of animals and plants are arranged in an ascending scale. Humans, being the highest organisms in this scale, had the capacity to modify nature, and indeed to improve upon it. The ideal of a pleasant and harmonious relationship between humankind and nature developed out of this kind of thinking, which is discussed by Aristotle, and later by Cicero, Seneca, and Pliny [41].

Theophrastus, a student of Aristotle, was especially interested in the interrelationships and interdependencies in the natural world. He noticed, for example, how human activities sometimes brought about certain changes in climate. In fact he gathered information on the actual temperature changes during his own lifetime in Greece caused by the draining of marshes, the alteration of the course of a river, and deforestation (see note 38).

The Romans were an intensely practical people, and their attitudes towards nature were 'distinctly utilitarian'. The gods which they invented and their interactions with them were designed to increase their control over nature, especially in relation to food production and the avoidance of disease. They were interested in knowledge mainly in so far as it could be applied to improving their ecological situation. They were less interested in 'knowledge for its own sake'.

Artefacts

Discussion on the man-made environment in the early urban phase will be brief. Basically, we can recognize four classes of artefact as follows: *buildings; tools and machines; aesthetic or ornamental objects; symbolic objects* (e.g. for storing and communicating information; money).

The *buildings* of the early cities had the following main functions: to provide dwellings for people; to provide homes for gods and places for their worship; to provide storage for foodstuffs and other commodities; to provide shelter for human beings engaged in certain societal activities, such as administration and commercial transactions. In the early days of Mesopotamia, these four functions were basically performed by two kinds of building—the residential buildings where people lived and stored their own personal belongings, and the temples, which were at once places of worship and of storage and centres of administration and commercial activities. According to Lewis Mumford:

... the general appearance of these ancient Mesopotamian cities must have been ... very much like that of a walled North African city today: the same network of narrow streets or rather, alleys, perhaps no more than eight feet wide, with the same one, two, three storey houses, the same usable rooftops, the same inner courts, and finally, the steep pyramid of the ziggurat dominating it all, as the towers of the mosque now dominate the Moslem city. Beyond the walled but spacious temple precinct, spread a series of more or less coherent neighbourhoods in which smaller shrines and temples serve for the householder [42].

Indeed, the residential quarters in some of the earliest cities in Sumer had some features which are found in modern cities, but which disappeared in later cities. These included bathrooms, inside latrines, pottery pipes, and brick-lined drainage channels in the streets. On the whole, however, standards of hygiene in early urban communities were extraordinarily low (see below).

The main temple was an extremely important component of the early cities. In Ur, for example, the temple complex, known as the Temenos, covered an area of a little over four hectares and was raised above the level of the rest of the city. At one end of this complex, and on a higher terrace, a staged tower, the Ziggurat, rose 68 feet into the air. The terraces of the Ziggurat were planted with trees, and at the top there was a shrine for the Moon God. At a lower level and beside the Ziggurat there were store-rooms and offices to which the tenants who farmed the God's lands brought rents and sacrifices. The palace of the King, the earthly representative of the Moon God, was built against a wall of the Temenos and at a slightly lower level than the shrine of the Moon God. Thus we see in Ur an early example of a psychological device which was to be used again and again throughout the history of civilization; that is, the use of mere size and height of buildings, frequently combined with an appropriate setting on the top of a hill, to inspire in the community a sense of awe and respect for the establishment. Some later examples are the Parthenon in Athens, the Capitol in Washington DC, and the Kremlin in Moscow. The tallest building at the Australian National University is the administration block, and the Vice-Chancellor is located on the next-to-top floor. On the top floor is the large room where the ultimate decision-making body, the University Council, meets.

While there is little evidence in the very early cities of major disparities in residential accommodation, this situation did not persist. Indeed, such variability became an outstanding feature of early urban societies. In the case of Rome around the time of Julius Caesar, for instance, about 1800 families of patricians lived in huge private mansions, usually with big gardens and additional buildings for servants and slaves, while the great mass of the population lived in extremely overcrowded tenement buildings under appalling sanitary conditions [43]. Tenement buildings such as these were not seen again until the 17th century AD, when multi-storey buildings were constructed in a number of cities of Europe to accommodate the rapidly growing urban populations. Paris provides an example. In 1684, the Chief of Police in Paris referred to the 'frightful misery that afflicts the greater part of the population of this great city' [44]. This was at the same time as the ruling classes were creating the magnificent palaces of the baroque period and the great boulevards, parades, and vistas which serve today to provide us with a somewhat unbalanced impression of Parisian life at that time.

Most of the *tools and machines* which were characteristic of the early urban phase of human history as a whole had already been introduced by around 3000 BC. These included the ox-drawn plough, wheeled vehicles, sailing ships, and the potter's wheel. By the time of the Roman Empire, the basic technology for the period was well established; it included not only the above, but also pulleys, treadles, the Archimedean screw, lathes, bellows, and hand tools made out of metal, including, for example, hammers, chisels, and files. The various machines were designed mainly to make certain kinds of work easier to carry out; but, with a few exceptions, they all depended for their functioning on the somatic energy of human beings or animals. The important exceptions were the sailing ship, the windmill, and the water-mill—devices driven by extrasomatic energy in the form of wind or running water. Another exception late in the period was the steam engine, a form of which was first put to use around 1700 for pumping water out of a mine, and which was later to play so crucial a part in the transition of Western society into the high-energy phase.

Perhaps the most influential technological innovations made in the later part of the period were gunpowder, the mechanical clock, the compass, and the printing press. Some of the inhabitants of the early cities, as in the case of all other human societies, devoted their efforts to making objects of no practical value whatsoever—objects created simply because people enjoyed making them or because they enjoyed looking at them. Such *aesthetic* or *ornamental objects* are to be found today in almost every home, and art galleries contain nothing else. Let it be noted, however, that many of the objects made for practical purposes, such as buildings and vases, also have an aesthetic component, and are shaped or decorated to please the eye.

It is worth drawing attention to the fact that there are two different aspects of artistic endeavour which are relevant to the quality of human experience. First, there is the value of beautiful objects to the individuals who contemplate them, which is due to the important contribution that passive aesthetic experience makes to human health and well-being. This applies to any object of beauty whether or not it is made by humans. Second, there is the question of the value, to the creator of the object, of the act of creation itself. While comments on this question are necessarily subjective, I have no doubt myself that personal creative behaviour contributes positively to the health and well-being of the individual who engages in it. It is therefore important, I suggest, in attempts to assess the 'quality of life' in a population, to take into account the proportion of people who participate in creative activities. In hunter-gatherer times, this proportion would have been high: most, if not all, members of the population were involved. It would also seem that a high proportion of the population in early urban society regularly shaped things with their hands, although their experience differed from that of hunter-gatherers in that many of them were confined to a single spot and practised the same creative act for many hours of each day.

For the sake of completeness, mention must be made of *symbols* as the all-important fourth category of human-made components of the early urban environment. It includes all writing, whether on walls, clay tablets, papyrus, or paper, and it also includes money. These artefacts are discussed elsewhere.

HUMAN EXPERIENCE

Infectious diseases

It is a well established epidemiological principle that when animal populations become significantly more dense than is usual in the evolutionary environment of the species, disturbances occur in host–parasite relationships. The overcrowded conditions provide greatly increased opportunities for the spread of pathogenic microbes and other parasites from one host to another.

An intestinal roundworm, for example, is likely, through natural selection, to have developed a rate of egg production such that its eggs will be picked up by potential hosts in numbers which, while ensuring the perpetuation of the particular species of roundworm, will not interfere significantly with the health of the host under normal conditions. If the ecological conditions deviate from those typical of the evolutionary situation in such a way that the potential hosts pick up vastly larger numbers of eggs, then pathological manifestations of the parasitic infection are likely to occur. In some crowded cities in developing countries today, 5 per cent of the body weight of some children is attributable to the parasite content of their gastro-intestinal tract. Similarly, in the case of bacterial and viral parasites, crowding increases both the sources of infection and likely size of infective doses.

The concentration of large numbers of people in a small area can be expected to facilitate the spread of potentially pathogenic organisms in various ways. One of the most important of these is spread by touch—either when non-infected persons touch infected persons or when they touch, for example, handles of tools or doors, bed-clothes, and so on which have been touched by infected persons. Clearly, the more people there are around, the greater are the possibilities of rapid transfer of infectious agents by this means.

Many pathogenic organisms, such as those responsible for typhoid, cholera, infantile diarrhoea, and dysentery are spread mainly by contact, direct or indirect, with human excreta. In the primeval situation, infections of this kind are unlikely to have been a major problem. Excreta were deposited well away from the camp, and people did not usually stay at one site long enough for large concentrations to accumulate in the area. The situation was very different in the early cities, in many of which accumulating masses of human excreta were a constant menace to health, especially when they contaminated water supplies. Mumford has written:

In this respect, Rome, for all its engineering skill and wealth, failed miserably in the rudiments of municipal hygiene. As a result, the danger of having a chamber-pot emptied on one's head was as great, again, as in Edinburgh (Gardy-lo!) ... . In sum, in the great feats of engineering where Rome stood supreme, in the aqueducts, the underground sewers, and the paved ways, their total application was absurdly spotty and inefficient. By its very bigness and its rapacity, Rome defeated itself and never caught up with its own needs. There seems little doubt that the smaller provincial cities were better managed in these departments, just because they had not overpassed the human measure [45].

Another reason why big increases in the size and density of human populations resulted in greater prevalence of infectious diseases is associated with the fact that many pathogenic micro-organisms require a large contiguous population in order to exist. This applies particularly to those viruses which, if they do not kill their host, give rise to a state of long-lasting immunity. It has been calculated, for example, that the measles virus requires a contiguous human population of at least 300 000 to keep it going. If by some means the measles virus had come into existence in primeval times and had infected a member of a hunter-gatherer group, it would have been passed on to the other members of the group, some of whom might have died, while others would have recovered and become immune: and that would have been the end of the line for the virus [46]. This would also be true of the large number of mild acute virus infections which are circulating in the urban populations of the world today. They are all products of civilization.

Thus as soon as cultural developments resulted in large numbers of people living crowded together in cities, a potentially hazardous situation was created with respect to infectious disease. Although we know little about disease in the early Mesopotamian cities, we learn from the Bible that epidemics were well known to the Israelites. We read of the 'ten plagues of Egypt' and of how 'the angel of the Lord went out, and smote in the camp of the Assyrians an hundred four-score and five thousand: and when they arose early in the morning, behold they were all dead corpses' [47].

The first recorded outbreak of infectious disease in Rome occurred in 707 BC, only 44 years after the supposed foundation of the city, and many more epidemics were recorded thereafter [48]. Perhaps the best documented epidemic of antiquity was that which hit Athens at the time of the Peloponnesian Wars [49]. Political events had created a situation conducive to infectious disease; in the summer of 430 BC, large armies were camped around Athens, and the population of the countryside swarmed into the city, which became extremely overcrowded. The disease, which was probably smallpox, struck suddenly and its symptoms were distinctly unpleasant: severe headache, redness of the eyes, followed by inflammation of the tongue and pharynx, with sneezing and coughing followed by vomiting, diarrhoea and excessive thirst. The individuals that died usually did so between the seventh and ninth days. At the height of the fever, the body was covered with reddish spots, some of which ulcerated. Severe cases which recovered often suffered from necrosis of the fingers, toes and genitals: some lost their eyesight and in many cases there was complete loss of memory. The epidemic had a major impact on political events, and it essentially brought the current hostilities to an end. The Peloponnesians left the region in a hurry for fear of the disease: and in Athens itself people became completely demoralized and a period of lawlessness resulted. The statesman Pericles died in this epidemic, as did two of his sons and his sister.

Infectious disease thus became firmly established as a feature of Western civilization. Sometimes it took the form of terrible epidemics; but it was also endemic, as in the case of tuberculosis and infantile diarrhoea and other causes of death in childhood. The case of the children of Queen Anne (1665–1714) is often mentioned in this context. There were sixteen of them in all, and of these the only one to survive infancy died at the age of eleven [50].

The most important of the infectious diseases throughout the early urban phase were dysentery, enteric fever, typhoid, typhus, the plague, smallpox, tuberculosis, and, late in the period, cholera.

Pestilence then, was, in large part a biological response to the high population densities in and around cities and to the generally poor standards of hygiene of the times, but it was exacerbated by another evodeviant situation, namely, the large and mobile concentrations of adult males taking part in military expeditions [51]. As numerous authors have pointed out, pathogenic micro-organisms often had more influence on the outcome of battles and of wars than did military commanders. In fact it would be impossible to exaggerate the part played by infectious diseases in human affairs throughout the early urban phase and, indeed, well into the period of transition into the high-energy ecological phase of human society. The disease due to infection with the bacterium known as *Pasteurella pestis* deserves special mention in the annals of culture–nature interplay. It takes one of two forms in human beings, known respectively as bubonic plague and pneumonic plague. The most characteristic symptom of the former is the presence of extremely painful 'buboes' in the groin, the armpit, or behind the ears. While this form of the disease was frequently fatal, if the buboes suppurated the patient had a good chance of recovery. The pneumonic form of the disease was, before the age of antibiotics, uniformly fatal.

The plague bacillus is a natural parasite of various species of burrowing rodents, in which it probably causes little trouble. It is transferred from one animal to another by means of fleas. Black rats can pick up the infection initially by being bitten by infected fleas from the wild rodents, but thereafter the infection is spread from rat to rat, also by fleas. Unlike the situation with the natural hosts, however, the bacillus is virulent for black rats, in which species it causes lethal epidemics as it does in humans.

Human beings can be infected by one of two means. The most common is through being bitten by fleas which have left rats that have died from the disease [52]. Infection by this means results in bubonic plague. However, infection can also pass directly from one human host to another when infected droplets are discharged into the air by coughing and sneezing or inhaled into the lungs of a healthy person. Infection by this means gives rise to the pneumonic form of plague.

The first definite outbreak of plague in Europe was the epidemic which played such a big role in bringing an end to Justinian's attempts to re--establish the Roman Empire. It is said that at one stage, in AD 542, ten thousand people died each day in Justinian's capital, Constantinople. After this, the Empire collapsed 'like a pricked bladder' [53].

It is not known when the black rat first established residence in northern Europe, although this must have occurred at sometime between the 5th and 12th centuries AD, possibly associated with the mass movements into Europe of people from the East. Certainly, rats were firmly established on the continent by 1284, the date at which the story of the Pied Piper of Hamelin is supposed to be set. Soon after this, rats were also in England.

The epidemic of plague which came to be known as the Black Death appeared first at Caffa in the Crimea in 1345 and spread rapidly through the continent. Estimates of the death rate range from 25 per cent to 75 per cent of the total population [54]. Between 1348 and 1374, the population of England fell from 3.8 to 2.1 million. The Black Death was not restricted to Europe; it eventually moved east and it is said to have reached China thirty or forty years later, where thirteen million people are estimated to have died.

Although this particular epidemic died out in Europe in 1351, many somewhat less devastating and less widespread epidemics of bubonic plague occurred sporadically in that continent during the following four hundred years. There were many outbreaks of the disease in London, the last of which was in 1665; and the last epidemic in France occurred in 1720 at Marseilles. After that, the plague withdrew eastwards, and there was a serious epidemic in Moscow in 1771. A re-activation of the disease occurred in China in 1894, and in Hong Kong, for example, there were several thousand deaths within a few months [55].

It has been suggested that the maintenance of plague for over three hundred years in England, and possibly also in other parts of Europe, was facilitated by the destruction of woodland for agricultural purposes. This brought the wild rodents, which are the natural reservoir for the plague bacillus, into close contact with the domestic rat population. It has been further suggested that the natural habitat for the wild rodents was eventually further diminished to the extent that it could no longer support the reservoir population of wild rodents, and that this development led in the end to the disappearance of the plague in rats and in humans [56]. Others have suggested that in essence the Black Death of 1348 was a manifestation of a fundamentally 'Malthusian crisis'—a response to overcrowding in some areas, undernutrition, and a general decline in health and resistance to disease [57].

With regard to the repercussions of plague, there occurred in Europe in the decades which followed the Black Death a series of serious societal disturbances, including many peasant uprisings. Some controversy exists on whether these developments were the consequence of the plague, or whether both the epidemic itself and the societal disturbances were the result of a common cause—the particular set of socio-environmental conditions prevailing at the time. Without becoming involved in this debate, let us note that it would be astonishing if the sudden loss of up to 75 per cent of the population did not throw society into a state of confusion for a while. Some historians favour the view that it was basically the shortage of labour resulting from the plague that precipitated the disturbances, since this meant that the noblemen and rich abbots had a much weaker hold over their serfs and tenants. In England, the restlessness of the peasants culminated in the assassination of their leader, Wat Tyler, by the Mayor of London in 1381. It now seems to be generally agreed that the reduction in labour supply which resulted from the Black Death in the 14th century accelerated a trend that was already underway in rural areas for serfdom to give way to tenancy [58].

H. G. Wells has interpreted the peasant uprisings as reflecting the emergence of a primitive communism and an attempt to reach back to the fundamentals of Christianity [59]. He cites the 'mad priest of Kent', John Ball, who for twenty years preached along the following line:

'Good people, things will never go well in England so long as goods be not in common, and so long as there be villeins and gentlemen'.

The possible reasons for the eventual disappearance of bubonic plague from Europe has been another area of conjecture and debate. Was it, as suggested above, because the rodent reservoir population disappeared? Was it the consequence of changes in the organization of society? Did the domestic cat population increase, and so interfere with the close relationship between black rats and human beings? Or was it because another species of rat, the brown rat, *Rattus norvegicus*, which is resistant to the plague, displaced the black rat throughout most of Europe?

Whatever may have been the reason or reasons for the disappearance of plague from Europe, it certainly was not because of any trend towards smaller or less dense human populations, or because of any major improvements in sanitation and the life conditions of people. It is true that complex societal, political, and economic changes were taking place in Europe from the 15th century onwards; but these changes were not associated with improved conditions of life for the masses.

Cultural adaptation to infectious diseases

With respect to the great epidemic diseases, it is clear that the first prerequisite for successful cultural adaptation was satisfied (page 24); that is to say, the existence of an undesirable state was well recognized. Indeed, people were at times living in almost constant terror of being struck by one or other of the severe infectious diseases.

Successful cultural adaptation was impeded, however, by lack of satisfaction of the second prerequisite. Knowledge was not available on the nature of the causes of the illness, nor had trial-and-error approaches produced any effective cures. The Anglo-Saxon belief that pestilence was due to 'elf-shots'—that is, to arrows fired by malevolent elves—was not useful in promoting effective counter-measures. It is not surprising, therefore, that people turned to religion. Romans, for example, appealed for help to the Goddess of Fever. And in England in 1666, after the Great Plague and the Fire, there was widespread feeling that these calamities were divine punishment for neglect of religious matters by the community, and a bill was introduced in Parliament designed to curb atheism. Although the bill was eventually dropped, Parliament continued to be influenced in its actions by this general attitude. Thus among the cultural responses aimed at preventing a recurrence of the plague, we can note the refusal of the authorities to allow the English philosopher, Thomas Hobbes, whose views were frequently at variance with those of the Church, to publish any of his writings on religion or morality [60].

In the light of our present knowledge of the nature of infectious disease, the most rational of the cultural adaptive responses of the early urban period was the introduction of isolation and quarantine procedures. The books of Leviticus and Deuteronomy in the Bible give a fairly complete picture of these procedures in relation to leprosy, and in medieval times many European cities established quarantine procedures for people passing in and out of their precincts. Although resented by the travellers, the practice was based on sound empirical observations, 'erring only on the side of caution, by allowing for almost three times the necessary incubation period' [61].

Diet

There is no aspect of human experience in which the importance of the interplay between culture and nature is more evident than in the field of food and nutrition. This has been especially so for urban populations, and it is very apparent in our own society today.

The quantity and quality of what people eat and drink is clearly an important influence on their biopsychic state. It affects their development in childhood, their physical health, mental state, behaviour, and susceptibility to infectious disease. In turn, what people eat is determined by a wide range of environmental factors, ranging from the properties of local ecosystems to various cultural influences: it is affected by regional climate, soil quality, and farming practices, by intersocietal trade patterns, by food technology, by culinary conventions, by economic forces, and by societal patterns of redistribution. Other influences include beliefs about and attitudes towards different foods, and knowledge relating to their effects on health and well-being.

In considering the effects of cultural and societal developments on diet and nutrition, it is useful to recognize three kinds of effect. These are effects on: the *quantity* of food consumed by individuals (this may be above or below that which is desirable for optimum health); the *quality* of the food consumed (essential nutrients may be lacking, or harmful contaminants or additives may be present); the *feeding pattern* (for example, how often people eat, how regularly, and at what times of day).

Let us consider each of these three kinds of effect in relation to the early urban situation.

Quantity of food

We have already noted that our hunter-gatherer ancestors were likely, most of the time, to have been well fed. There is no biological or anthropological justification for the view that 'Palaeolithic people hardly ever had enough to eat' [62].

As discussed in the last chapter, when people became dependent for most of their calories on a harvest, often of a monoculture, their food supply was more vulnerable to the potentially devastating effects of abnormal weather or disease. The Old Testament is full of references to famine, but the most vividly described are those of medieval Europe and India. France seems to have been especially susceptible, and sixty famine years were recorded there between AD 970 and AD 1100, and in 1418 one third of the population of Paris is said to have died from starvation [63]. Famine also occurred sporadically in England and throughout the rest of Europe. The last widespread famine in western Europe was in 1817, except for the famine associated with potato blight in Ireland, Belgium, and Silesia in 1846. Famines were common in Russia, especially through the 19th century and again in 1911–12 and 1921–2; and over the centuries famines causing the death of hundreds of thousands of people have occurred from time to time in India and China.

Quality of food

While periodical food shortages, sometimes very severe, have been a feature of recorded human history, variations in the quality of the diet have also, from time to time, contributed to ill health and mortality. We noted above that the human organism is an unusually unspecialized eater and that the diet of people in primeval society was characteristically diverse. Clearly this lack of specialism with respect to diet is a distinct biological advantage, especially when, as a result for example of abnormal weather conditions, certain foodstuffs become scarce. There is, however, another side to the coin. It is apparent that the human organism is to a certain extent *dependent* on dietary diversity. The human species is, in fact, a specialized eater, in the sense that individuals *need* a wide variety of foodstuffs in order to remain healthy.

Thus an important kind of evodeviation which has been liable to occur during the course of civilization has been the tendency for populations to become overdependent on a single food source, or else to omit from the diet a certain major group of foodstuffs, such as fresh fruits and vegetables, leading to a state of nutritional imbalance.

Specific nutritional deficiencies include lack of certain amino acids and of vitamins in the diet. Of the twenty or so amino acids that are found in human tissues, eight cannot be synthesized in the body, and these are therefore essential components of the diet. Certain foodstuffs are, in terms of human health needs, deficient in one or other of these: maize (or 'corn'), for instance, is deficient in tryptophan, and plant foods in general, especially wheat protein, tend to be deficient in lysine. With respect to the vitamins, heavy dependence on white flour (if not artificially supplemented with vitamins) or on polished rice has resulted in thiamin and riboflavin deficiency with symptoms of beriberi, and a high maize diet has been associated with niacin deficiency and consequent maladjustment in the form of pellagra.

One of the best documented specific nutritional deficiency diseases in history is scurvy—due to lack of vitamin C. It is well known that scurvy was a common occupational hazard for seamen before the importance of fresh fruit or vegetables in the diet was recognized. The massing together of large numbers of men in armies was also associated with vitamin C deficiency. Zinsser has said that the history of the crusades reads like a series of diseases, with scurvy as important as infections: 'scurvy not infrequently became decisive in itself or so weakened large bodies of men that subsequent infectious disease found them without normal powers to resist' [64].

Another important disease in Europe in the early urban phase was rickets, which is associated with a deficiency of vitamin D. Evidence for the prevalence of rickets in the Low Countries and in northern Germany is provided by religious paintings from the 15th and 16th centuries, which often depict small children, for example the Christ child, with bent limbs, a swollen belly, and the squarish head typical of the disease [65]. This illustrates an important biosocial principle; namely, the fact that if disorders appear slowly and insidiously in a population, and if they are not especially painful nor obviously lethal, they may well come to be accepted by the people as 'normal'—which, indeed, they are under the prevailing conditions (see boiling frog principle, chapter 3, p. 25).

Rickets, which became a scourge of the towns at the beginning of the industrial transition, also illustrates in another way the complexity of the interrelationships between cultural and biological processes. Vitamin D can be provided by the diet, but it can also be synthesized within the body, provided that the skin is exposed to direct sunlight. The wearing of clothes, the habit of living indoors, as well as air pollution from industry all contribute to effective prevention of any such synthesis. The poorer classes at the beginning of the industrial era were unable to afford such luxuries as eggs and milk products—the only common foods containing vitamin D in quantity. It is estimated that as many as 75 per cent of the children in some European cities in the 19th century were seriously affected with rickets; and this 'disease of poverty and darkness' is not uncommon among children in some European cities even today.

Cultural adaptation to nutritional deficiencies   The notion that what one eats affects one's health and well-being is a very old one, and it almost certainly predates agriculture and urbanization. Certainly, in the days of classical Greece, there was no shortage of opinions about the relationships between diet and health. In particular, Hippocrates and his school had much to say on the subject. In *A Regimen for Health*, for example, advice is given on what to eat at different times of the year [66].

The teachings of Hippocrates and those of Galen in the 2nd century AD, and indeed all medical teaching throughout the Middle Ages, was based on the idea that the natural world is made up of the *four elements*, air, fire, water, and earth. Each of these had a characteristic *quality*: air was cold, fire was hot, water was moist, and earth was dry. A combination of any two elements produced a *complexion*. Four complexions were recognized, and for each one there was an appropriate 'humour'. The relationships between the complexions, the qualities (each corresponding to two elements), and humour were seen as follows:

Complexion	Qualities	Humour
sanguine	hot and moist	blood
phlegmatic	cold and moist	phlegm
choleric	hot and dry	yellow or green bile
melancholic	cold and dry	black bile

All foodstuffs were considered to be made up of a combination of the four elements, and their nutritive value was thought to be determined by the corresponding qualities. Lettuce, for instance, was considered to be cold and moist, while cabbage was hot and dry.

While this theory of nutrition and health clearly had intellectual appeal, we now know it to be based on entirely false premises. It must have been of little value in preventing or in curing malnutrition. For instance, the *Regimen for Health* advocates that the ordinary man should eat as few vegetables as possible during winter-time; and Galen believed that fruits could give rise to fevers and reported that his own father lived to be a hundred years old as a result of not having eaten them. This suspicion of fruits was still common in Europe in the Middle Ages.

The history of the cultural adaptive response to scurvy is an interesting one [67]. As mentioned above, the disease was a particularly serious problem among seamen who went on long voyages. Many different medical authorities in Europe participated in the discussions and debates concerning both the nature of the condition and the effectiveness of the whole range of different remedies. Scurvy at sea was generally considered to be a different disease from the scurvy which occurred on land, and most authorities believed that it was in some way the result of the seamen being exposed to too much salt. Among the numerous treatments which were advocated, fresh fruit, and particularly juice of lemons, was frequently cropping up. For example, in the year 1600, four ships of the East India Company left Woolwich for the East Indies. In three of the ships many men fell sick with scurvy; but in the ship of the Commander, Master James Lancaster, there were very few cases indeed; and it was noted that 'he brought to sea with him certaine Bottles of the Juice of Limons, which hee gave to each one, as long as it would last, three spoonfuls every morning fasting: not suffering them to eate any thing after it till noone' (Drummond and Wilbraham 1958, p. 140). The East India Company then arranged for a supply of lemon-water for all its ships. However, the practice was not universally accepted. Some physicians, having noted that it was particularly the more acid fruits that seemed to be beneficial, came to the view that it was the acidity itself that was effective. Consequently oil of vitriol came to be recommended by many of them, along with a whole range of other remedies.

An enormous amount of literature about 'sea-scurvy' appeared during the 18th century. One of the most important contributions was written by Johannes Bachstrom and published in Leyden in 1734. It was a small book and apparently attracted little attention at the time. However, in the words of Drummond and Wilbraham:

With a few clear arguments and examples the author demolished all the complicated differentiations between land-scurvy, sea-scurvy and all the other forms of scurvy ... and showed that there is, in truth, only one disease meriting this name. He was equally clear and definite about its primary cause, which he held to be a deficiency of fresh vegetables in the diet [68].

In 1753, James Lind published his famous work, *A Treatise of the Scurvy*. He produced a comprehensive review of the opinions of earlier authorities who had found oranges and lemons to be the most effective means of dealing with scurvy. He also carried out a comparative trial himself, the results of which fully confirmed this view.

Lind recommended to the Admiralty that orange and lemon juice be supplied to all men at sea. The Admiralty, apparently mainly for economic reasons, did not accept this recommendation, advocating instead the use of wort—the infusion of malt which forms the first stage in the preparation of beer. It was not until 1795 that the Admiralty finally decided to adopt lemon juice as the principle anti-scorbutic agent. The result was dramatic. While there had been 1054 cases of scurvy in the naval hospital at Haslar in 1760, in 1806 there was only one.

In summary, the cultural adaptive response to scurvy was a slow process, hindered at first by lack of knowledge both of the general concept of specific deficiency disease and of the specific preventive and curative power of fresh fruit and vegetables. Success, when it came, was the outcome of a great deal of trial and error in which people were seeking a specific 'cure' to a disease which was seen as a reaction to some undesirable agent associated with voyages at sea. Even when clear evidence of effective preventive measures was produced and presented in the reports of Backstrom and Lind, the authorities were reluctant to accept it. Also of relevance is the role of economic considerations in blocking a rational response to this evidence.

Additives and contaminants   There are basically two ways by which potentially harmful chemical compounds or micro-organisms can become incorporated into drinks and foodstuffs: they may be introduced inadvertently or they may be deliberately added.

The case of lead in the diet of the inhabitants of Rome at the beginning of the first millennium AD provides an example of inadvertent contamination. It was the practice, especially in the case of the nobility, to store fruit juice and other drinks in lead containers, and we can assume that many people were therefore regularly consuming small amounts of lead. The analysis of human bones from this period confirm that this was the case.

According to the *British Medical Journal*:

Mild symptoms and signs of lead poisoning include: tiredness; lassitude; constipation; slight abdominal discomfort or pain; anorexia; altered sleep; irritability; anaemia; pallor; and, less frequently, diarrhoea and nausea [69].

The nature of these symptoms and signs is such that the affected individuals in Rome might well have come to take them for granted, assuming perhaps that they were merely the natural consequences of growing older; this would be another example of the boiling frog principle. In fact, the first prerequisite of successful cultural adaptation would not have been satisfied, with the consequence that no cultural adaptive response would have been brought into play. Nevertheless, if most of the members of the ruling class of the population, and possibly many others as well, were suffering from symptoms such as those listed above, the impact on the society as a whole might well have been considerable.

The following quotations provide some idea of the problems of contamination, accidental and deliberate, of foodstuffs in the London area around the year 1770. The first was written by a visitor to London from the Continent, and the other, hopefully somewhat exaggerated, by the Scottish poet, T. G. Smollett:

All that grow in the country about London, cabbage, radishes, and spinnage being impregnated with the smoke of sea-coal, which fills the atmosphere of that town, have a very disagreeable taste. I ate nothing good of this sort in London, but some asparagus.

But the milk itself should not pass unanalysed, the produce of faded cabbage-leaves and sour draff, lowered with hot water, frothed with bruised snails; carried through the streets in open pails, exposed to foul rinsings discharged from doors and windows, spittle, snot, and tobacco quids from foot-passengers; overflowings from mud carts, spatterings from coach-wheels, dirt and trash chucked into it by roguish boys for the joke's sake; the spewings of infants, who have slabbered in the tin-measure, which is thrown back in that condition among the milk, for the benefit of the next customer; and, finally, the vermin that drops from the rags of the nasty drab that vends this precious mixture, under the respectable denomination of milkmaid [70].

Feeding patterns

Culture has a powerful influence on when, and to some extent on how, food is eaten by human beings. Thus it is culture that determines whether food is consumed in a squatting position on the ground or in the sitting position at a table, and whether it is conveyed to the mouth by means of the fingers or by means of artificial extensions of the fingers in the form of spoons, forks, or chopsticks. These variations are not particularly important biologically, and other influences of culture on feeding behaviours seem more relevant. For instance, it was noted above that in primeval societies people tend to eat at any time of day when food is available and when they are hungry; in contrast, in most civilized societies special times are set aside for eating. At these times, most people in the community have food placed before them, regardless of whether or not they are actually hungry. This has become especially so since the rise of the clock as an organizer of human behaviour. Mealtimes are presumably influenced to some extent by biological requirements, but even more so by tradition and by the demands of the social machine. In the late-17th century, fashionable people in England ate dinner, their main meal, at 5 or 6 p.m. Later, in the 19th century, this meal was eaten at 7 p.m. or later. This change led to the introduction of an extra meal in the middle of the day—luncheon; at first this was no more than a very light snack, perhaps a glass of wine and a biscuit, but it gradually grew into something more substantial. It is easy to see how changes in customs of this sort can influence the total amount of food consumed each day.

One particularly aberrant form of feeding behaviour, representing an extravagant deviation from primeval conditions, occurred in the case of the upper classes in ancient Rome. In the spacious homes of the patricians, place was found, beside the dining-room, for a special chamber known as the *vomitorium*. The Latin name for this room leaves no doubt about its function. According to custom, the diners, having consumed as much food as they were able, would retire to the vomitorium for the express purpose of regurgitating their meal, so that they could return to the dining-room to further indulge. Slaves would sometimes assist the process by tickling the throats of the patricians with feathers.

The wearing of clothes

The wearing of clothes represents a significant culturally induced modification in the biological conditions of life of human beings. There was a time in the past, of course, when our ancestors were totally unclothed at all times;

but already by about five hundred thousand years ago some *Homo erectus* individuals were wrapping themselves in animal skins for protection against the elements, and in the course of time the use of clothing became an important cultural adaptive response to cold conditions and contributed importantly, no doubt, to the ability of human populations to survive in colder regions of the earth. Early farming people in south-west Asia were probably usually lightly clothed. Clothing was certainly worn in the first cities in Sumer and, even if very scantily, in early Egyptian civilization; and it has been worn most of the time by most of the people in the Western world ever since.

It is not surprising to find that the custom of wearing clothes resulted in new cultural attitudes towards the exposure of the body, which in many Western societies was totally unclothed only for certain specific activities, such as washing and sexual activity. According to Herodotus, there was already, by 450 BC, wide variability among human populations in attitudes towards the exposure of the body [71]. Disapproving attitudes towards the exposure of the body probably became more pronounced in Europe during the Middle Ages, although many etchings from the period illustrating the use of public baths in Europe suggest that attitudes at the time to nudity were considerably more liberal than they are even at the present day [72].

These developments eventually led to a set of social norms relating to the wearing of clothes that our hunter-gatherer ancestors would indeed have found hard to comprehend—a situation in which, in Victorian England, for example, the mere exposure of a female ankle was regarded as a sexual invitation. Although more recently in Western society there has been some swing away from these extremes, the situation is still such that a section of the community is able to make a living by marketing photographs of the nude, or almost nude, female subject.

There were other important biological implications of the wearing of clothes. In medieval Europe, most of which was a good deal colder than Mesopotamia, Egypt, Athens, and Rome, the natural tendency of people to seek maximum comfort resulted in the donning of many layers of protective cloth, which were seldom washed or cleaned. This evodeviationary custom provided a biologically novel and ideal environment for the protection of small external animal parasites—especially fleas and lice [73]. A good illustration of this is provided by the story of Thomas á Becket's funeral, quoted by Zinsser:

The Archibishop was murdered in Canterbury Cathedral on the evening of the twenty-ninth of December. The body lay in the Cathedral all night, and was prepared for burial on the following day. The Archbishop was dressed in an extraordinary collection of clothes. He had on a large brown mantle; under it, a white surplice; below that, a lamb's wool coat; then another woollen coat; and a third woollen coat below this; under this, there was the black, cowled robe of the Benedictine Order; under this, a shirt; and next to the body a curious hair cloth, covered with linen. As the body grew cold, the vermin that were living in this multiple covering started to crawl out, and (in the words of the chronicler): 'the vermin boiled over like water in a simmering cauldron, and the onlookers burst into alternate weeping and laughter' [74].

Perhaps this was an exceptional case—and, in any case, it was mid-winter. But the description illustrates well the principle that clothing provides a haven for fleas and lice which can play so important a role as vectors of devastating epidemic diseases.

Another extremely important aspect of the wearing of clothes in the early urban phase of human existence, and also in modern high-energy society, has been the use of clothes to signify status, position, and role in society. Different designs of clothing have become associated with different professions and with position in the social hierarchy. Thus the occupation and rank of a soldier, a railway worker, or a member of a religious order can be seen at a glance. This use of clothing has played, and continues to play, a major role in reinforcing the hierarchical structure of society.

COMMENT

It is worth noting that certain aspects of human experience in the early urban phase probably remained much the same for most people as they had been in primeval and early farming times. For example, the extended family was still the basic economic unit of society, and it probably provided an effective emotional support network for the majority of individuals. It is also likely that early urban society, like hunter-gatherer and early farming societies, was characterized by a fairly strong sense of community on the local level. With regard to the various postulated intangible health-promoting experiences (pages 37, 40, and 79), there is likely to have been considerable variation from time to time, from place to place, and from one section of the community to another. In general, however, we are left with the impression that the life conditions of most people were conducive to a sense of personal involvement, a sense of belonging, a sense of interest, a sense of responsibility, and a sense of comradeship.

On the negative side, it appears that the degree of deliberate cruelty which human beings displayed to other human beings was, by modern Western standards, extraordinarily high in some regions during the early urban period. This state of affairs is further evidence for the view that the human species does not possess any innate concern for the well-being of people who are not members of the individual's own in-group. This cruelty persisted in spite of legislation designed to maintain order in society and to protect the weak from the powerful, and in spite of (and sometimes in the name of) religious teachings. Indeed, the law itself, as well as some of the religious codes, prescribed punishments for antisocial behaviour which, to many of us today, seem unbelievably callous and cruel [75].

NOTES

1. For discussion on the concept of the *city*, various uses of this term, and of the words *urban* and *urbanization*, see Davis 1965; Petersen 1969, chapter 13; Whitehouse 1977, pp. 1–7.
2. This period is referred to as the 'formative period' by some authors. See, for example, Service 1962.
3. For good accounts of these early cities, see Woolley 1963; Whitehouse 1977.
4. Sjoberg 1965; Keesing 1976, chapter 3.
5. See Jacobs 1969; Whitehouse 1977; Redman 1978.
6 Redman 1978, p. 298.
7. There is some difference of opinion among authors about the occupational structure of the early Sumerian cities. Some believe that *all* people in the city were farmers at least on a part-time basis, and worked in the field at least at some times of the year. See the discussions by Frankfort 1951, p. 63; Whitehouse 1977, p. 55.
8. The phenomenon does occur, however, among social insects. In bee colonies, for instance, the queen and the drones do not participate in the acquisition of food.
9. Woolley 1963; Petersen 1969, p. 377; Whitehouse 1977.
10. Cary 1962, p. 454; Petersen 1969, p. 377.
11. Derry and Williams 1960, p. 25.
12. Mumford 1966, p. 300; Petersen 1969, p. 377.
13. Mumford 1966, p. 68.
14. Boyden, Millar, Newcombe, and O'Neill 1981.
15. Petersen 1969, p. 378.
16. Woolley 1963, p. 467.
17. For a more complete list, see Hunter 1971, pp. 221–2.
18. Whitehouse 1977, pp. 78–81.
19. Whitehouse 1977, p. 187; but see Service 1962, p. 221.
20. Wilkinson 1973, p. 81.
21. This concept is well described by James 1983.
22. Woolley 1963, p. 475.
23. Galbraith 1975, p. 15.
24. Grigg 1974, p. 162; Wiet, Elisseeff, Wolff, and Navdov, 1975, pp. 953–4.
25. Woolley 1963, p. 482.
26. Micah, chapter 4, verse 3.
27. Woolley 1963, p. 481.
28. For a background to these terms, see Nisbet 1970; Dunphy 1972.
29. Nisbet 1970, pp. 107–8.
30. Ferguson 1767.
31. Woolley 1963, p. 574.
32. Kramer 1967, p. 81.
33. Kramer 1967, p. 83.
34. Woolley 1963, p. 93.
35. Schmandt-Besserat 1978.
36. Woolley 1963, p. 637. Note that this simplification of writing involved, on average, the removal of two signs from the total system each year.

37. Woolley 1963, pp. 660 ff.
38. The comments in this section are derived mainly from Hughes 1975.
39. Hughes 1975, pp. 149–50.
40. Hughes 1975, pp. 43, 45.
41. Glacken 1967, p. 72.
42. Mumford 1966, pp. 91, 92.
43. Mumford 1966, p. 254.
44. Mumford 1966, p. 493.
45. Mumford 1966, pp. 251–2.
46. Fenner 1970.
47. Second Book of Kings, chapter 19, verse 35.
48. Hare 1954, p. 71. According to another author, Rome was swept by bubonic plague in AD 68, AD 79, AD 125, and AD 164, and this last epidemic raged for 16 years, killing 10 000 people (Leavesley 1984).
49. Zinsser 1935.
50. Young 1971, p. 336.
51. For a detailed and interesting account of the interplay between infectious diseases and human society in history—an interplay of immense significance in human affairs—see McNeill 1976. The influence of contagious diseases on the outcome of military campaigns is discussed by Zinsser 1935 and by Hare 1954.
52. It was previously thought that humans were infected as a result of being bitten by rat fleas. Recently, however, the remains of a human flea have been found on the remains of a 15th century rat in London, raising the possibility that human fleas may have played a role in the spread of plague from rats to humans.
53. Wells 1961, p. 562.
54. Estimates of the death rate for Europe as a whole range from 25 million out of a population of 105 million to 75 per cent of the population (see Hare 1954, p. 91; Langer 1964).
55. Boyden *et al* 1981, p. 75.
56. Alland 1968.
57. See Alland 1968 and Herlihy 1980, p. 113.
58. Wiet *et al* 1975, p. 953. See also Ziegler 1969, chapter 15.
59. Wells 1961, p. 740.
60. Russell 1946, pp. 203, 570.
61. Mumford 1966, p. 341.
62. Lorenz 1966, p. 218.
63. Banks 1969.
64. Zinsser, 1935, p. 156.
65. Drummond and Wilbraham 1958, p. 151.
66. Drummond and Wilbraham 1958, chapter 4; Phillips 1973; Lloyd 1978.
67. The information presented here on the history of scurvy and the various quotations are mainly taken from Drummond and Wilbraham 1958.
68. Drummond and Wilbraham 1958, pp. 140, 260.
69. Diagnosis of Inorganic Lead Poisoning: A statement. *British Medical Journal* (1968) **4**, p. 501.
70. Drummond and Wilbraham 1958, pp. 193–4.
71. Herodotus 1921.
72. Trevelyan 1950.

73. One specific biological consequence of this state of affairs was the evolution, through natural selection, of a new kind of louse—the body louse—quite distinct from the head louse which had been a companion of humankind relatively unchanged for tens of thousands of years. Mummified head lice are found associated with the human remains in the tombs of Ancient Egypt, but there is no sign of body lice.
74. Zinsser 1935, p. 185.
75. Hibbert 1963.

7

THE TRANSITION TO THE
HIGH-ENERGY PHASE AND THE
PUBLIC HEALTH MOVEMENT

BACKGROUND TO THE TRANSITION IN ENGLAND

A great deal has been written about the various factors that different authors believe lay behind the fundamental societal changes which led to, or became part of, the 'industrial revolution'. No attempt will be made here to summarize this literature or to present a detailed assessment of the different viewpoints on the causes of these changes. We will be more concerned with the *impacts* of the transition from an agricultural to an industrial economy, on the interrelationships between human society and the biosphere, and on the life experience of human beings. Nevertheless, some brief comment on the background to the situation is appropriate.

Broadly, we can recognize two kinds of development which preceded the processes of industrialization and the associated changes that took place between around 1750 and 1850 and which have been seen by various authors as contributing to these changes. One set is essentially cultural and relates to the activities of intellectuals, and the other is basically ecological.

First, with respect to events in the cultural arena, much weight has been put on the writings of Francis Bacon at the beginning of the 17th century. Bacon was a great exponent of the inductive method of reasoning and had supreme confidence in the power of scientific knowledge and its technological application to 'the effecting of all things possible' [1]. René Dubos, for example, has written:

... the scientific revolution which began yielding its fruits during the seventeenth century can be traced in large part to the writings of one person: Francis Bacon. His place in the history of science is unique because his influence was exerted through words rather than deeds. He did not add to knowledge, but became the prophet of scientific civilization [2].

At around the same time, and also reflecting a new spirit or mood among thinkers, René Descartes, the 'founder of modern philosophy', was trying to shake off the intellectual shackles of past centuries and to take a new look at reality [3]. He, too, had great faith in the usefulness of science for improving the lot of humankind.

However, despite the confident assertions of authors like Dubos, it is difficult to be sure to what extent either Bacon or Descartes, or other philosophers of the 17th and 18th centuries, influenced what actually happened. Would the developments in science and technology of the 18th, 19th, and 20th centuries really have been any different if these men had never existed? The old conundrum, whether writers of this kind have a strong influence on the nature and pace of societal change, or whether they merely reflect an overall cultural mood, remains unanswered.

A similar question can be asked with respect to the philosopher–economists of the early 19th century. Thomas Malthus developed Condorcet's idea that population always grows to the limits of the food supply until checked by starvation—although Condorcet's emphasis on birth control as a cultural adaptive means of overcoming the human distress which this concept implied was missing from Malthus's writings [4]. Then there was David Ricardo's economic theory, with his stress on the naturalness or, at least, inevitability of the law of the survival of the fittest. It is said that the works of these two men were 'taken as gospels by the most serious and even pious men, who used them to justify actions they would never have thought of defending on human grounds' [5]; but again, we ask, would things have turned out any differently if these two individuals had never been born?

Whatever the causes, it became increasingly fashionable for members of a small privileged section of the community, who had the time and the resources, to indulge in scientific experimentation and technological innovation. Eventually a positive feedback mechanism came into play, and technological advance became a self-perpetuating and self-accelerating process.

Turning to the more basic socio-ecological dimension, a number of important developments were taking place towards the second half of the 18th century. Probably the most significant in the present context was an acute land shortage associated with a rapidly growing human population [6]. The total population of England more than doubled between 1700 and 1800, and 4 or 5 million people had to be provided with homes. Compared with previous centuries, this was a big increase, as indicated in Table 7.1 [7].

There were also some important societal changes taking place in the countryside during the second half of the 18th century. As a result of a series of Acts of Parliament, the common land of villages was transferred to the private ownership of wealthy families which claimed some ancient right to it. This *enclosure of the commons* hit the small farmers, the cottagers, and the squatters very severely and their impoverishment contributed to a general movement of people away from rural areas to towns and cities in search of something better [8]. On the positive side, however, the 'age of enclosure' did result in some important improvements in agricultural methods, leading to a considerable overall increase in primary production [9].

Table 7.1. *Population of England and Wales, 1086–1841.*

Date	Population (in millions)
1086	1.1
1348	3.7
1377	2.2
1545	3.2
1695	4.8–5.5
1801	9.2
1811	10.2
1821	13.9
1831	2.0
1841	15.9

Another relevant societal trend was the continuation of the baroque rise of capitalism and the interplay between this factor and the advances which were taking place in manufacturing technology. Small factories began to spring up all over the country, offering employment to the displaced people from country areas; and capitalism began to shift its emphasis from mercantilism to industrialism. The distribution of factories at this time was decentralized, in that they appeared in small towns as well as in big ones. The factory employees were often expected to work on dull and monotonous tasks for very long hours for very low wages; and child labour was widely used. Thus the patterns and conditions of work which were characteristic of the industrial transition were already well established before the introduction of the steam engine and of fossil fuel as a source of power [10].

EARLY BIOSOCIAL IMPACTS OF INDUSTRIALIZATION

The introduction of steam power for manufacturing purposes towards the end of the 18th century had far-reaching biological, social, and economic repercussions. Probably the most important initial effect was the change it brought about in the distribution of the industrial work-force. Because the new technology depended on the use and supply of coal, industry became concentrated in towns which were not far from coal-fields or, later, which were along the main-line railways and canals; and it was in these towns that massive increases in population occurred. Before long, 80 per cent of the population was living in communities consisting of more than 2500 individuals. A typical example of a new industrial city was Manchester, the population of which increased from 30 000 or 40 000 in 1760 to over 303 000 in 1851 [11].

The population movements were necessarily accompanied by building programmes on a massive scale; but these were haphazard and uncontrolled, and builders and landowners were free to determine the standards to suit their own financial interests. 'Improvements' in the form of gas lighting, gas cooking, running water, stationary baths, and proper water-closets became available in the 1830s, but only the middle and upper classes could afford them [12]. The mass of the urban population lived under conditions of extreme squalor, in overcrowded, badly ventilated buildings which, like the tenements of ancient Rome, were devoid of any form of sanitation. The lavatories were usually in the cellars and 'foul beyond description'. In Liverpool, for example, 20 000 people lived in cellars, with no sanitation whatsoever. In a street in Leeds, three toilets were the only sanitation available to 386 people, and for 7055 people living in 645 houses in a part of London only 33 toilets were available [13]. Whereas in the 16th century it had been an offence in many English towns to throw garbage into the street, in the early 19th century this practice was the regular means of disposal of household refuse and human excreta; and there it remained until a sufficiently large volume had accumulated to make it worthwhile for someone to come and take it away for use on the land.

The overall result, then, of the complex interacting societal and economic processes was a massive influx of people into the big industrial centres, surpassed only by similar mass movements of people into urban areas in some of the developing regions of the world today.

Under these conditions of high population density and absence of plumbing, drains, and washing facilities, a high prevalence of severe infectious disease is predictable from basic epidemiological principles. Although information is not available on the actual incidence of different diseases during this period, there is evidence that the overall death rates in cities were much higher than in rural areas. For instance, for the period 1831-9 the mortality in country districts was reported to be 18.2 per thousand, compared with 26.2 per thousand in town districts. In 1840 childhood diseases were reported to be twice as fatal in towns as in the country [14].

Typhus seems to have been ever-present, sometimes smouldering inconspicuously and sometimes flaring up into violent epidemics. This disease was probably the major factor that stirred the consciences of the small group of humanitarian reformers who initiated the cultural adaptive responses of society that eventually led to the Public Health Movement of 19th century Britain. This movement illustrates some important principles of cultural adaptive processes, and the remainder of this chapter will be given up to a discussion of its main features.

THE PUBLIC HEALTH MOVEMENT

The appalling environmental conditions experienced by the mass of the working class in the big industrial cities early in the 19th century eventually prompted a societal response, the end result of which was that, for the first time, government on both national and local levels accepted a degree of responsibility for the health of the human population. This cultural adaptive response was painfully slow, but at least it was progressive and it led in the long run to increasing governmental control of the design of the built environment, in so far as it might affect the health and well-being of the people who live in it.

Governmental concern in Britain for the welfare of the less fortunate members of the community can be said to go back to the 16th century. In the last chapter we discussed how, with the advent of civilization, occupational specialization became a feature of urban societies. Eventually there emerged a class of people who, for one reason or another, had no occupation, and consequently often no source of livelihood. The numbers in this *unemployed* group varied from time to time according to economic circumstances. During the Middle Ages unemployed people who had no family to rely upon had to turn to the Christian Church, which was considered to have a moral and ecclesiastical responsibility to give alms to the poor. For a number of reasons, this system was proving inadequate in the early part of the 16th century. At that time, bands of 'sturdy beggars' were wandering the countryside and terrorizing its inhabitants. They came from many sources—'the ordinary unemployed, the unemployable, soldiers discharged after French wars and the Wars of the Roses ... serving-men set adrift by impecunious lords and gentry, Robin Hood bands driven from their woodland lairs by deforestation and by the better enforcement of the King's peace, ploughmen put out of work by enclosures for pastures, and tramps who prudently pretended to belong to that much commiserated class' [15]. The fact that the Crown showed some interest in the destitute was probably attributable rather to politics than to compassion: 'Vagrants spread disaffection and might provide raw material for a rebel army' [16].

The first statute dealing specifically with poor relief was that of 1531, which gave local justices the power to license aged and impotent persons to beg within their own neighbourhood. In 1535 revolutionary draft proposals were produced, suggesting, for example, that the able-bodied poor should be employed on public works and that both central and local administrative machinery should be set up to make this possible. This did not, however, become law, and the statute of 1536 concentrated largely on the organization of voluntary funds for the relief of those who could not work due to age or infirmity. This statute and that of 1531 are often regarded as the first English 'Poor Law'. In 1547 the city of London found that voluntary contributions were insufficient, and the local government decided to impose compulsory rates to provide funds for this purpose. This has been described as a momentous step forward, in that it introduced a new administrative device and also broke away from the mediaeval idea of charity as a religious duty only. In 1572 a compulsory rate was imposed on a national scale. Other important developments were the introduction of the Speenhamland System in 1796 which declared that wages below what was considered to be an absolute minimum should be supplemented by local parishes up to the appropriate level, according to the number of children that an individual had and the price of bread. In 1834 this system gave way to the New Poor Law in which poor relief was granted to the able-bodied poor and their dependents only in well regulated *work-houses*, under conditions which were intended to be inferior to those of the humblest labourers outside.

The reformers

As discussed in chapter 2, the first prerequisite for successful cultural adaptation is awareness that an undesirable state exists. This was important because the operative classes in Britain—that is, the middle and upper classes—were largely oblivious of the conditions experienced by the working people. The task of creating awareness of the problem and of the urgent need for reform was initially taken up mainly by members of the medical profession. As early as 1806, for example, some of these *first-order reformers*—a group of doctors—investigated conditions in Plunket Street, Dublin, and reported that:

... a great proportion of the lower classes live in lanes and back yards. The houses through the Liberty in general are unprovided with privies, or the privies are choked up. The lane, therefore, is frequently the deposit of all the filth of the adjacent buildings. If the attention of the scavengers is seldom directed to the streets of the Liberty, still more neglected are those recesses, which in fact, are hardly ever cleansed; the constant respiration of air thus tainted, must gradually weaken the powers of life; and if diseases be not the immediate consequence, the system is at least fitted for the reception of contagion whenever it presents itself [17].

Many other doctors, especially those working for the Poor Law Commission, played a vital role. In the words of Flinn:

It is hardly possible to over-estimate the value of the work done by the medical officers in the service of the poor law in the 1830s. Almost every page of the *Sanitary Report* bears ample testimony to their hard work, conscientiousness, experience, medical commonsense, and compassion [18].

Especially important in the 1830s was the 'great trio' of reforming doctors— James Kay, Neil Arnott, and Southwood Smith. It was the reports of these three men on the conditions of life in certain districts of London which initiated the wider enquiry described in the *Sanitary Report* (see below). In the long run, it was probably Southwood Smith who contributed most to the cause of sanitary reform. His three papers published in 1838 and 1839 'presented a vivid and horrifying proof of the interconnection of insanitation and disease in London's East End' [19].

Voluntary organizations

We have noted in a previous chapter the important role of the Society of Friends and other voluntary organizations in the societal response in England to the Irish potato famine. Voluntary groupings played an even more important role in the Public Health Movement, and in particular are said to have had a marked influence upon the course of sanitary legislation and administration during the period from 1844 to 1849. One of the organizations was the Health of Towns Association in London—'born of the reformer's intense zeal for sanitary reform' [20]. It was originally established by Southwood Smith. This association included among its sponsors and supporters such individuals as the Duke of Cambridge, the Duke of Norfolk, the Marquis of Normanby, Earl Grey, the Earl of Ellesmere, the Earl of Granville, the Earl of Derby, Earl Fortescue, and other high-ranking personages. The fact that this kind of backing was necessary for the organization to have any influence on public affairs reflects a principle which is not irrelevant to reform movements in modern high-energy society.

Other similar organizations were later established in many of the larger towns in England and Scotland. The chief object of the Health of Towns Association was stated to be:

To diffuse among the people the valuable information elicited by recent inquiries, and the advancement of science, as to the physical and moral evils that result from the present defective sewerage, drainage, supply of water, air, and light, and construction of dwelling-houses [21].

The role of governmental bureaucracy

It was a government body, the Poor Law Commission, which, in the 1830s, secured the services of the three doctors, James Kay, Neil Arnott, and Southwood Smith, to enquire into the prevalence and causation of preventable disease in the metropolis. Following receipt of these reports, the Government instructed the Poor Law Commission to make inquiries in regard to the health of the working population throughout England and Wales, and later these inquiries were extended to Scotland. During the following three years, an enormous amount of evidence was collected and this formed the basis of the Commission's *Report on the Sanitary Conditions of the Labouring Population of Great Britain*, published in 1842. This report was laid before Parliament and consisted of three volumes, the first two being local reports covering, respectively, England and Scotland, while the third was a synoptical volume, entitled *General Report on the Sanitary Condition of the Labouring Population of Great Britain*. This last volume was the work of Edwin Chadwick (see note 23).

Chadwick was a government employee, and had been Secretary to the Poor Law Commission. Between 1832 and 1838 he had been occupied with factory and Poor Law reform. But in 1838 he embraced the public health idea, and this development represented 'an important turning point in the history of the Public Health Movement'. Chadwick was a born leader, and he effectively brought together the separate strands of the Public Health Movement and gave it the impetus which it needed. According to Flinn, 'That Chadwick was the ultimate instrument of success was due in a large measure to his rugged determination, to his humanitarianism, and to his skill as a sociologist' [22]. Another factor contributing to his enthusiasm was his passionate hatred of all forms of vested interests.

Some of the conclusions which Chadwick considered to be established by the evidence contained in the first two volumes of the Report are as follows [23]:

That the various forms of epidemic, endemic and other disease caused, or aggravated, or propagated chiefly amongst the labouring classes by atmospheric impurities produced by decomposing animal and vegetable substances by damp and filth, and close and overcrowded dwellings, prevail amongst the population in every part of the kingdom, whether dwelling in separate houses, in rural villages, in small towns, in the larger towns—as they have been found to prevail in the lowest districts of the metropolis.

That such disease, wherever its attacks are frequent, is always found in connection with the physical circumstances above specified, and that where those circumstances are removed by drainage, proper cleansing, better ventilation and other means of diminishing atmospheric impurity, the frequency and intensity of such disease is abated; and where the removal of the noxious agencies appears to be complete, such disease almost entirely disappears... .

That the formation of all habits of cleanliness is obstructed by defective supplies of water.

That the annual loss of life from filth and bad ventilation are greater than the loss from death or wounds in any wars in which the country has been engaged in modern times... .

That the ravages of epidemics and other diseases do not diminish but tend to increase the pressure of population.

That in the districts where the mortality is the greatest the births are not only sufficient to replace the numbers removed by death, but to add to the population.

That the younger population, bred up under noxious physical agencies, is inferior in physical organisation and general health to a population preserved from the presence of such agencies... .

That these adverse circumstances tend to produce an adult population short-lived, improvident, reckless, and intemperate, and with habitual avidity for sensual gratifications.

With respect to the means by which the sanitary condition of labouring classes might be improved, Chadwick states that:

The primary and most important measures and, at the same time the most practicable, and within the recognised province of public administration, are drainage, the removal of all refuse of habitations, streets and roads, and the improvement of the supplies of water... .

That the expense of public drainage, of supplies of water laid on in houses, and of means of improved cleansing would be a pecuniary gain, by diminishing the existing charges attendant on sickness and premature mortality.

That for the protection of the labouring classes and of the ratepayers against inefficiency and waste in all new structural arrangements for the protection of the public health, and to ensure public confidence that the expenditure will be beneficial, securities should be taken that all new local public works are devised and conducted by responsible officers qualified by the possession of the science and skill of civil engineers.

It has been pointed out that these recommendations state the principles on which sanitary reform during the next fifty years was based. The Report played an immensely important role in the reform process. It provided the solid base of information which could be used effectively by those in Parliament and elsewhere who believed that the Government should take action to improve the conditions of life of the working people.

The part played by Chadwick himself in the whole process nicely illustrates a point of considerable importance in cultural adaptation—that is, that so much depends on whether the right person is at the right place at the right time. It also shows the advantage in reform processes of the existence of sympathetic and vigorous individuals in influential positions in the governmental bureaucracy.

It is noteworthy that, in 1889, the year before his death, this great sanitary reformer, Edwin Chadwick, was created a Knight Commander of the Bath.

The anti-reformers

As in the case of all reform movements, those who advocated reform were countered by others who denied the existence of a problem and who referred to the reformers by such disparaging terms as alarmists and fanatics. These anti-reformers were mainly individuals (or representatives of groups) whose financial interests, or whose influence and power, were likely to be threatened by government action aimed at improving the situation. Referring to the early mid-19th century in Britain, Flinn has written:

A very wide range of social and economic issues were raised, debated, and made the subject of legislation by the parliaments of this period. Those whose interests were likely to be protected by these measures gave enthusiastic support: those whose interests were threatened, opposed them, and if, in doing so, they invoked the 'principle' of *laissez-faire*, they were only grasping at a perfectly legitimate straw in the circumstances. Thus the campaign for sanitary reform was not opposed by an immutable and unchallengable principle; it was faced instead with a powerful opposition whose economic and political interests might be threatened by measures likely to reduce some incomes or diminish local autonomy. Chadwick and his supporters had to arm themselves, therefore, against the spurious use of economic and political theory which was merely the first line of defence of a group of opponents very well aware of the real nature of the threat [24].

Anti-reformers representing vested interests and playing precisely this same role in precisely the same way can be recognized in any reform movement of the 19th or 20th centuries.

Government action

On receipt of Chadwick's Report, the Prime Minister, Sir Robert Peel, and the Home Secretary, Sir James Graham, arranged for the problem to be placed before a Royal Commission, which was set up in May 1843. This Commission not only confirmed the broad thrust of Chadwick's report, but also came up with firm recommendations. One of the most important of the proposals was that the Government should have power to inspect and supervise the execution of all general measures for the sanitary regulation of large towns. This, of course, would involve the setting up of a new government department. For a variety of reasons, including opposition within Parliament, no legislation was introduced until 1848 when the first Public Health Act was finally passed.

The Public Health Act of 1848 was in essence a compromise, and was a far cry from the suggestions put forward by Chadwick in the *Sanitary Report*. In fact, some of the Report's main recommendations were not included in the Act. For instance, there was no comprehensive national system of sanitary, sewage, or public health commissions. Instead there were to be local public health boards which were to be compulsory only in places where the death rate exceeded 23 per 1000. Elsewhere, local boards could be established only as a result of petitions signed by not less than one-tenth of the ratepayers. Nor were the local boards required to appoint medical officers—they were merely permitted to do so. The Act set up a General Board of Health with three members, who turned out to be Chadwick, Southwood Smith, and Lord Ashley. In 1854 Chadwick and Southwood Smith were both removed from the Board—a step which essentially marked the end of Chadwick's official career. Nevertheless, in the words of Flinn, the Act of 1848:

... constituted a tentative and uncertain start to government action in a major field ... it had put a foot through a door which had hitherto defied all attempts at opening, and although the detailed administrative arrangements it laid down were scrapped within half a dozen years, its principle of state responsibility was not discarded. It was this principle which the *Sanitary Report* had sought to establish [25].

The next fifty years saw, step by step, increasing government involvement in public health matters. For example, in 1875 another Public Health Act was passed by Parliament. This Act established a governmental *Department of Health*—a development which some see as the most important step forward up to that time in the history of the public health movement. This Act of 1875 represented a consolidation of the law pertaining to public health. It was very comprehensive, and included law pertaining to such matters as local government responsibilities, sewerage and drainage, offensive trades, unsound food, infectious diseases, hospitals, the prevention of epidemics, markets and slaughter houses, highways and streets, the purchase of land, the making of local by-laws, and so on. It covered the whole of the legislative needs in the field of sanitation at that time [26].

The concern of local authorities as well as parliamentarians about public health received a considerable boost from the outbreaks of cholera which occurred in 1831, 1848, 1854, and 1861. It was a new and alarming disease and directly threatened the wealthier classes, and it 'stirred even the moribund, degraded, unreformed municipal corporations into fits of unwonted sanitary activity' [27]. At the other extreme was tuberculosis, which probably killed more people than any other disease; but there was a tendency to regard it as an act of God, affecting both rich and poor alike, against which little action was possible.

The role of knowledge

In chapter 2 we discussed four prerequisites for successful cultural adaptation. The second of these was *knowledge* about the causes of an undesirable state, or at least of effective remedial procedures. It was pointed out, however, that cultural adaptation is occasionally successful in the absence of such knowledge or in situations in which the cultural adaptive responses are based on a false or incomplete understanding of the nature of the problem. This was so in the case of the Public Health Movement.

The dominant theory around the middle of the 19th century concerning the cause of infectious disease was the so-called *miasmatic theory*, according to which disease was seen to be generated by unpleasant odours. The following quotations illustrate this theory:

I think it tolerably evident that the contagion may be propagated by an impression on the olfactory nerves [28]. The immediate, or the exciting cause of fever is a poison formed by the corruption or the decomposition of organic matter. Vegetable and animal matter, during the process of putrefaction, give off a principle, or give origin to a new compound, which, when applied to the human body, produces the phenomena constituting fever [29]. This disease-mist, arising from the breath of two millions of people, from open sewers and cesspools, graves and slaughter-houses, continually kept up and undergoing changes; in one season it is pervaded by cholera, in another by influenza; at the one time it bears smallpox, measles, scarlatina and whooping-cough among your children; at another it carries fever on its wings. Like an angel of death, it has hovered for centuries over London. But it may be driven away by legislation ... . The poisonous vapour may yet clear away from London—and from all the other towns of the kingdom:- some of the sunshine, pure water, fresh air and health of the country may be given to the grateful inhabitants of the towns by the parting voice of Legislature [30].

Although the miasmatic theory was widely held, not all authorities embraced it fully. For example, around 1840 Professor Alison in Edinburgh put much more emphasis on the enfeebling effect of poverty on the human organism 'by deficient nourishment, by insufficient protection against cold, by mental depression, by occasional intemperance, and by crowding in small ill-aired rooms' [31]. And John Snow's thorough investigations in the London cholera epidemic of 1854 traced the outbreak around Broad Street, St James' Parish, to the Broad Street pump, and showed clearly that cholera could be transmitted by means of water [32]. Nevertheless, the miasmatic theory remained dominant well into the second half of the 19th century, and even later. It was embraced by Chadwick himself and by John Simon and Southwood Smith, and it was supported by the General Board of Health and the Privy Council. It was not until the discovery of the role of micro-organisms in infectious disease by Louis Pasteur and other bacteriologists in the 1870s and 1880s that the miasmatic theory eventually collapsed.

The point of interest is the fact that, although the proposals of the sanitary reformers were based on an erroneous theory about the transmission of communicable diseases, the cultural adaptive processes which were finally put into action were essentially the same as they would have been had the reformers and the professional advisers of government been aware of the role of pathogenic microbes in the disease process.

COMMENT

Kenneth Clark has written [33]:

The early reformers' struggle with industrialised society illustrates what I believe to be the greatest civilising achievement of the 19th century, humanitarianism. We are so much accustomed to the humanitarian outlook that we forget how little it counted in earlier ages of civilization.

He then goes on to suggest that the concept of 'kindness', which is important to us today, was quite unimportant for our ancestors before the 19th century. He says: 'Our ancestors didn't use the word, and they did not greatly value the quality'.

Clark's point is an important one, and most relevant to the theme of this book. His statements, however, need some qualification. Concern for the well-being of other human beings has always been a characteristic of our species at the level of the primary or in-group. Nevertheless, as discussed in an earlier chapter, there seems to be no evidence for any innate tendency in humankind to feel such concern for the well-being of members of out-groups. It was this 'deficiency' in the human character which led to the necessity, when many different in-groups came to live in close proximity with each other in the early cities, of introducing cultural adaptive measures, in the form of legislation, to protect individuals from the undesirable consequences of out-group antagonism. However, moves aimed at preventing conflict and aggression between out-groups are distinct from efforts to encourage mutual aid among them, and cannot be regarded as manifestations of kindness. It is true that the importance of showing kindness to members of out-groups was inherent in the teachings of Jesus Christ, but this aspect of his teachings has not always featured prominently among motivating values of professed Christians. Nevertheless, the Church itself in the Middle Ages was seen to have an obligation to help the destitute, and this can be viewed as an expression of kindness on the institutional level.

The new development in the 19th century was *secular humanitarianism;* that is, the idea that those who are better off in society, and indeed the government itself, have a responsibility to take steps to alleviate suffering among less fortunate sections of the community. This idea found expression in a range of reform movements, including those aimed at bringing an end to slavery, preventing exploitation of children in factories, improving the conditions of the prisoners, providing sustenance for the very poor, alleviating suffering in Ireland at the time of the famine, and, as discussed in this chapter, improving the living conditions of the working class. In all these instances, the situation was characterized by the existence of a relatively small group of vocal reformers, and an equally vocal group of anti-reformers, most of whom represented, or in some other way were closely associated with, vested interests which could be disadvantaged by the proposed reforms.

NOTES

1. See Passmore 1980, p. 19. For general discussions on this theme see Dubos 1961 and Medawar 1982, pp. 324–39.
2. Dubos 1961, p. 19.
3. Russell 1946, p. 580.
4. Russell 1946, p. 750.
5. Clark 1969, p. 327.
6. Wilkinson (1973) takes the view that technological progress is almost always a response to ecological pressure, and in particular suggests that the industrial transition in Britain was a natural consequence of population growth (see his chapter 6).
7. From Petersen 1969.
8. Some authors consider that historians have exaggerated the degree of poverty in England during this period (see Petersen 1969, pp. 406–8; Grigg 1980, p. 163; Wiet, Elisseeff, Wolff, and Navdov 1975, part 2, p. 957).

9. Petersen 1969, p. 164; Grigg 1974, pp. 164–77. This increase in primary production is sometimes referred to as the 'second agricultural revolution', the 'first agricultural revolution' being associated with the introduction of annual clover (see Nicol 1967, p. 62).
10. Mumford 1966, pp. 518–9. There is some debate about the severity of the working conditions in factories at this time—see, for example, Petersen 1969, pp. 409–13.
11. Mumford 1966, p. 518.
12. Mumford 1966, pp. 529–30.
13. Hare 1954, p. 62.
14. Flinn 1965, p. 13.
15. Trevelyan 1951, p. 106.
16. *Encyclopaedia Britannica* 1962, volume 18, p. 215.
17. Flinn 1965, p. 24.
18. Flinn 1965, p. 32.
19. Flinn 1965, p. 34.
20. Frazer 1950, p. 33.
21. Frazer 1950, p. 33.
22. Flinn 1965, pp. 35–6.
23. Flinn 1965, pp. 422, 423, 424.
24. Flinn 1965, p. 42.
25. Flinn 1965, p. 73.
26. Frazer 1950, p. 120.
27. Flinn 1965, p. 8.
28. John Ferriar, quoted by Flinn 1965, p. 62.
29. Southwood Smith, quoted by Flinn 1965, p. 62.
30. William Farr, quoted by Flinn 1965, p. 29.
31. Flinn 1965, pp. 63–4.
32. Frazer 1950, pp. 65–70.
33. Clark 1969, p. 329.

8

INTERRELATIONSHIPS BETWEEN SOCIETY
AND THE BIOSPHERE IN THE
HIGH-ENERGY PHASE

INTRODUCTION

The two most important ecological characteristics of the high-energy phase of human society have been technological innovations resulting in the widespread use of machines driven by extrasomatic energy and changes in human life conditions which have dramatically reduced mortality rates in human populations. There have also been important changes in the quality of human experience. Many of the major causes of distress characteristic of the early urban societies have been largely removed through cultural adaptive processes, and the average life expectancy of human beings has greatly increased. Advances in the science and practice of agriculture have brought about big increases in food production, and this development, along with progress in nutritional science, has meant that malnutrition due to deficiencies of specific nutrients, and undernutrition associated with low calorie intake, are now much less common in populations of the developed world than they were in earlier urban situations.

On the negative side, however, some of the changes associated with the fourth ecological phase of human history are patently undesirable. First and foremost among such changes are the advances in the technology of warfare, with the invention, development, and production of weapons of mass destruction which are capable of devastating the processes of life on earth and of bringing an end to the human species. Other characteristics of the modern world which are causes for concern include rapid population growth, and the accelerating use of both renewable and non-renewable resources and the discharge of wastes from industrial activities on a scale which many ecologists believe constitutes a threat to the integrity of the biosphere [1]. Certainly, these particular characteristics of the high-energy societies are ecologically unsustainable in the long run; the only room for debate is on the question: How much longer can the processes of expansion proceed before irreversible damage is done to the biosphere?

This chapter, and the two which follow it, will focus mainly on different aspects of ecological phase four society in its most advanced form, as it occurs in the developed Western countries. I will not attempt an exhaustive and comprehensive account of the ecology of modern society; much excellent material on this theme has been written by other authors [2]. The intention is rather to highlight the salient ecological and biosocial aspects of the current situation and to discuss them in terms of the principles and concepts introduced in earlier chapters. The present chapter consists of an overview of the intensifying metabolism of modern high-energy society.

POPULATION

A salient feature of the human ecological situation since the 19th century has been the spectacular rate in the growth of the world population. The unprecedented acceleration in population growth which has occurred during the past fifty to one hundred years is depicted in the graph in Fig. 8.1 [3]. At the time of writing, the doubling time for the world population is estimated to be about thirty-five years. During the early farming and early urban phases, the average doubling time was about 1500 years. At present, more than 70 million people are added each year to the global population: two hundred years ago the figure would have been about 7 million, and one thousand years ago it would have been about 50 000. As Singh has put it: 'Today's situation is unique in that the largest mass of humanity that has ever existed is growing at the highest rate in its experience' [4].

The explanation of the increasingly rapid rate of population growth in the modern world lies mainly in the effects of cultural developments on death rates. In an earlier chapter we discussed how the potential demographic consequences of the protection afforded by early farming societies from the main causes of death in the primeval habitat were to a considerable

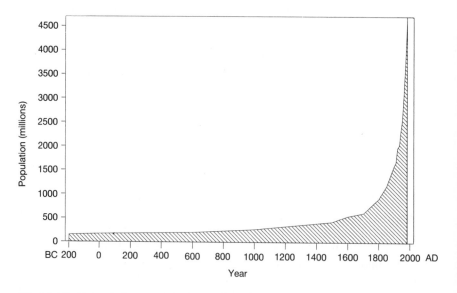

Fig. 8.1. World population growth in the past 2000 years.

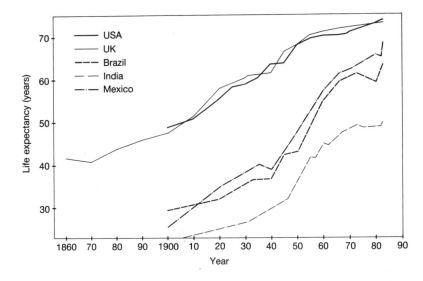

Fig. 8.2. Life expectancy at birth in selected countries (1960–80).

extent offset by the introduction of new causes of ill health and death. These causes were mainly infectious disease, malnutrition, and starvation. The situation in the high-energy societies is entirely different. The main causes of death in early farming and early urban societies have been largely overcome through improvements in life conditions and various cultural adaptive processes, and as a consequence there has been a great increase in life expectancy. A far higher proportion of babies born now reach reproductive age and thus contribute to population growth, not only by surviving longer themselves but also by adding their own offspring to the population pool. Some typical figures for changing life expectancy at birth for a number of countries is shown in Fig. 8.2 [5].

There has been a tendency to over-emphasize the role that chemotherapy and the use of antibiotics has played in this lowering of death rates. Certainly these cultural adaptive measures have played a part, but a far greater contribution has been made by the introduction of new standards of public and private hygiene—standards which are necessary to counteract the effects of high population density on the susceptibility of human pupulations to contagious diseases. Improved nutrition has also played an important part [6].

The other factor affecting the rate of population growth or decline is fertility rate [7]. In the early days of the high-energy phase, birth rates were high and this fact, along with the declining death rates, resulted in relatively rapid population growth. The crude birth rate in England and

Wales in the first decade of this century, for example, was about 30 per 1000. However, this pattern of high birth rates and low death rates with consequent rapid population growth has not persisted in most of the developed regions of the world. It has been replaced by a situation characterized by low death rates, low fertility rates, and only small annual increases in population. The crude birth rate in England and Wales in 1976 was 11.9 per 1000 [8].

The major factor determining fertility rates in such societies, in sharp contrast to the situation with respect to mortality rates, is *personal choice*— the desire, that is, to have or not to have children. Deliberate family planning has been made somewhat easier through the use of hormonal contraceptive pills, although it is noteworthy that fertility rates dropped dramatically in Western populations well before their introduction. In fact, the fall in fertility rates in England in the early part of this century is said to have been due mainly to the practice of *coitus interruptus*.

The drop in birth rates in high-energy societies has been associated with an increase in material standard of living, improved nutrition, and healthier working conditions [9].

The situation is very different in many developing regions of the world, where death rates have fallen markedly as a result of certain influences of the high-energy societies, but where birth rates remain high. This difference between developed and developing regions is resulting in an increase in the proportion of the world's population living in the latter regions. The population of the world as a whole in 1982 was growing at a rate of 1.7 per cent per year. The rate in developed regions was 0.6 per cent and in the developing areas it was 2 per cent or more. It has been estimated that by the year 2000, four-fifths of the world's population will be living in today's developing countries [10].

These differences in fertility rates are also reflected in the age-structure of human populations in developed as compared with developing regions and are depicted in the population pyramids for Sweden and the Philippines in 1980 shown in Fig. 8.3 [11]. The much higher proportion of young people in the developing regions is due in part to the higher birth rates and in part to the fact that a significant increase in life expectancy in these regions has been a relatively recent development. An important implication of the present age-structure in the developing areas is the fact that the high proportion of young people will result in an increase in the number of births in the population when they reach maturity, even if by that time they are planning to keep their families small.

It is self-evident that the population of the human species on earth cannot continue to grow indefinitely. At some time it must cease to increase, and it is to be hoped that this change will come about deliberately, and not as a consequence of large-scale catastrophe. There are hopeful indications that in some parts of the developing world the rate of population

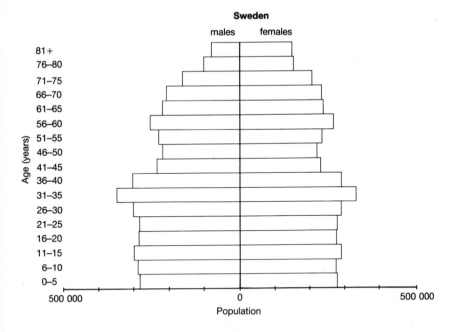

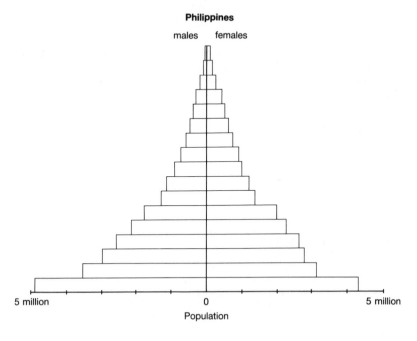

Fig. 8.3. Population by sex-and age-groups in Sweden and the Philippines, 1980.

growth is slowing down [12]. However, unless there are catastrophic rises in mortality rates, population growth up to at least eight thousand million persons seems to be inevitable.

FARMING

Introduction

In chapter 5 we noted how, during the early urban phase of human existence, the proportion of the population engaged in food production was much higher than the proportion engaged in other pursuits. In the Middle Ages in Europe, for instance, about 90 per cent of the population consisted of farmers and their families. The situation in modern industrial countries is reversed, and the farming sector usually comprises 10 to 15 per cent, providing food for the other 85 to 90 per cent of the population— as well, in some cases, as a surplus for export. The proportion of the population of the USA now engaged in farming is about one-eighth of what it was in 1920, and half of those now classified as farmers also have other jobs. During this same period, the number of man-hours devoted to farming has halved, while the output (food calories) per man-hour has increased four-fold. In the 1970s one farm-worker in the USA produced sufficient food for 50 people [13]. Australia is an extreme case, where one farmer now produces food for 125 people, 83 of whom live overseas [14].

This new situation is characteristic of high-energy societies and has come about through the introduction of new farming methods, particularly those involving the use of farm machinery powered by fossil fuels. In terms of man-hours of farm labour, the transition can be illustrated by comparing the figures for the yields of grain crops per man-hour of farm labour in typical phase four situations with those of farming systems in developing countries, as shown in Table 8.1 [15].

Table 8.1. *Yield (MJ/man-hour) in different farming systems.* (Note: one adult human requires on average about 10 MJ per day.)

Crop and place	Output
Rice (USA)	2800
Cereals (UK)	3040
Maize (USA)	3800
Subsistence rice (tropics)	11–19
Subsistence maize (tropics)	25–30

The overall production of food the world over has been increasing slowly but steadily as a consequence of the spread of more intensive farming methods. Indeed, this has been very necessary, with 70 million or more extra people to feed each year. Between 1971 and 1978 cereal production rose from 1315 million tonnes to 1596 million tonnes, and there were also increases in the production of pulses, fruits, nuts, meat, and milk (increases equivalent to about 0.6 tonnes of grain per year per additional person) [16].

The biomass harvested by humankind today for food, fuel, fibre, and timber has been estimated to represent about 7 per cent of the energy trapped by photosynthesis over the surface of land as a whole (i.e. about 7 per cent of the global nett primary production). The proportion of the energy trapped by photosynthesis which is actually harvested for eating by humans is around 1.0 per cent [17].

The Green Revolution

The figures given above indicate the huge differences which exist in primary productivity between developed and certain developing regions, and yet it is in the latter areas that the population is increasing most rapidly and where more food is desperately needed.

Nevertheless, in some regions of the Third World large increases in yield have been achieved as a consequence of the so-called Green Revolution. This began in the 1960s and it continued to have an important influence on agricultural trends during the 1970s. It was the result of scientific advances in the high-energy societies and was based on the development of high-response varieties of wheat and rice, and also to some extent of maize and millet, as well as the increasing use of artificial fertilizers and pesticides. In the case of wheat, new high-yielding 'dwarf' varieties developed in Japan were introduced into various developing countries. Big doses of fertilizer, which in the conventional strains of wheat had produced heads heavier than the plants could support, resulted in similar heavy heads of grain in the dwarf varieties: however, in the latter case these heads could be supported by the shorter, sturdier stems. Following the establishment of the International Rice Research Institute in the Philippines, similar new varieties of rice were produced.

The use of the new varieties resulted in an increase in the Asian food supply of around 16 million tonnes in the mid-1960s, enough to feed 90 million people. In India, for instance, there was a doubling of the wheat crop in a six-year period; and there was also a remarkable increase in Pakistan, and some increase in the Philippines, Sri Lanka, Indonesia, and Malaysia. Another country which was much affected was Mexico, where the Japanese dwarf varieties of wheat were first tried out.

The Green Revolution has provided, however, only temporary relief to the problem of food shortage in these developing countries where populations are rapidly increasing. Indeed, countries like India and the Philippines are now once again importing large quantities of wheat [18].

Another problem is associated with the fact that the changes in farming practices in these developing areas have tended to benefit the rich rather than the poor. Although the use of high-yielding varieties could well be adapted to labour-intensive conditions, in practice it has often resulted in a boost in capital-intensive and energy-intensive mechanization, giving rise to widespread unemployment. In many situations the introduction of the high-response varieties has made the rich landlords richer and the poor peasants poorer [19].

Artificial fertilizers

Research in the past hundred years on the nutritional requirements of plants has resulted in the large-scale redistribution of certain essential plant nutrients, through the use of artificial fertilizers. The overall global use of artificial nitrogenous, phosphate, and potash fertilizers has increased more than ten-fold since 1945, and it is safe to say that the present world population could not be supported were it not for the redeployment of chemical compounds in this way [20]. It is an outstanding example both of effective societal adaptation and of technoaddiction (see page 89).

The rate of use of artificial fertilizers has been increasing in all parts of the world; the total use increased from 13.5 million tonnes in 1950 to 121 million tonnes in 1984. The increase has been much higher in typical high-energy societies. The following figures for rates of fertilizer use (nitrogen, phosphorus, and potassium) in Kg/Ha indicate the degree of variability among countries: USA, 82.0; Japan, 439.8; Taiwan, 292.6; China, 39.6; Burma, 1.8; Pakistan, 13.7; India, 16.4 [21].

A variety of unintended ecological repercussions, especially affecting waterways, have followed the increased use of artificial fertilizers. In many instances over-fertilization, or *cultural eutrophication*, as it has been called, has resulted in heavy growths of algae in streams and rivers. In other situations nitrogen fertilizers have resulted in toxic concentrations of nitrates in waterways [22].

The energetics of food production

Changes in the technology of farming in the high-energy societies, and especially the shift from labour-intensive to machine-intensive methods, have had major effects on the energetics of food production. The energy budget of modern farming is very different from that of pre-industrial, early farming societies. The difference lies partly in the fact that human (and animal) somatic energy has to a large extent been replaced by extrasomatic energy, mainly in the form of fossil fuels. There is also, for instance, an important energy cost inherent in the use of artificial fertilizers when account is taken of the energy which is used in their extraction or manufacture, in their transport, sometimes thousands of kilometres, to the farm, and in their eventual distribution on the land by mechanical means.

In the 19th century the ratio of energy output (yield) to energy input in farming was around 15. In the USA, UK, Holland, and Israel today this ratio ranges from around 0.5 to 0.7. In Australia it is about 2.8. The discrepancy between Australia and the other high-energy societies is due mainly to the greater reliance in Australia on legume nitrogen rather than on bag nitrogen [23].

By the time the food reaches the home of the consumer further energy has been spent in transport, in wrapping, and, in some cases, in the preparation of the food for the consumer. In Australia, the output–input ratio for food reaching the retail store is about 0.35, and in the USA it is about 0.2 [24].

The fact that farming in many regions of the world has become so heavily dependent on machines which use petroleum as their energy source renders the whole process extremely vulnerable to the scarcities and increases in oil prices which are bound to occur in the future.

Things that go wrong and cultural adaptation

Food shortages

Whereas shortage of food, at least on a per capita basis, is now rare in high-energy industrialized society, this is not the case for the rest of the world. In fact distress and death due to poor nutrition are widespread in many developing regions. It is not easy to assess precisely the degree of ill health and mortality resulting from starvation and malnutrition, but it has been suggested that in 1970 at least 500 million people were malnourished. This represents a shortfall, in terms of cereals, of about 20 million tonnes [25]. Children are particularly badly affected. It is said that at least 10 thousand people die every day from starvation or malnutrition. Indeed, in a bad year, 10 million people may die from these causes [26].

In the ten years from 1972 to 1982 there was a substantial increase in the average food supply of industrial countries, in both absolute and per capita terms. In developing regions, the increase was more modest and only just kept pace with the growth of population. According to Holdgate and his colleagues, the 'main problem in 1980 was the unequal distribution of food within and between countries, and the situation was least satisfactory in Africa and the Far East' [27].

Populations which are near the limits of human tolerance with respect to nutritional intake are clearly especially vulnerable to the undesirable effects of bad seasons. Moreover, the effects of unusually unfavourable weather are much more damaging in the ecosystems which are already under stress as a consquence of unsatisfactory farming practices. These two factors partly account for the devastating effects of droughts during the past fifteen years on human populations in many parts of Africa [28].

Degradation of ecosystems

The main causes of loss of productivity in ecosystems resulting from societal activities are the same in the modern world as they were in early farming and early urban situations, although nowadays it is very often a case of *informed abuse*, rather than of *ignorant abuse of nature*.

Thus serious degradation is occurring in many parts of the world as a result of erosion, chemicalization (e.g. salinization and alkalinization), and desertification. However, one additional factor has come to be important; it is the displacement of agricultural land by human-made artefacts, mainly in the form of buildings and roads. It has been estimated that if trends underway between 1975 and 1980 were to continue, by the year 2000 the annual losses of agricultural land would be about 15 million hectares—made up of 8 million hectares lost through displacement to non-agricultural use, 3 million hectares through erosion, 2 million hectares through desertification, and 2 million hectares through chemicalization [29].

It should be noted that problems involving soil degradation are not restricted to developing countries. One authority predicts, for example, that if effective counter-measures are not soon taken to halt erosion, many soils in the USA will last only another 35 years. The situation is more serious in Australia, where the depth of fertile topsoil is often only about 15 cm as compared with several metres in North America. A great deal of land has already been rendered useless in Australia through erosion, and in other parts of the country the rate of disappearance of soil is the cause of serious concern [30].

Considering the world as a whole, it has been estimated that the present rate of soil loss to the oceans resulting from human activities is about four-times the 'natural' average rate for the last 200 million years [31]. The rate of soil loss on cropland is said to exceed the rate of formation of new soil the world over by 25 400 million tonnes each year, and it has been estimated that topsoil reserves will disappear in around 150 years if appropriate steps are not taken to protect them [32].

Salinization also continues to be an important cause of land degradation in many regions of the world. According to Holdgate and his colleagues, between 1972 and 1982 salinization caused the abandonment of about the same area as was being reclaimed and irrigated. It was particularly acute in semi-arid and arid regions. During the same period, desertification also continued on a grand scale and apparently at an increasing rate. About 60 000 km^2 of land was destroyed or impaired annually as a result of severe and recurrent drought and human exploitation [33].

Pests and disease

The high-energy phase of human society has seen major advances, not only in the understanding of the causes of infectious and parasitic disease of plants and farm animals but also in various means of control. The resulting cultural adaptive measures fall broadly into the following classes: (a) The use of *biocides*; that is, the use of chemical compounds or certain biological products (e.g. antibiotics) which are toxic for the parasite but not for the host. (b) *Biological control*—this is used especially in the case of animal parasites which attack plants, and involves the introduction of other living organisms which prey on, or cause disease in, the parasites. (c) *Vaccination*— this approach has been widely used in the case of micro-organisms which cause disease in domestic animals. It involves the induction of a state of specific immunity in the animals through the injection of products or components of the disease-producing organism or of living micro-organisms related to the infectious agent, but not themselves capable of causing disease. (d) The *selective breeding or cultivation* of genetic variants which are relatively resistant to particular infective agents.

Of these four cultural adaptive approaches aimed at countering the undesirable effects of parasites and pathogens on food production, biological control, vaccination, and selective breeding or cultivation have been relatively free of undesirable side-effects. In contrast, the use of biocides has frequently had far-reaching repercussions, including, in some instances, ill health and death in human beings.

The Western world was alerted to some of the undesirable effects of the widespread use of synthetic chemical biocides for protecting food plants from parasites by the extraordinary book of Rachel Carson, *Silent Spring*, which was published in the USA in 1962 [34]. This book drew attention to the wide-ranging disturbances in ecosystems, sometimes involving the virtual extermination of many organisms other than the target arthropod pests. Carson was a *first-order reformer* in the best tradition (see page 00). She was in the forefront of those who alert society to the existence of an undesirable state of affairs. She was, like all reformers, ridiculed and attacked by anti-reformers who called her a fanatic and an alarmist, and who criticized her for overstating her case and for being emotional. Most of the anti-reformers represented vested interests; but unlike the situation in the Public Health Movement discussed in chapter 7, the vested interests in this case were large corporate organizations, rather than individual landowners or businessmen. Carson was described by the president of a large chemical corporation as 'a fanatic defender of the cult of the balance of nature' (see note 34).

Carson aroused a great deal of interest in the community at large, and she was soon joined by many others—and slowly but surely legislation was introduced in the USA, and eventually in other Western countries, controlling the use of DDT and other biocides. The sequence of events in this episode is summarized in Fig. 8.4.

It is noteworthy, however, that although the use of DDT and various

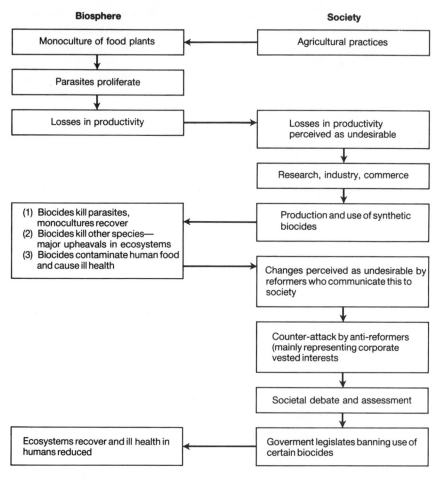

Fig. 8.4. Cultural adaptation to undesirable side-effects of synthetic biocides.

other biocides is now banned in the USA, the manufacturers are still permitted to export them in large quantities to Third World countries, where their use is not yet prohibited by legislation and where they are often used indiscriminately and without adequate precautions to protect human health [35].

Predictions, prophecies, and pious hopes

There has been a good deal of discussion and debate during recent decades on the potential carrying capacity of the earth. What is the maximum number of people that could possibly be supported by the biosphere? Will it be possible to provide food for the predicted world population in, say, fifty years time?

Views range from those of authors who argue that the present population of over 4000 million is already above that which could be fed on a typical North American diet, to those of Colin Clark (1967) who considers that 47 000 million people could be adequately fed [36]. Another economist, Heinz Arndt, has written: 'There are no technical or economic reasons for doubting that the world's capacity to produce food can grow ahead of the world's needs for food for many generations' [37]. A more common view, however, is that it might be possible to double food production, but probably not to treble it [38].

Certainly, if the population predicted for the year AD 2000 is to be fed even to the same level on a per capita basis as the world population today, extraordinary increases in production will be necessary. The Food and Agricultural Organization (FAO) has estimated that a 60 per cent increase in food, fish and forest production will be necessary to maintain the current pattern of consumption, assuming that the world population rises to 6300 million by AD 2000. The *Global 2000 Report* (Council of Environmental Quality 1980) optimistically suggests that an increase of 90 per cent in food production could be achieved, using 2.5 per cent more land than was used in 1978. Some authors believe that the situation will be relieved by advances in biotechnology, such as hydroponics, greater use of micro-organisms as nitrogen fixers and as agents of organic synthesis, and new forms of cereals which fix nitrogen and which have a higher capacity for photosynthesis than existing forms [39].

Apart from the question of food supply, there are two other important factors to be taken into account when considering the possibilities for the future with respect to population size on a global level. First, attention must be paid to the technological and economic bases of society; that is to say, account must be taken of both the biometabolism and the technometabolism of the human population. Whittaker and Likens have estimated, for instance, that an 'agricultural world', in which most human beings are peasants, should be able to support 5000 to 7000 million people, probably more if the large agricultural population were supported by industry-promoting agricultural activity. In contrast, they suggest that a reasonable estimate for an industrialized world society at the present North American material standard of living would be 1000 million. At the more frugal European standard of living, 2000 to 3000 million would be possible [40].

Forests

For several reasons, concern has been expressed over recent years about trends affecting the world's forests. Behind all these reasons lies the simple fact that more trees are destroyed every year than are planted—the annual deficit is about 11 million hectares of forest [41].

As discussed in earlier chapters, extensive deforestation began in Europe around 1100 and continued until the 14th century, when it declined after the Black Death. Deforestation began again about 1500. Eighty per cent of central Europe had been covered with forest around AD 900, and only 25 per cent in AD 1900. The European forests reached a minimum around the time of the First World War. By this time the forests of North America had declined by about 60 per cent—that is, by 1.5 million square kilometres. Reforestation has taken place since then, covering about one-third of this area.

Similar trends have been taking place in many other parts of the world. Most of North Africa and the Middle East, as well as most of continental Asia, Central America, and the Andean regions of South America are now virtually treeless. Problems associated with deforestation are also looming large in parts of Central Africa and on the Indian subcontinent.

Figures for the per capita consumption of wood are interesting. In the developing areas, 80 per cent of all wood consumed is used as fire-wood. The annual per capita consumption for this purpose varies between 0.35 cubic metres in India to 1.3 cubic metres in forest-rich countries like Indonesia. The per capita consumption of wood in Britain each year is 0.72 cubic metres. The average person in the developed countries uses about 127 kg of paper each year, as compared with 5 kg for every individual in developing regions. The developed countries, with 30 per cent of the world's population, consume roughly 88 per cent of all industrial wood. At present most of the timber and paper used in the developed countries is provided by the temperate forests of Canada, Finland, Norway, Sweden, and the USSR. However, according to some reports from the FAO and the Centre for Agricultural Strategy, the wood consumption of the world is likely to have doubled by the year 2000, and this will result in a severe shortfall in supply.

Tropical forests at present account for about 69 per cent of forest productivity in the world. It is claimed that about 200 000 square kilometres of tropical forest is damaged, if not destroyed, each year by farmers— mainly in south-east and southern Asia, in tropical Africa, and in parts of tropical Latin America. In the words of Brunig, this forest is:

... widely considered by politicians, industrial entrepreneurs and pioneer farmers to be an unproductive asset that is going to waste, one that should be mined, if possible, and replaced by more productive agri- and silvicultural ecosystems. There are serious socioeconomic and ecological objections to this view... [42].

The forests play a number of important ecological roles relevant to human society. These include energy exchanges, water vapour exchanges, the fixing and release of carbon dioxide, the release, dilution, and absorption of aerosols, gases, and solutes, the production of goods, as well as various intangible cultural services.

Some ecologists have pointed out that the delicate nature of the various balances involved in global climatic processes make it likely that ecological changes in the tropics of the kind that would follow further extensive deforestation could give rise to important climatic changes in other regions of the world.

In recent years serious concern has been expressed about the damage to forests in the industrialized nations due to *acid rain*. This issue will be discussed later in this chapter.

WEAPONS OF MASS DESTRUCTION

As noted in earlier chapters, various technological advances in human history have led to a progressive increase in the lethality and destructive capacity of weapons used in warfare. The invention and use of gunpowder, although at first not very effectual, was perhaps the most significant development in this regard. Cultural processes in the high-energy societies have resulted in further highly significant developments in the destructive power of weapons manufactured and used by human beings for the purpose of killing other members of the species.

The situation is well summarized in the following paragraph taken from UNESCO's *History of Mankind*:

The application of scientific knowledge to the means of destruction during the first half of the twentieth century radically changed the character of warfare and created threats to the very survival of mankind. In a succession of stages it transformed warfare from the specialized activity of a limited military group, first into the total mobilization of industrial resources to produce weapons which would overwhelm military objectives, then into the total involvement of entire industrial societies in the process of hurling concentrated, mechanized force against military or civilian targets anywhere on the globe, and finally into a scientific contest to develop weapons of annihilation and the means of delivering them to destroy the opponent's total society. At each stage the forces of destruction were increased many times, the complexity of organization was very greatly enhanced, the moral and political stability of the home front as well as the armed forces became a more critical factor, and dependence on scientific development became more complete... . By the opening of the century weapons designed by the application of scientific technology had already given the industrialized nations an overwhelming advantage against the non-industrialized peoples and had provided a basis for the political and economic imperialism of the western powers. The machine-gun had given the unskilled foot soldier a formidable weapon and had stepped up the destructive power of a given fighting force [43].

World War I marked an important transition in the history of warfare. With the use of rifles, machine-guns, tanks, howitzers, and the occasional aircraft, the face-to-face aspect of combat was beginning to disappear. This fundamental change in the nature of armed conflict had proceeded further by the time of World War II, when bombing by air became important and long-range weapons, including rockets of various kinds, were introduced. The situation now exists that deliberate hostile action by a very small number of individuals many thousands of miles away from their targets can result in the deaths of millions of human beings.

Among the range of highly sophisticated instruments of mass destruction which exist today in the arsenals of powerful nations, nuclear weapons are by far the most destructive. These have come into being as a consequence of massive programmes of research on radioactivity which had its beginnings in the work of Pierre and Marie Curie and Ernest Rutherford at the end of the last century and the beginning of this one.

During the last part of World War II, Germany and the USA were urgently competing to be the first to produce nuclear weapons. At 8.15 a.m. on the 6th August 1945, three months after the capitulation of Germany, a bomb was dropped out of the sky onto the city of Hiroshima in Japan. At least 140 000 people, that is about 40 per cent of the population of the city, were killed or died soon afterwards [44]. The buildings of the city were flattened over an area of 13 square kilometres. Three days later, another nuclear bomb was dropped on the Japanese city of Nagasaki, and 26 per cent of its population of about 280 000 people was killed outright.

Since that time, the governments of opposing ideologies have channelled immense financial resources and human effort into research on and development of nuclear weapons, so that bombs now exist with an explosive power one thousand times that of the bomb which was dropped on Hiroshima. A single US B-52 strategic nuclear bomber can now carry more explosive power on one trip than has been used in all the wars in human history.

At the time of writing, the USA, the USSR, the UK, France, and China are known to possess nuclear weapons. India has detonated a nuclear explosion, and Israel and South Africa are believed to have nuclear arms. Between 1970 and 1980, 469 nuclear devices were exploded, 41 of these in the atmosphere [45]. The arsenals of the USA and the USSR are gigantic, and they completely overshadow those of the other nations. Together, these two powers possess weapons with explosive power equivalent to 12 000 million tonnes of TNT (12 000 megatonnes); that is, they possess weapons which are equivalent to one million of the bombs that were dropped on Hiroshima. This means that the explosive power of the nuclear arsenals of the world is equivalent to at least three tonnes of TNT per man, woman, and child of the world population. These figures are so astronomical that it is difficult to grasp their full meaning. The individual weapons range in strength from the equivalent of around one hundred tonnes to twenty million tonnes of TNT, depending on the particular use for which they are designed.

The growth in the explosive power of bombs during this century can be illustrated by the following analogy. If we imagine the explosive power of the biggest bombs in World War I to be represented by a pea, then the biggest weapons (other than the atomic bombs used at Hiroshima, Nagasaki) used in World War II would equal the size of a large plum. The Hiroshima bomb would be equivalent to a sphere of about 50 centimetres across, and the most powerful bombs now ready for use would have a diameter of 5 metres.

The likely effects of a nuclear war on the people of the world and on its ecosystems, as far as can be gauged on the basis of existing knowledge, may be summarized by the following two quotations:

In a nuclear world war, not only would a fair part of the urban population in the Northern Hemisphere be killed by fire and blast, and most of the survivors by radiation, but much of the rural population would be killed by radiation from fallout. Many millions in the Southern Hemisphere would also be killed by radiation from fallout [46].

Although the impact of the nuclear war ... would be widespread and terrible, there would probably be survivors. Their fate, however, is extremely uncertain. The human and social environment in which they will have to live will be changed far beyond our comprehension. In addition to wartime destruction and poisoning, the natural enviroment might suffer such grave long-term changes as to severely threaten the survivors' fight for recovery. In any case societies as we know them today will most certainly cease to exist [47].

It is now widely believed that a nuclear war would, as a result of smoke and soot released into the atmosphere from fires started by nuclear explosions, produce a 'nuclear winter'. Temperatures would fall to extremely low levels and the productivity of crops would, due to the cold and the lack of sunlight, almost cease [48].

Thus, very recently, and for the first time in the history of life on earth, a single species, through its unique and peculiar capacity for culture, has developed the means to destroy most, if not all, of its kind within a few days, and to cause devastating damage to the biosphere as a whole.

The arms race is clearly an escalatory societal process which is at present totally out of control. While this process is to some extent fuelled by, or at least made possible by, such innate human behavioural tendencies as suspicion of and lack of concern for out-groups, it is at root the inexorable consequence of the promotive behaviour of corporate organizations— industrial, military, and governmental. The leaders of the superpowers appear as little more than role-playing puppets in this context. It is true that a vigorous movement is underway in the Western world today at the community level drawing attention to the tragic absurdity of the situation and calling for the abolition of nuclear arms; but so far the movement has been without effect.

Mention must also be made here of the immense amount of effort and of resources which have been devoted in modern high-energy societies to the development of other sophisticated and extremely potent methods of killing people. Thus, apart from the advances in nuclear armaments, great progress has been made in the development and production of chemical and biological weapons. We will not, however, discuss these weapons here since, horrific though they may be, their impact on civilization and the human species and on other life forms would be small in comparison with that of a nuclear war.

Human society as a whole at present spends over US $1 million per minute on the development and manufacture of homicidal devices. In six hours more money is spent on the manufacture of arms than was spent by the world community to bring about the eradication of smallpox from the face of the earth.

THE TECHNOMETABOLISM OF MODERN INDUSTRIAL SOCIETY

Since the introduction and spread of metallurgy a few thousand years ago, and up until the time of the industrial transition, the level of use of non-renewable resources and of extrasomatic energy increased at a rate which ran roughly parallel to the rate of increase of the human population. That is to say, until the 19th century, the intensity of the technometabolism of human society remained more or less constant. With the transition to the high-energy ecological phase, two significant changes took place. First, as discussed earlier in this chapter, the human population began to grow at an unprecedented rate. Second, the per capita use of non-renewable resources and of extrasomatic energy started to increase; that is, the technometabolism of society began to intensify.

It is an ecological truism that no organism or population of organisms can grow and expand its metabolism in a finite environment indefinitely. One of three factors will bring the process to an end: (a) the organism or population will run out of resources; (b) it will 'choke' in or be poisoned by its own wastes; or (c) it will reduce or destroy, either through the production of waste material or through other impacts, the productivity of the ecosystem of which it is a part and on which it depends. In the current human situation, the first of these possibilities would be by far the most desirable, because it would allow for some adjustment, after which human beings could continue to exist and to enjoy life (assuming that the scarcity of resources does not provoke devastating international warfare).

Whether the fundamental principle that there exist ecological limits to growth should be regarded as a basis for concern in the human context depends, of course, on the scale of human activity in the biosphere. If this activity accounted for only a small fraction of the overall metabolism of the system, as reflected, for example, in the amount of energy used by the human species, then it might be reasonable to postpone consideration of the implications of these ecological constraints for the time being. However, the facts which will be discussed in this section show that this is not the case. The societal metabolism of the human species already represents a very significant component of the metabolism of the biosphere as a whole.

Resource extraction and energy use

Minerals

For various reasons, it is not easy to provide an accurate picture of the changing rate of use of different mineral resources either by humankind as a whole or by separate societies. However, it stands to reason that increasing industrial productivity involves increasing use of resources.

The figures available for the per capita production of pig-iron the world over show a steady increase during the past 120 years, from around 6 kg per person for the year in 1860 to 122 kg per person in 1978. In 1972 in the USA the per capita use of iron and steel was 545 kg, and of total metals 609 kg. In Australia the per capita use of iron amounts to nearly 500 kg per year (this estimate does not include imports, for example, of motor vehicles, but it does take account of the fact that two-thirds of the iron produced in Australia is exported) [49].

With respect to the use of metals in general, although relatively sophisticated metallurgy was practised five thousand years ago, more primary metal has been consumed by society during the past twenty-five years than during the whole of previous history [50].

Needless to say, the control and use of material resources, and the material benefits derived therefrom, are far from evenly distributed within the populations of most high-energy societies. Even more striking, however, are the disparities which exist in the intensities of resource use between populations in the developed and the developing regions of the world. The industrially developed world, which contains about one-third of the total human population, uses 90 per cent of the non-renewable resources produced. This means that the populations in these countries are using about twenty times as much non-renewable resources per capita as are populations in the Third World. The USA, which has 5 per cent of the world's population, uses 27 per cent of the materials extracted. This is thirty-six times more, per capita, than in the developing world [51]. To bring the developing world up to the same level of resource use as the USA would require a five-fold increase in global resource production. The estimates which have been made for the 'lifetimes' of various raw materials would have to be divided by five.

Clearly, a critical ecological question is: How long will the deposits in the earth's crust of the important minerals used in the technometabolism of the industrialized world last, either at the present rate of use or at the present rate of increase in use? The answer to this question is not easily determined and much caution is necessary in interpreting estimates of existing reserves. Among other factors to be taken into account is the

Table 8.2. *Per cent of consumption recycled.*

Metal	USA, 1974	UK, 1978
Iron	24	69
Lead	40	65
Copper	23	37
Aluminium	8	29

extent to which resources are used, or might be used over again. In fact, recycling is already an important feature of the flow path of some minerals in developed countries. The percentage rate of recycling for four metals in the USA in 1974, and in the UK in 1978, is shown in Table 8.2 [52].

Another major cause of uncertainty is the degree of accuracy in the various estimates of remaining resources of different minerals. In some cases different estimates vary by as much as 600 times. The range of recent estimates for the 'lifetimes' of six important metal resources is as follows: aluminium 53–146 years; copper 30–61 years; lead 29–50 years; zinc 18–32 years; nickel 42–92 years; tin 32–82 years. These figures are taken from a table presented by Holdgate and his colleagues, who write:

The 'lifetimes' shown ... are rather low, all but one being less than a century. This creates an initial impression that there may be shortages of some mineral raw materials in the coming decades, if they have not already begun during the 1970's, but ... such a presumption is subject to many qualifications [53].

Opinions differ about the seriousness of the problem of depletion of non-renewable resources, as indicated by the following statement from an OECD report:

The concern so often expressed about the exhaustion of resources, based on the undeniable but sweeping assumption that our world is limited, does not for the moment justify any change in present policies [54].

Extrasomatic energy

Almost all aspects of the use of non-renewable resources, including their extraction from the earth's crust, the extraction of minerals from ores, the transport of raw materials, the manufacture of commodities, and often their use by consumers, involve the utilization of extrasomatic energy. The changing pattern of energy use is, in fact, the best single indicator of the overall scale of human industrial activity and of the expanding technometabolism of society. As indicated above, the use of extrasomatic energy, mainly from the combustion of fossil fuels, for driving machines is one of the outstanding ecological differences between ecological phase four and the earlier phases. The rate of energy use by human society as a whole at the time of the domestic transition was about 100 terajoules per day, and by the beginning of ecological phase four around 180 years ago it had increased to around 20–30 000 terajoules. By 1984 it had reached about 800 000 terajoules. The extraordinary increase in overall energy use during the past 180 years is in part the direct consequence of greater use of somatic energy resulting from population growth, but it is mainly due to the increasing use of extrasomatic energy in machines. In the case of coal as an energy source, for example, mining of this commodity started about 800 years ago; and half the total amount extracted by 1976 had been taken in the preceding thirty years [55].

At present the human species is using about one-fifteenth as much energy as is fixed daily by photosynthesis. If energy use by human society continues to increase at the same rate as it has increased over the past two decades, the human species will, by the year 2050, be using as much energy as all land animals and plants put together. One does not have to be an ecologist to appreciate that the introduction of a new force of this magnitude into the biosphere is likely to have far-reaching consequences for the system as a whole, or to appreciate that this rate of increase cannot possibly go on indefinitely.

The changing pattern of use of *extrasomatic energy* by humankind between 1800 and 1975 is shown in Fig. 8.5 [56]. As a consequence of technological developments the rate of use of extrasomatic energy used by human society in 1984 was about 300×10^6 terajoules per day, which is about 8–9000 times the figure for the time of the domestic transition when the only extrasomatic energy used by the human population of around five million was in the form of fire.

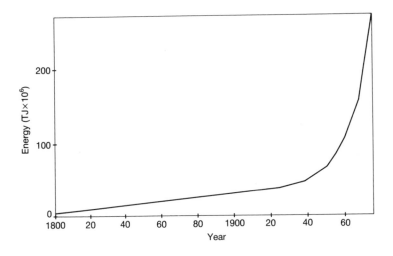

Fig. 8.5. World extrasomatic energy use per year (1800–1975).

At present the intensity of energy use, like the intensity of natural resource use, is much greater in the developed countries than it is in the rest of the world. In the USA, the daily use per capita of energy is around 1000 megajoules; that is, each individual has the equivalent of 100 'energy slaves' working 24 hours a day for him or for her. In Australia the figure would be 46 energy slaves and in the UK it would be 60 energy slaves. In some developing countries, the rate of energy use is less than the equivalent of one energy slave per person, and in this situation most of the extrasomatic energy is used, not in machines, but in the burning of wood or other organic material. The figures in Table 8.3 indicate the trends in extrasomatic energy consumption per capita in three parts of the world since 1925.

Table 8.3. *Per capita extrasomatic energy consumption per day (MJ).*

Place	1925	1950	1960	1972
USA	500	650	720	1000
Latin America	20	40	67	87
China	6	6	30	46
Total world	64	90	120	170

Since 1973 the extrasomatic energy use by human society as a whole has, in fact, been rising much more slowly than in earlier post-war years. Indeed, mainly as a result of improvements in efficiency of use, it actually declined very slightly between 1979 and 1982, but has since been increasing again. Moreover, the contribution of oil to the extrasomatic energy budget also declined, and in 1981 it was lower than in 1975 [57].

Sources of extrasomatic energy

Throughout the high-energy phase of society, the main sources of extrasomatic energy have been fossil fuels, although the relative contributions of coals, oil, and natural gas have changed considerably over the past sixty years. Hydroelectricity has increased slowly during the period, and very recently nuclear energy has made a contribution in some countries. The changing pattern on a global level is reflected in the figures in Table 8.4 [58].

Energy debates

Since the energy crisis of 1973, when there occurred a substantial increase in oil prices as well as some temporary interruptions in the flow of oil into industrialized countries, energy has become the topic of a great deal of discussion and debate in economic, political, and ecological arenas, and the volume of recent literature on energy use by society and on possible options for the future is now massive. Most of this literature is concerned

Table 8.4. *World energy consumption by source, 1925, 1950, 1979 (percentages).*

Source	1925	1950	1979
Solid fuels	82.9	61.0	28.6
Liquid fuels	13.3	27.7	45.2
Natural gas	3.2	9.7	18.0
Hydroelectricity	0.7	1.7	6.0
Nuclear energy	—	—	2.2

with economic and engineering aspects of energy use: but social and environmental aspects have also received considerable attention [59].

The so-called 'energy debate' consists, in fact, of a number of different, although obviously interrelated, energy debates, and it is important to appreciate the distinctions between them. Broadly there are three main areas of concern, and these are discussed below.

Problems associated with the fact that the supply of oil, and later other fossil fuels, will eventually be exhausted   The main current source of extrasomatic energy, petroleum, is predicted, at the present rate of use, to last another 20–40 years [60]. Coal is likely to last considerably longer and, at the present rate of use, *proved reserves* will last about 230 years; and it has been suggested that *assumed recoverable reserves* will last 1800 years. However, some estimates predict that rises in consumption will occur and that consequently the proven reserves will be used up by the year 2070. Debates concerning alternative sources of energy range from those which take account only of economic and political aspects to those which are concerned with the side-effects on the biosphere of different energy forms (see below).

Problems associated with the side-effects of energy use on the integrity of the life-supporting processes of the biosphere   The side-effects of extrasomatic energy use depend on the source of the energy. In the case of the fossil fuels, the most discussed side-effect is the predicted impact of the release of carbon dioxide on the global climate (see page 204).

In the case of nuclear power, the main cause for concern is the production of radioactive wastes which are harmful to living organisms. Authorities disagree on whether or not these can be disposed of in a completely safe way. There has also been much discussion on social and political aspects, relating especially to the possibilities of nuclear material being used for non-peaceful purposes and of the risks of sabotage [61]. The debate about the advantages versus the disadvantages of nuclear power appears to have had an influence on the policies of governments in a number of countries, as reflected in the figures for the number of nuclear reactors ordered each year, from 3 in 1960 and in 1961, fluctuating between 30 and 76 between 1966 and 1975, and dropping to 5 in 1979 [62]. The most important aspect of the present situation is well summed up in the following quotation from the *New Scientist* in 1979, referring to a meeting of Environmental Ministers from more than twenty countries: 'The most significant point to emerge was that the Ministers feel they do not know enough about nuclear power to say whether it is environmentally good or bad' [63].

All energy used by human society ends up as heat, and the use of fossil fuels and of nuclear power results in some heating up of the biosphere [64]. Such effects are at present only discernible locally—for example, in cities where the intensity of energy use is high, and in aquatic systems affected by discharges from the cooling systems of power plants. With respect to the long-term implications of increasing heat production resulting from increasing use of fossil fuels and nuclear power, some authors consider that if present trends in energy use continue there will be noticeable warming on a global scale in 100–125 years. Some estimates suggest that heat from extrasomatic energy use would render the earth unsuitable for human habitation as early as 150 years from now.

Problems arising out of heat production do not apply to hydroelectric power, wind power, tidal power, and biomass as sources of energy, since the energy derived from them is part of that naturally flowing through global ecosystems and is consequently destined to finish up as heat in any case. The situation with respect to 'direct' solar power is more complicated, in that a part of the energy trapped by such energy-converting devices as photovoltaic cells would, in fact, have been reflected back into the sky as light energy. Only about half would temporarily remain in the biosphere in the form of heat (the actual proportion depending on the nature of the earth's surface at the point of contact).

Another ecological advantage of the renewable energy sources is the fact that they do not give rise to chemical or radioactive products which might adversely affect the viability of the earth's ecosystems. For these reasons, some authors feel strongly that society should move towards the adoption of the renewable energy options.

Concern about the question whether increasing levels of extrasomatic energy use, or the present high levels in phase four societies, are really necessary, or indeed desirable, for a good society   The dominant view at present, and the one accepted by all governments in the high-energy societies, is that continuing increase in energy use, associated with increasing industrial production, is necessary for the welfare of people. It is considered that the best approach to poverty and discontent is to increase the material wealth of the whole society; and, in the present economic situation, this requires increasing use of extrasomatic energy. Nevertheless it is also generally agreed that many technologies are unnecessarily wasteful of energy and that there is ample scope for energy conservation within the present economic arrangement, involving improvement in the design of equipment and in the organization of industrial and commercial processes.

Some authors, however, go much further and take the view that the real needs of people, in terms of their health and enjoyment of life, could be met just as well as at present, perhaps better, by greatly reducing the current levels of resource, energy, and machine use [65]. This, of course, would involve important changes in the economic system and in the occupational structure of society. Let it be noted, however, that this aspect of the energy debate is much less in evidence than the controversies about the first two themes discussed above.

Waste products

From the ecological standpoint, the possibility of scarcity of resources is less critical than that of civilization 'choking in its own wastes' or severely damaging the productive processes of the biosphere. As indicated above, scarcity of resources which are not actually essential for human survival, health, and well-being simply calls for, or would result in, changes in societal organization. The human species would survive, and there is little reason to doubt that the people could live in a state of physical health and would find ways of enjoying themselves. But collapse of the life-support systems due to abuse and pollution of the ecosystems on which people depend would be a very much more serious matter.

Broadly, the wastes produced by ecological high-energy society are of two kinds. First, there are the organic waste products which are an aspect of the biometabolism of society. They include sewage and unused food, as well as other discarded products of primary production, such as timber and organic fibres. This aspect of societal metabolism is not, of course, peculiar to the high-energy societies. The main differences between the phase four situation and earlier societies lie in the quantities of organic wastes produced and in the fact that in many modern situations little of the organic waste material is returned to the soil. The natural biogeochemical cycles are thus interrupted. An extreme example is the case of Hong Kong where, in 1971, between six and seven thousand tonnes of organic solids in sewage were discharged into the ocean everyday, while artificial fertilizers for use on the land were imported from the other side of the world [66].

The other kind of waste produced is that which results from industrial activities and the use of machines—an aspect of the technometabolism of society. It includes the by-products of extrasomatic energy use itself, such as carbon dioxide, carbon monoxide, and radioactive wastes, as well as other waste products of industrial processes and the residues of various commodities, chemical and otherwise, which are used in society. There is now an enormous literature on this subject, and we will only deal here with a few salient examples which draw attention to some of the key ecological issues of our time. The following topics have been selected for brief comment: acid rain; carbon dioxide in the atmosphere; threats to the ozone layer; and chemicalization of the biosphere. First, however, let us indulge in a brief historical interlude.

An historical perspective

The recognition of pollution of the environment as undesirable and the application of cultural adaptive measures aimed at rectifying the problem is not new and is by no means an exclusive feature of modern high-energy societies. Indeed, it is clear that pollution was recognized and considered undesirable very early on in the history of cities. As discussed in chapter 6, the Hittite Code from 15th or 16th century BC imposed a fine for individuals who polluted pots or tanks; and air pollution was well recognized as undesirable in England around AD 1300, when King Edward I is said to have decreed: 'Be it known to all within the sound of my voice, whosoever shall be found guilty of burning coal shall suffer the loss of his head' [67].

Nearly a century later, Richard II introduced milder measures, in the form of taxation, to control the use of coal, presumably because of its undesirable effects on the atmosphere; and early the following century Henry V set up a Commission to oversee the use of coal in the city of London [68].

In 1661, John Evelyn, who later became one of the founding members of the Royal Society, wrote his famous and extraordinarily interesting pamphlet on air pollution in London. This work nicely illustrates some important characteristics of environmental reform processes. Evelyn, whose paper was entitled 'Fumifugium: or the inconvenience of the aer and Smoake of London dissipated' [69], played the role of a typical *first order reformer*. He was clearly deeply concerned about the environmental changes which he perceived as undesirable:

That this Glorious and Antient City ... which commands the Proud Ocean to the *Indies*, and reaches the farthest *Antipodes*, should wrap her stately head in Clowds of Smoake and Sulphur, so full of Stink and Darknesse, I deplore with just Indignation.

Evelyn proposes that:

... by an *Act* of this present *Parliament*, this infernal *Nuisence* be reformed; enjoyning, that all those *Works* (i.e. Brewers, Diers, Sope and Salt-boylers, Lime burners and the like) be removed five or six miles distant from *London* below the River of Thames.

He also advocates a programme of planting trees, shrubs and flowers, suggesting that: 'The City and environs about it, might be rendered one of the most pleasant and agreeable places in the world'.

Evelyn's paper is remarkable because it draws attention to an environmental consequence of societal developments which he perceived to be undesirable, because it reflects understanding of the links between pollution and health, and because it proposes a number of clear and positive steps that government and society could take to rectify the situation. It is also interesting because of the sycophantic approach that the reformer takes, in the hope of eliciting appropriate action by the authorities or rather , in this case, the authority—in the form of Charles II, who had been newly restored to the throne. The following excerpt illustrates this feature of his approach:

It was one day, as I was Walking in your MAJESTY'S Palace at WHITE-HALL (where I have sometimes the honour to refresh myself with the Sight of Your Illustrious Presence, which is the Joy of Your Peoples hearts) that a presumptuous Smoake issuing from one or two Tunnels near Northumberland-house, and not far from Scotland-yard, did so invade the Court; that all the Rooms, Galleries, and Places about it were filled and infested with it; and that to such a degree, as Men could hardly discern one another for the Clowd, and none could support, without manifest Inconveniency. It was not this which did first suggest to me what I had long since conceived against this pernicious Accident, upon frequent observation; But it was this alone, and the trouble that it must needs procure to Your Sacred Majesty, as well as hazzard to Your Health, which kindled this Indignation of mine, against it, and was the occasion of what it has produced in these Papers.Your Majesty, who is a Lover of noble Buildings, Gardens, Pictures and all Royal Magnificences, must needs desire to be freed from this prodigious annoyance ... .

Regrettably, Evelyn's efforts were to no avail. It was not until the notorious smog episode of December 1952, which killed four thousand people, that the government was finally galvanized into introducing effective legislation banning the burning of coal in open fires in the London area [70].

The history of air pollution in London illustrates an unfortunate characteristic of so many cultural or societal adaptive patterns—the fact that effective regulatory action is not taken until a really dramatic and frightening event occurs, involving, in this instance, the death of a large number of people over a short period of time. We have noted elsewhere how it was the outbreak of an acute epidemic of cholera that finally gave the necessary impetus to public health reform in Britain in the mid-19th century, although reformers had been calling attention to the chronic ill health and high mortality rates of the working class in the manufacturing cities of that country for at least fifty years.

As a consequence of more efficient methods of combustion, the use of 'clean' fuels, and the installation of smoke-remover equipment and of taller chimneys, the smoke and sulphur dioxide concentrations in the atmosphere have declined in London, and no serious acute pollution episodes have been reported since the early 1960s. The notorious 'pea-soupers' of that city have become a thing of the past. Similar trends have resulted from the introduction of regulations aimed at controlling air pollution in many other cities in the developed world. In other places, however, especially in the developing countries, the atmospheric concentrations of sulphur dioxide and sulphurous particulate matter are increasing. The total global emissions of sulphur dioxide by human society grew at a rate of around 5 per cent per year in the 1970s and reached around 200 million tonnes a year at the end of that decade [71].

Motor-car engines are mainly responsible for *photochemical oxidant smog*, which is due to the release into the atmosphere of nitrogen oxides which react with certain hydrocarbon combounds to form various noxious substances, especially peroxyacetyl nitrate [72]. This substance is toxic for human beings and plants, and when photochemical smog was at its worst in Los Angeles, citrus fruits could not be grown within 50 km of the city. Photochemical smog is especially important in towns with high traffic densities and warm sunny climates, but it also occurs in Canadian and northern European cities. Controls have been brought into effect in some North American cities, such as San Francisco and Los Angeles, and in Japan, reducing the oxidant concentrations by about 40 per cent.

The undesirable consequences of air pollution so far discussed and which have prompted various societal adaptive measures occur locally in the vicinity of the site of release of the noxious agents. The scale of the problem, however, is now such that ecologically harmful effects are sometimes observable many hundreds of kilometres away from the origin of the pollutants.

Acid rain

On the regional level, the problem which has caused most concern in recent years has been the impact on the structure and productivity of ecosystems of *acid rain* caused by the release into the atmosphere of the products of combustion of fossil fuels. The effect is due to the conversion of the oxides of sulphur and nitrogen to strong mineral acids, and their eventual precipitation. These pollutants may be transported in the atmosphere for 1000 km or more before they reach the ground.

Acid precipitation is recognized as a major global environmental problem [73] and the phenomenon has led to extensive regional acidification of water courses in southern Scandinavia, in parts of the eastern side of North America, and in many other areas. There is also concern that such acidification is interfering with the productivity in terrestrial ecosystems. In particular, acid rain and other pollutants are causing severe and progressive damage to forests over a wide area of Europe. It has been reported, for example, that the area of forest showing serious damage in West Germany increased from 562 000 Ha in 1982 to 2 545 000 Ha in 1983

[74]. Similar damage has recently been reported in forests high up in the Swiss Alps.

Carbon dioxide

The rest of this section on air pollution will be concerned with societally induced changes in the atmosphere which are seen as important on the global level. First, let us summarize the situation with respect to carbon dioxide.

Carbon dioxide is a normal component of the atmosphere and of the oceans. It is an essential reactant in the process of photosynthesis, which results in the incorporation of carbon, along with hydrogen and oxygen, into the energy-bearing organic molecules on which life depends. The carbon thus incorporated in the living components of the biosphere is eventually released again to the atmosphere in the form of carbon dioxide through respiration and other metabolic processes in plants themselves and in animals. It is also released in the processes of decomposition of organic matter by micro-organisms and its combustion by fire. This is the carbon cycle, and its integrity is clearly essential for the maintenance of animal and plant life on earth.

The carbon in the biosphere is distributed in a series of pools or reservoirs, and the interrelationships between the main carbon reservoirs are summarized in Fig. 8.6.

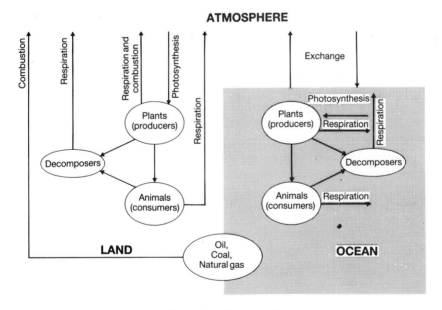

Fig. 8.6. Interrelationships between the main carbon reservoirs.

Cultural developments characteristic of the high-energy phase of human history have resulted in important modifications in the carbon cycle, brought about mainly through the use of machines powered by the combustion of coal, oil, and natural gas. These fossil fuel sources contain carbon which was fixed by photosynthesis over many tens of millions of years; it is now, as a consequence of industrial processes, being released into the atmosphere in what is, relatively speaking, an extremely short period of time. Further increase in carbon dioxide production is the result of the clearing and burning of forests.

The changing pattern of release of carbon dioxide into the atmosphere by human society during the high-energy phase is depicted graphically in Fig. 8.7 [75], which indicates a fifty-fold increase between 1860 and 1975. This development has resulted in an overall increase in the carbon dioxide concentration in the atmosphere since 1860 of about 17 per cent—that is, from around 290 parts per million to 340 parts per million in 1982. The annual increase in the rate of use of fossil fuels, and hence in carbon dioxide production, has hovered around 4.3 per cent, except for the Depression in the 1930s and the two World Wars. However, it is noteworthy that following economic pressures and some consequent reduction in oil use, as well as some improvements in efficiency of use, a fall of 10 per cent in carbon dioxide emissions has been reported for the period 1979–85 [76].

Predictions of future changes in the level of carbon dioxide in the atmosphere depend on the assumptions made with respect to the rate of increase in use of fossil fuels in years to come. Assuming, conservatively, an average growth rate of two per cent in the future, it is estimated that carbon dioxide will reach 450 parts per million by the year 2025, and 600 parts per million by 2030 (that is, double the pre-industrial level). In the words of a leading article in *Nature* in 1979: 'The release of carbon dioxide to the atmosphere by the burning of fossil fuels is, conceivably, the most important environmental issue today' [77].

The reason for this concern lies in the notion of the *greenhouse effect*. Like the glass of a greenhouse, carbon dioxide in the atmosphere is relatively transparent to incoming solar radiation in the wave lengths of visible light, but it is relatively opaque to infra-red or heat irradiation. Much of the sun's light energy is reflected back from the crust of the earth into the atmosphere in the form of infra-red irradiation. The higher the concentration of atmospheric carbon dioxide, the more of this outgoing infra-red is blocked and sent back to the earth's surface. The overall effect is a warming of the biosphere.

While there is agreement among scientists that this effect will be manifest, considerable uncertainty exists about the precise degree of warming that might be anticipated for a given increase in carbon dioxide concentration. In general, simulation models suggest that, with a carbon dioxide concentration

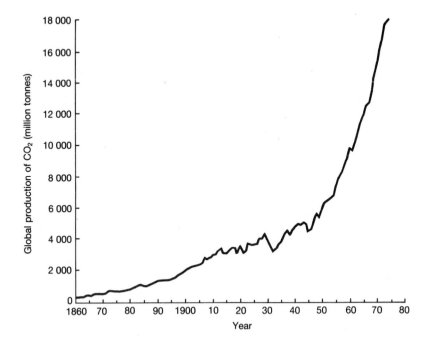

Fig. 8.7. Annual global production of carbon dioxide from industrial activities.

of 600 parts per million (a conservative prediction for the year 2050), the average global temperature rise would be 3°C $\pm$ 1.5°C. Temperatures up to 7°C higher than those at present prevailing would be anticipated in the polar regions. (As a rule of thumb, each rise of 100 parts per million in carbon dioxide concentration will be accompanied by an average temperature rise of about 1°C). Mention should also be made here of recent work which indicates that the concentration of methane in the atmosphere is also increasing and the possibility that this, along with other carbon-containing trace gases, is likely to contribute to the greenhouse effect [78].

A global temperature rise of 3°C would be unprecedented in human evolutionary history. It would exceed temperatures prevailing in the antethermal period around 10 000 years ago, as well as the previous (Eemian) interglacial period of 125 000 years ago. It would approach the warmth of the Mesozoic period—the age of the dinosaurs some 200 million years ago.

There has been much discussion about the possible consequences for the ecosystems of the world and for humankind of an increase in global temperature of this order. It is generally agreed, for instance, that one of the effects would be melting of ice in the Arctic and Antarctic, and this, together with the swelling of the oceans due to increase in temperature, would result in a rise in the sea-level the world over of up to 5 or 7 metres. It is also expected, on the basis of comparison with such earlier periods and of recent improvements in climatic models, that the temperature increase would give rise to considerable changes in global precipitation patterns. This, in turn, would have severe effects on ecosystem dynamics, on agricultural productivity, and consequently on the viability of human societal, economic, and political institutions.

There is still uncertainty and scientific debate about several aspects of the carbon dioxide budget—relating, for instance, to the absorptive capacity of the oceans, to the extent to which increasing concentrations of carbon dioxide in the atmosphere may promote additional photosynthesis, and to whether or not deforestation is resulting in a significant net increase in the flow of carbon dioxide from the biomass to the atmosphere.

While there exist these uncertainties, it is worth noting two points on which there appears to be little or no disagreement among the experts: (a) The concentration of carbon dioxide in the atmosphere has increased as the result of societal activities, and in particular through the combustion of fossil fuels, and it is continuing to increase. (b) If this trend continues there will be a significant increase in the average global temperature, and this will cause ecological and societal upheaval on an immense scale but of unpredictable kinds.

No government in the world is taking these two facts into account in the formulation of its policies.

In terms of the prerequisites for successful adaptation to threats to natural systems (see pages 24-9), it is evident that the problem is recognized to exist by a section of the community in Western societies, but not by the majority. Thus the first prerequisite is only partially satisfied. With respect to the second prerequisite, knowledge and understanding exist about the cause of the problem, and it is known that it could be overcome by drastically reducing the use of fossil fuels. Clearly, those in power in society are not motivated to take remedial action. Perhaps this is partly a function of ignorance, but other factors, such as the psychological defence mechanism of denial of information, may also be playing a blocking role.

The situation with respect to the third prerequisite—the means or resources for successful cultural adaptation—is a more complicated one. If industrial productivity of the kind which has characterized the last fifty years is to be maintained and to grow, then alternative sources of energy must be introduced; but it would be extremely difficult, if not impossible, to achieve this on the scale that would be required. An alternative approach would be a reorganization of society and the economic system in such a way that the health and well-being needs of humans could be satisfied with a different pattern of industrial productivity which uses less extrasomatic energy.

The ozone layer
Concern has been expressed during the last decade about the predicted effect

of the chlorofluorocarbons released into the atmosphere on the ozone in the stratosphere. These substances have been used as propellents in spray cans, replacing vinyl chloride monomer which was found to be carcinogenic, as well as for other purposes, such as in the manufacture of foam plastics and in air conditioners, refrigerators, and freezers. Considerable uncertainty exists about the degree of effect that the chlorofluorocarbons are likely to have on the ozone layer. According to a report of the United States National Academy of Sciences, continual release of these substances at the 1977 rate would eventually result in depletion of the ozone layer by 16.5 per cent, while a report of the United Nations Environmental Programme (UNEP) Co-ordinating Committee on the Ozone Layer estimates a drop of 10 per cent [79].

The ozone in the stratosphere absorbs solar ultraviolet irradiation, and this provides some heat which helps to maintain the stability of the stratosphere—one of the ultimate controls on the global climate [80]. This absorption process also shields the earth's surface from the most biologically damaging wavelength of ultraviolet irradiation, UV-B. It is estimated that a one per cent reduction in the ozone layer would increase the amount of UV-B reaching the ground by about two per cent, which would result in an increase of four per cent in skin cancer in human beings. The model used by the UNEP Committee calculated that a one per cent depletion of the ozone layer has already taken place.

In summary, while there is considerable uncertainty and indeed controversy about the seriousness of the threat to the ozone layer posed by the chlorofluorocarbons [81], the body of scientific opinion expressing concern is sufficient to warrant prompt and effective societal action, at least until uncertainties are resolved. The argument that there is as yet 'no positive proof' is surely a dangerous and meaningless one. By the time there is positive proof it will be too late.

Chemicalization of the biosphere

The chemical waste products of modern industrial society discussed above are those which have given rise to most concern, controversy, and, in some cases, counter-activity. They represent, however, only a very small fraction of the chemical compounds produced by high-energy societies and released into the environment. It has been estimated that about 30 000 chemicals are made in quantities greater than one tonne per annum, and 1500 of these are produced in amounts in excess of 50 000 tonnes per annum. On the basis of predictions of future industrial growth, an Organization of Economic Cooperation and Development (OECD) report estimates an increase of 20–25 per cent in emissions of major conventional pollutants between 1978 and 1985 within OECD countries, and a greater increase for less controlled substances [82]. These chemical substances may be transported across the world by various means, and it is clear that already a significant proportion of the world's human and terrestrial animal populations are exposed to the cumulative effects of low levels of chemical pollutants from a wide range of different and distant sources. The implications of the artificial chemicalization of the oceans is also a subject for serious discussion.

A biologically significant aspect of the technometabolism of modern industrial society is that a high proportion of the chemical waste products of civilization are newly synthesized compounds with no natural counterparts, although they frequently have effects on the natural metabolic processes of plants and animals.

COMMENT

From the facts discussed in this chapter it is abundantly clear that human cultural processes have brought about, during the past fifty or so years, a new and critical ecological situation on the earth's surface, and the future of humanity and the biosphere hang in the balance. The threats inherent in the present situation are associated with the manufacture and deployment of technological devices of tremendous destructive potential and with major changes in the patterns of flow of energy and of key elements in biogeochemical cycles [83]. While there is uncertainty about the precise effects on the biosphere of a continuation of current trends in societal technometabolism, there can be no doubt that the present pattern of change is ecologically unsustainable in the long run. It is a matter for conjecture whether humankind will, through the processes of cultural adaptation, successfully overcome these various threats to its existence.

NOTES

1. 'There are three sets of restraints imposed by the finiteness of the planet; the limits on the earth's non-renewable resources, such as fossil fuels and minerals, the limits of production of renewable resources, such as food and timber, and third, the pollution absorption capacity of the plant' (Birch 1980, p. 3).
2. See, for example, Ward and Dubos 1972; Eckholm 1976, 1977, 1982; Ehrlich, Ehrlich, and Holdren 1977; Brown *et al* 1984, 1985; Myers 1985.
3. Sources: Carr-Saunders 1965 and United Nations *Demographic Yearbooks* 1967, 1980, 1985.
4. Singh 1978, p. 35.
5. Sources: United Nations 1967; Arriaga 1968; Glick 1975; Unesco 1975; Tuve 1976; Clark 1977; United Nations 1980; United Nations 1981*a*.
6. McKeown and Brown 1955; McKeown and Record 1962.
7. The term 'fertility' is used here in the demographer's sense: 'In 1934 the Population Association of America officially endorsed the distinction between *fecundity*, the physiological ability to reproduce; and *fertility*, the actual birth performance as measured by the number of offspring' (see Petersen, 1969, pp.

81, 173, 183). The *general fertility rate* is 'the number of births per 1000 women in the fecund ages' (Petersen 1969, p. 81). In physiology the meanings of the two terms are reversed.

8. Williams 1978.
9. United Nations 1978, p. 22. It is also likely that the 'zero population growth' movement and such books as *Too Many Americans* by Day and Day (1964) have had some impact on the fertility rates in the high-energy societies.
10. Holdgate, Kassas, and White 1982, p. 304. There are also common differences in fertility rates between rural and urban areas, especially in the developing countries. Usually, fertility rates are considerably higher in the rural areas (see Boyden 1972).
11. United Nations *Demographic Yearbook* 1985.
12. Some demographers are more optimistic about the implications of this slight trend than are others—see, for example, Borrie 1972.
13. Skogerboe 1974, p. 127; Starr 1979; Steinhart and Steinhart 1974, p. 310; Rasmussen 1982.
14. This statement is based on the following figures published by the Australian Bureau of Statistics (1980) from their *Annual Report* for the year 1976: population of Australia = 13 991 200; population rurally employed = 336 300. It takes account also of the fact that two-thirds of the food (by energy value) produced in Australia is exported.
15. Leach 1976, p. 9.
16. In 1979, the overall cereal production dropped as a result of crop failure due to adverse weather in the USSR, southern Asia, and many African countries (Holdgate *et al.* 1982, p. 256).
17. Gifford 1982, p. 463. For general information about bioproductivity in the biosphere, see Lieth and Whittaker 1975.
18. Brown 1978, p. 148.
19. Barbour 1980; Bull 1982.
20. Brown 1970; Eyre 1978; Brown *et al.* 1985; United Nations 1981*a*.
21. Ehrlich *et al.* 1977, p. 338.
22. *Eutrophication* is a natural process in lakes which takes place as a result of the inflow of nutrients in streams draining the surrounding area; it involves enrichment of the organic content of the water of the lake and the deposition of dead organic matter on the floor of the lake. *Cultural eutrophication* is the term used by some authors for over-fertilization of water due to culturally induced societal activities (see Ehrlich *et al.* 1977, p. 666 and Holdgate *et al.* 1982, pp. 140, 278).
23. Leach 1976; Gifford 1982.
24. Gifford and Millington 1975.
25. Gabor, Colombo, King, and Galli 1978, pp. 199–200; Holdgate *et al.* 1982, pp. 250, 386–7.
26. Barbour 1980, pp. 245–6.
27. Holdgate *et al.* 1982, p. 387.
28. Eckholm 1982, pp. 59, 183.
29. Buringh (1981) cited by Holdgate *et al.* 1982, p. 255.
30. Seymour 1980.
31. Gregor 1970; Garrels and Lerman 1977.

32. Brown *et al.* 1985, p. 8.
33. Holdgate *et al.* 1982, p. 271.
34. Carson 1962. See Ehrlich *et al.* 1977, pp. 854–6 for an interesting account of the societal responses to Carson's book.
35. Bull 1982.
36. See the discussion in Freedman and Berelson 1974; Pimental 1977; and Eyre 1978.
37. Arndt 1978, p. 144.
38. Barbour 1980, p. 250.
39. Holdgate *et al.* 1982, pp. 251–2.
40. Whittaker and Likens 1975, p. 317.
41. The information presented in this section is taken from the following: Brunig 1977; Gribbin 1979; Sagan, Toon, and Pollack 1979; Grainger 1981; Myers, 1981.
42. Brunig 1977, p. 187.
43. Ware, Panikkar, and Romein 1966, p. 337.
44. Most of the information in this section is taken from Barnaby and Rotblat 1982; Barnaby, Kristofferson, Rodhe, Rotblat, and Prawitz 1982.
45. Holdgate 1983.
46. Barnaby and Rotblat 1982, p. 93.
47. Barnaby *et al.* 1982, p. 162.
48. Crutzen and Birks 1982; Turco *et al.* 1983; Harwell 1984.
49. See Alexandersson and Klevebring 1978; United Nations 1981*a*; Holdgate *et al.* 1982, p. 194; Australian Bureau of Mineral Resources 1984.
50. Eyre 1978, p. 77.
51. Leiss 1978, p. 5.
52. Holdgate *et al.* 1982, p. 192.
53. Holdgate *et al.* 1982, p. 195.
54. OECD 1979, p. 47.
55. Singh 1978, p. 45.
56. Sources for this figure and other information in this section: Tuve 1976; Ehrlich *et al.* 1977; Alexandersson and Klevebring 1978; Ridker and Watson 1980; United Nations 1981*b*.
57. International Energy Agency 1981, p. 17. For a discussion on the impact of increasing efficiency of energy use, see Brown *et al.* 1986, p. 85.
58. Darmstadter 1971; Taher 1982.
59. For a useful annotated bibliography on this topic (to 1979), see Crossley 1979.
60. Mesarovic and Pestel 1975; Holdgate *et al.* 1982, chapter 12.
61. Schumacher 1973, pp. 124–35; Surrey and Huggett 1976; Patterson 1976; Elliott 1977; Nelkin and Fallows 1978.
62. Rotblat 1979.
63. Lloyd 1979, p. 524.
64. Singer 1970; Cole 1971; Steinhart and Steinhart 1974; Lovins 1975; Lovins and Price 1975; Oke 1978.
65. Illich 1974.
66. Boyden, Millar, Newcombe, and O'Neill 1981, chapter 8.
67. Perkins 1974, p. 5.
68. Chambers 1976.

69. Evelyn 1661.
70. Chambers 1976, p. 8.
71. Holdgate *et al.* 1982, Chapter 2.
72. Holdgate *et al.* 1982 pp. 36–8; Howard and Perley 1982.
73. Sagan *et al.* 1979; Howard and Perley 1982; Wetstone and Rosencranz 1983; Torrens 1984.
74. Postel 1985, pp. 104–5.
75. Figure adapted from Holdgate and White (1977) and Brown *et al.* (1985, p. 28). See also Keeling 1973; Rotty 1973, 1979; Rotty and Weinberg 1977.
76. Brown *et al.* 1985, p. 29.
77. *Nature* (1979) **279** (5708), p. 1.
78. Fraser, Khalil, Rasmussen, and Steele 1984; Ramanathan, Cicerone, Singh, and Kiehl 1985.
79. Holdgate *et al.* 1982, chapter 2.
80. Isaksen and Stordal 1981.
81. Allaby and Lovelock 1980; Isaksen and Stordal 1981.
82. Sors 1980.
83. Anyone doubting that the biosphere is experiencing ecological overload as a consequence of human activities should consult Brown *et al.* 1985 and Myers 1985.

9

SOCIETAL DEVELOPMENTS IN THE HIGH-ENERGY PHASE

INTRODUCTION

In this chapter we will consider the features of modern high-energy society which are of particular significance in the biohistorical context. The focus will be especially on those societal processes which lie behind the important culturally induced changes in biological systems, and we will consider briefly the debates which are taking place about the desirability or otherwise, from the ecological standpoint, of the economic systems of high-energy societies.

First, however, some comment is called for on one of the most characteristic features of the human situation in the modern world, namely, the extreme unevenness in the distribution of goods and commodities, both from one society to another and within societies. This is one of the changes in societal conditions that came about initially with early urbanization. It has not only persisted, but the differences in material wealth are even greater than in the past, and they are the cause of a great deal of discontent the world over. In the case of food and nutrition, for example, the majority of people living in the affluent high-energy societies have sufficient to eat: malnutrition among them, when it occurs, is usually a reflection of ignorance or of pressures of lifestyle (see next chapter); but in the developing world many millions of people are forced to go hungry and are suffering from various undesirable consequences of malnutrition or undernutrition [1].

It is a depressing reflection of the state of human society that, despite all the knowledge and affluence that exists in the world today, this situation persists. Certainly, improved distribution would involve difficult problems of organization; but when we witness the organizational skills that are devoted in developed countries to industrial production, commercial activities, space programmes, preparation for warfare, and even staging the Olympic games, it is apparent that the continued maldistribution of foodstuffs throughout the world is not due to any innate lack of organizing ability in the human species. It is true that many valiant aid programmes exist and that they do much to alleviate immediate distress. It should also be recognized that such measures, in which cultural adaptive processes are brought into play in certain societies aimed at overcoming threats to communities living in far-away places, are relatively new developments: and it is pertinent to note the vital role of the United Nations and its various agencies, including the Food and Agricultural Organization and the World Health Organization, in this transsocietal cultural adaptive process. Nevertheless, the financial and human resources which are devoted to humane activities of this kind are infinitesimal when compared with those which go into the development, manufacture, and deployment of weapons of mass destruction [2].

The uneven distribution of resources, including food, is not restricted to the differences between populations in the developed and developing world. Severe poverty exists within even some of the most affluent countries. Heilbroner has remarked:

I think the problem is that there is not one American Economic Republic but two. One is inhabited by those on whom fortune has smiled. It begins at the Pan Am building and ends on Park Avenue and 96th Street. It is the view from the top, the view in which things work, in which flaws seem small and relatively unimportant, in which reasonable men, lunching comfortably, can come to reasonable agreement about how to decide reasonable things, in which the impotence of the unlettered, the unimportant, the unskilled is never experienced at first hand, in which—with full recognition of the exceptions which must, of course, be borne in mind—all's right with the world. But there is also another republic, which begins at 96th Street and continues north into the wynds of Harlem and the dreariness of the Bronx and Yonkers, and from this republic there is another view, more constricted, less Olympian, meaner, poorer, unhappier. We fortunate citizens may never have to visit, much less live in, the commonwealth of the less fortunate, but a book which aims to describe and to judge the American economic republic as a whole must give just as much weight to each miserable and unprivileged citizen as to each contented and favored one, if it is to be, in fact, a description of the way things are, and not merely the way we hope they are [3].

Australia is often seen as an egalitarian society. Nevertheless, in 1978, 1 per cent of the adult population in that country owned 22 per cent of the personal wealth, the top 5 per cent owned 46 per cent, and the top 10 per cent owned almost 60 per cent. One half of the Australian population owned less than 8 per cent of the personal wealth, and the top 5 per cent owned more than the bottom 90 per cent put together. This pattern had remained largely unchanged during the previous 60 years, although recent statistics show a further widening of the differentials [4]. In Britain in 1982, 1 per cent of the population owned 17 per cent of the nation's personal wealth, and the wealthiest 10 per cent owned 47 per cent [5].

Once the essential material needs for human health and survival have been met, the differentials with respect to non-essential goods become important. On the one hand, it is self-evident that people who are free from hunger, relatively free from disease, who live in a reasonable cultural setting, and who are not overwhelmed by tedious work, but who in other respects have a low material standard of living, may enjoy life at least as much as those who live in plush homes equipped with the ultimate in sophisticated electronic gadgetry and material comforts. On the other hand, an important factor that can severely interfere with enjoyment of life is the chronic *feeling of deprivation* which occurs when individuals are aware that others in their society have a much higher material standard of living than they have themselves. This sense of deprivation can lead, in turn, to a sense of chronic frustration and mental or physical ill health, and to patterns of interpersonal behaviour which interfere with the well-being of the affected individuals and of those around them.

These differentials in material wealth have important implications, not only for the quality of human experience but also for the ecological relationship between society and the biosphere. The *driving force of differential disadvantage* comes into play, resulting in a continual striving for more goods at all levels of material wealth, and consequently a progressive increase in the use of resources and the production of technological wastes. As Miller has written: 'needs' stem not so much from what we lack as from what our neighbours have ... . So long as there are people who have more, others will 'need more' [6].

THE ECONOMIC ARRANGEMENT

The economic arrangements in modern Western industrial society are unique in the history of humankind, and their characteristics are of great ecological and biosocial significance. We have noted how one of the early ecological effects of human culture, even in the primeval phase, but more so in the early farming and early urban phases, was the extraction from the environment of non-renewable resources for the production of artefacts of various kinds. The economic system was such, however, that the actual rate of use of these non-renewable resources, and thus the intensity of societal technometabolism, differed very little from generation to generation. In general, society was based on a *steady state economy*, at least with respect to resource use and waste production.

Certainly, there was, throughout the early urban phase, a progressive increase in the amount of goods exchanged between different societies— often over long distances. In the latter part of this period the numerous princely states of Europe regarded economic improvement as a major objective, and the 16th and 17th centuries saw the emergence of the capitalist economic system, with its emphasis on the accumulation of more money and goods, 'not only for the satisfaction of a burgeoning array of material needs but increasingly for the further expansion of productive enterprises and the pursuit of an even-higher standard of living' [7]. There were occasional economic disruptions and perturbations following periods of warfare, epidemic disease, and the influx of gold and silver into Europe as a result of the exploitation of Meso-American societies by European adventurers [8]. On the whole, however, the changes which occurred in the overall metabolism of society were mainly the consequence of changes in population size or of occasional natural disasters. The bigger the population, the greater the level of use of resources, both renewable and non-renewable. On a per capita basis, societal metabolism remained more or less steady. Indeed, even in the 19th century leading economic theorists, while advocating economic growth, still considered that the mass of the population was destined for all time to live in a state of semi-depression, with their survival needs only barely satisfied [9].

During the 19th century industrialization, based on the use of machines powered by fossil fuel and the accumulation of capital through industrial activities, became the outstanding feature of the economic scene in Europe and the USA. Concurrently, changes occurred in the prevailing value system of society, and religious movements, in particular Calvinism, 'cast a halo of sanctification' around the conventional vice of making money [10]. And, late in the century, the spurious ideas of the Social Darwinists supposedly provided scientific support for the new system.

These developments eventually resulted in the entirely new pattern of technometabolism of the high-energy phase of human history, involving an ever-increasing per capita use of non-renewable resources and energy and an ever-increasing production of technological wastes. In this 'high-intensity market setting', the modern consumer society, the production of goods and services is ever-growing, and the tempo of exchange is steadily increasing [11]. Between 1950 and 1973 the industrial output of Western nations more than quadrupled—a rate four-times as fast as the rate of increase in real economic activity that occurred between 1900 and 1950 [12].

It is widely assumed in all Western countries, and accepted by all their governments, that economic growth based on increasing industrial productivity is desirable, if not essential, in order to satisfy the needs of the people. Thus in 1980 a Committee appointed by the Australian Government to report on technological change wrote:

The view of the Committee is that the Australian community currently has substantial unsatisfied needs for private and public goods and services, more leisure and better work, and that these could be met by increased national income through economic growth [13].

According to Herman Daly:

Growth is *not* seen as a temporary *means* of attaining some optimum level, but as an *end* in itself. Why? Perhaps because, as Henry C. Wallich so bluntly put it in defending growth, 'Growth is a substitute for equality of income. So long as there is growth there is hope, and that makes large income differentials tolerable'.... . We are addicted to growth because we are addicted to large inequalities in income and wealth [14].

Associated with these assumptions is the view put by Lewis and shared by many others that 'The advantage of economic growth is not that wealth increases happiness, but that it increases the range of human choice' [15]. The assumption that increasing the range of choice is good for people is seldom seriously questioned.

Another argument for economic growth is reflected in the following words of Walter Heller, a chairman of the President's Council of Economic Advisers in the USA:

We need expansion to fulfill our nation's aspirations. In a fully employed, high-growth economy you have a better chance to free public and private resources to fight the battle of land, air, water and noise pollution than in a low-growth economy [16].

In 1970, US President Richard Nixon said, in his state of the Union address, 'Continued vigorous economic growth provides us with the means to enrich life itself and to enhance our planet as a place hospitable to man' [17].

Similar views are shared by the majority of the business community in the Western world, as reflected, for example, in the following words by the chairman of a large pharmaceutical firm: 'A nation as well as a business that does not grow will go back to the Dark Ages. The price of lethargy is slavery' [18].

Since the above statements were made, the economies of Western countries experienced a temporary recession associated with both high unemployment and high inflation. The solution to such problems is seen by the governments of all affected countries to lie in increasing the rate of industrial production and commodity consumption.

Apart from these assumptions about the desirability of economic growth, mention must also be made of the prevailing attitudes about the market system of exchange. First, there is the common view that the market in some way automatically and satisfactorily performs a number of essential functions which, in other societal systems, such as those which operate in the planned economies, are performed by governmental institutions. These functions include the allocation and distribution of resources, the establishment of social priorities, and the development of programmes for the production of goods and services. Linked with this view is the attitude that we do not need to plan for the future, because the market economy itself will dictate 'its shape and content', and that the outcome will be to the benefit of humankind [19].

From the biohistorical standpoint, such assertions appear naive. There is no sign as yet, for instance, that the 'invisible hand' provides protection against threats to the biosphere on a global level. Indeed, even at the local level, state control has already, of necessity, become increasingly important as a tool of cultural adaptation against undesirable externalities arising from various economic activities.

Expressions of concern

The two fundamental causes for ecological concern inherent in economic growth—depletion of resources and threats to the life supporting processes of the biosphere—have received unequal attention in the literature. Most authors have concentrated on the former, despite the fact, commented on in the last chapter, that successful adaptation to scarcity of non-essential goods appears to be a more feasible proposition than adaptation to an ailing global ecosystem.

A particularly renowned work in this area is *The limits to growth* produced for the Club of Rome by Meadows and others in 1972 [20]. This study, which has given rise to a great deal of controversy, consisted of the application of a 'systems dynamics' approach to the economic and ecological situation. The authors concluded that, if present trends in resource use continue, 'the limits to growth ... will be reached sometime within the next one hundred years. The most probable result will be a rather sudden and uncontrollable decline in both population and industrial capacity'. The authors suggest that this collapse of the system could be avoided if deliberate steps were taken to 'establish a condition of ecological and economic stability that is sustainable far into the future. The state of global equilibrium could be designed so that the basic material needs of each person on earth are satisfied and each person has an equal opportunity to realize his individual human potential' [21].

The limits to growth promoted a flurry of activity among systems analysts, and the Meadows team was criticized for various weaknesses in their model, such as failure to take into consideration economic stabilizing forces acting through the price mechanism and for not having taken account of possible technological progress in the future [22]. In fairness, the theme of the Meadows report was that *if we continue as at present*, and if no deliberate action is taken, then collapse will occur. The various criticisms of the model used in *The limits to growth* programme have been seized upon by the supporters of economic growth as evidence that the work of the Meadows team is to be discredited and that there is therefore no cause for concern.

The link between the perceived threats to the biosphere inherent in continuing industrial growth and consumerism has been thoroughly developed by William Leiss in his book *The limits to satisfaction*. He writes:

It is the unforgivable squandering of resources in the developed countries and the environmental degradation resulting therefrom that currently determine the general direction of the global 'political economy' and that constitute the source of potential future disasters for the entire human population... one might describe our existing practices as a massive auto-experiment of the industrialized societies, the ultimate consequences of which may jeopardize not only the biological future of the human species, but that of many other living things as well [23].

Economic growth has also come under attack for other reasons, especially by those who take the view that it is not satisfactory socially [24]. Joan Robinson, for instance, has written as follows:

Not only subjective poverty is never overcome by growth, but absolute poverty is increased by it. Growth requires technical progress and technical progress alters the composition of the labor force, making more places for educated workers and fewer for uneducated, but opportunities to acquire qualifications are kept (with a few exceptions for exceptional talents) for those families who have them already [25].

More recently Fred Hirsch, in his book *The social limits to growth*, developed the view that some of the undesirable social problems associated with economic growth are more important, or at least more urgent, than the ecological ones [26]. Mishan has also questioned the assumed correlation between economic growth and human well-being:

Under such perverse conditions of growth men may continue, if they choose, to so juggle with words as to equate growth with 'enrichment', or 'civilization', or any other blessed word. But it is just not possible for the economist to establish a positive relationship between economic growth and social welfare [27].

Since the present economic arrangement, involving as it does continually intensifying societal metabolism, is incompatible with the long-term ecological sustainability of human society, it is apparent that, if civilization is to survive, at some time in the future a transition must be made to an alternative system. What would be the characteristics of this alternative system? How would it work? What kind of economic arrangement would be compatible with the long-term ecological sustainability of civilization? And how could this new arrangement be achieved? These are the essential questions, and it is disappointing to note that so few economists are seriously attempting to answer them. Of course, some may take the view that the imperatives of the economic system as it exists today are beyond our control. Others, like Coombs, are more optimistic:

There is nothing divinely ordained about the economic system: it is the product of human ingenuity, effort and capacity to organize and can therefore properly be questioned, criticised and, if a better alternative exists, rejected [28].

Perhaps another reason for the relative lack of interest among economists in this problem is that it is apparently much easier to envisage a smooth-running economy that is growing than one which is not. For one thing, there would be the difficult problem of disparities: 'If poverty is not alleviated by growth, then it must be alleviated by a redistribution of income' [29]. There are, however, a few economists who write about the necessity of changing to an alternative economic system, and who put forward ideas about how such a system might operate and how it might be achieved. Broadly, and rather obviously, the suggestions usually involve the replacement of the *flow economy* (Boulding's 'cow-boy economy'), in which industrial production and consumer spending are perpetually on the increase, with a steady state *stock economy* (Boulding's 'spaceman economy'), in which both production and consumption are minimized [30]. This suggestion then leads to further important questions. At what level of production and consumption should the economy be stabilized, so that it is indefinitely sustainable ecologically? If the USA maintained its present intensity of technometabolism, and the rest of the world were eventually to catch up, would this scenario be indefinitely sustainable? No one can answer this question with certainty on the basis of existing knowledge, but many ecologists would probably suspect that the answer is 'no'.

Various scenarios for an ecologically sustainable society have been suggested [31]. Apart from an emphasis on small-scale enterprises, on decentralization, on labour-intensive and energy-cheap industry, and on renewable energy sources, the suggestions include various institutional and economic devices for ensuring harmonious and fair distribution of wealth, for encouraging population control, and for discouraging the rate of use of non-renewable resources. Several authors have expressed the view that the proper control of the processes of economic growth would necessitate greater control by the state of large industrial and commercial corporate organizations.

All the specific suggestions put forward for moves towards an ecologically sustainable economy would require state intervention. In general, a shift in the balance of power between the state and the private sector in capitalist regimes is implied in all the recommendations—at least during the period of transition. However, discussion has also taken place on whether or not an ecologically sustainable economy would indeed be possible under a capitalist regime. Heilbroner writes:

The idea of a 'stationary' capitalism is, in Marxian eyes, a contradiction in terms, on a logical par with a democratic aristocracy or an industrial feudalism ... . But there is no doubt that the main avenue of traditional capitalist accumulation would have to be considerably constrained; that net investment in mining and manufacturing would likely decline; that the rate and kind of technological change would need to be supervised and probably greatly reduced; and that, as a consequence, the flow of profits would almost certainly fall.

Is this imaginable within a capitalist setting—that is, in a nation in which the business ideology permeates the views of nearly all groups and classes, and establishes the bounds of what is possible and natural, and what is not? Ordinarily I do not see how such a question could be answered in any way but negatively, for it is tantamount to asking a dominant class to acquiesce in the elimination of the very activities that sustain it. But this is an extraordinary challenge that may evoke an extraordinary response. Like the challenge posed by war, the ecological crisis affects all classes, and therefore may be sufficient to induce sociological changes that would be unthinkable in ordinary circumstances [32].

The manipulations and modifications suggested by such authors as Galbraith and Daly, and in the Gamma Report (Valaskakis *et al.* 1977) [33], while involving a shift in the balance and exercise of power between the state and the private sector in favour of the former, seem to assume the continuation of a capitalist economy. Other authors, however, totally reject this possibility. Roberts, for instance, has written:

Capitalist planning for profit which implies the massive shaping of consumer demand into precisely those channels most harmful to mankind's future, must be replaced by central planning of a far different kind: planning which adopts more far-sighted and rational criteria for production, taking fully into account the true overall social cost of a commodity—which includes the usage of limited resources and the deterioration of a limited environment [34].

Linked to this view is the common assumption that a steady state economy under a capitalist system inevitably leads to a progressive and inexorable decrease in the material standard of living. Let it be noted, however, that while the material standard of living in even the more affluent of the planned economies is well below that in most affluent capitalist states, the governments of socialist nations are also committed to the concept of economic growth based on increased industrial production [35].

The counterattack

As would be expected when an established societal pattern is criticized by a minority, there has been a strong counterattack, often vigorous and emotive, against the recent critics of economic growth and in defence of the conventional dogma [36]. As in all reform anti-reform patterns, the supporters of the status quo have seized especially on the more extreme and less rational of the would-be reformers and have made the most of inaccurate predictions and uncertainties. Also as in other anti-reform situations, some of the counterattack has been characterized by the use of ridicule, and the critics of industrial growth have been labelled with the usual disparaging terms used for reformers, such as 'prophets of doom', 'fanatics', and 'alarmists', as well as with some more specific expressions like 'ecofreaks' and 'ecodoomsters'.

The main thrust of the counter-movement has been aimed at the concerns which have been expressed about the *input* component of technometabolism—that is, about the depletion or exhaustion of resources. Because in the past when raw materials have come to be in short supply substitutes have often been found, it is assumed that this will always be the case in the future, and that this substitution, along with the application of recycling and other technological developments, will overcome the problems of depletion [37]. Goeller and Weinberg have presented an optimistic analysis of the prospects for substitution in a paper entitled 'The age of substitutability'. They see depletion of mineral resources to pose no problem 'provided man finds an inexhaustible, nonpolluting source of energy' [38].

As mentioned earlier, with respect to the depletion of non-renewable materials, the ecological problem is hardly one of survival of the human species, but rather a question of whether global society will be able to handle the situation in a way that is equitable and which avoids serious international conflict. The much more difficult issue—the threat to the productive capacities of the biosphere as a consequence of the massive growth of technological waste production—has received less attention by the pro-growth lobby and tends to be dismissed because of the prevailing uncertainties with respect to the specific changes which are likely to be critical [39].

Many economists are not convinced that the present economic system in any way represents a threat to the biosphere. The prophecies of Malthus, Ehrlich's future scenarios [40], and the predictions by the authors of *The limits to growth* are all seen by some as having been 'discredited'. Thus, from the perspective of economic theory, the response to expressions of concern about threats to the biosphere can be summed up in the words 'we have heard it all before'. As Heinz Arndt has written:

The one indubitably valid point made by the prophets of doom, that current rates of growth of population and per capita income cannot be maintained indefinitely, was neither new nor in itself very interesting. That man will not survive on earth for all time has long been known as a strong probability. If doom is millennia off, the responsibility of the present to the nth future generation is a fine point in moral philosophy. 'Why', as Arthur Lewis asked, 'should we stay poor so that the life of the human race may in some centuries to come be extended for a further century or so?' [41].

The issue here, however, is one of perspective. It is a fact, not a probability, that human beings will not be able to survive on earth indefinitely. But given an ecologically sustainable society, it may well be one hundred million years before, due to a fading or exploding sun, the earth becomes too cold or too hot to support humanity. *Homo sapiens sapiens* has so far been here only around fifty thousand years. The other relevant perspective is whether 'doom is millennia off'. Disregarding for the moment the possibility of nuclear annihilation, it should be clear from the facts relating to societal metabolism discussed in the last chapter that, despite all the uncertainties, it would be prudent to be thinking in terms of decades or centuries rather than millennia.

Arndt (1978, pp. 142-3) goes on to ask 'whether there are good reasons to fear that growth of per capita income, superimposed upon population growth, will outrun available resources in the near or foreseeable future'. He discusses the question at some length and concludes that this is unlikely, drawing attention to the fact that the concept of 'finite' resources is meaningless to an economist. As a resource becomes scarce, so does its price increase and its rate of use decline; and this process encourages the search for substitutes This is a well known economic argument, and it may well be valid in the case of most non-renewable resources. Exceptions include those resources, like phosphorus used in food production, for which there are no substitutes.

In this book on the economic growth–non-growth debate, Arndt has only one paragraph on the ecologically more serious problem—the toxification of the biosphere. It is worth quoting this paragraph in full because it nicely reflects, I believe, the majority opinion among economists today:

Aside from exhaustion of non-renewable resources, early limits to economic growth have been predicated on irreversible damage to the environment. On most of these dire predictions, of over-heating of the earth's atmosphere, destruction of the ionosphere and irremediable damage to ecological balance in other forms, the weight of reputable scientific opinion appears to have come down against the prophets of doom. But it is only fair to add that our ignorance about these matters is still great, and the prophets will have done good in making humanity aware of dangers that may need watching [42].

Apart from disagreeing with the ecological warning of the critics of economic growth, some pro-growth authors believe that a steady state economy would interfere seriously with human well-being and would be unfair to future generations. Lewis claimed, for instance, that economic growth is a necessary condition for full employment in a private enterprise economy and that 'the case for economic growth is that it gives man greater control over his environment, and thereby increases his freedom' [43]. This attitude is further reflected in the following statement of Barnett and Morse:

What we are saying is exceedingly simple: our debt to future generations will be discharged to the extent that we maintain a high rate of quantitative and qualitative progress, to the extent that we alter—in a direction favorable to human welfare generally—the conditions that determine the choices open to men when they are free to choose [44].

SOCIETAL ORGANIZATION

From the biosocial and ecological standpoints, two related developments in societal organization in the high-energy societies are of crucial significance. One of these was already noted late in the 19th century by Tonnies [45], who made the distinction between what he called *Gemeinschaft* and *Gesellschaft*, the former being characteristic of pre-industrial society and the latter of modern industrial society. In Gemeinschaft, social groups are united by common beliefs and family ties and are characterized by face-to-face relationships. In Gesellschaft, social groups are held together by practical concerns, and by formal and impersonal relationships. The implications of this change for the life experience of individual humans will be discussed in the next chapter.

The other development of major significance had been the organization of a substantial part of the population into large groupings or units with different and moderately well-defined societal objectives. In the words of Galbraith:

It is not to individuals but to organizations that power in the business enterprise and power in the society has passed. And modern economic society can only be understood as an effort, wholly successful, to synthesize by organization a group personality far superior *for its purposes* to a natural person and with the added advantage of immortality [46].

Although an increasing number of authors are expressing anxiety about the extraordinary power and potential for undesirable influence of these corporate organizations (private and governmental), the role of these organizations as significant ecological and biosocial forces has not received a great deal of attention in social science. Nevertheless, it is self-evident that many of them do play a very significant part in determining the ecological interrelationships between society and the ecosystems of the biosphere. The existence of nuclear weapons is in large part the outcome of the behaviour of corporations; and, through their role in the industrialization process, corporations make a major contribution to the technometabolism of society. Moreover, at the biosocial level, they have far-reaching effects on the life conditions, values, and aspirations of people.

This aspect of the human situation today is of great importance ecologically, but it is very complicated, and it is not possible to do it justice in this book [47].

TECHNOLOGICAL CHANGE

The most dramatic cultural feature of ecological phase four has been the explosive and accelerating advances in science and technology. These developments underlie most of the major impacts that human society has made on the biosphere (chapter 8), and they have also had profound effects on human experience (chapter 10).

The technological developments of the last 150 years have been associated with, and are largely responsible for, important changes in the organization and activities of society. In discussing some of these changes in the USA, Toffler has written:

Behind such prodigious economic facts lies that great growling engine of change— technology. This is not to say that technology is the only source of change in society. Social upheavals can be touched off by a change in the chemical composition of the atmosphere, by alterations in climate, by changes in fertility, and many other factors. Yet technology is indisputably a major force behind the accelerative thrust [48].

Why, we may ask, has there been this sudden 'leap forward', this sudden intensification, sophistication, and increase in complexity of technology? The idea that technological advance occurs mainly in response to new ecological needs of society [49] is surely not adequate to explain this explosion. Nor, of course, does the rapid rate of technological development in recent times reflect any change in the innate potentials and capacities of human beings. It is the consequence, rather, of a variety of interlocking cultural and societal processes. Among the contributing influences is that, as scientific knowledge and technological 'know-how' accumulate, so does the rate of technological change accelerate. To quote again words of Toffler:

Millennia or centuries go by, and then, in our own times, a sudden bursting of the limits, a fantastic spurt forward... . The reason for this is that technology feeds on itself. Technology makes more technology possible ... [50].

Another factor contributing to the acceleration of technological change is that about 90 per cent of scientists that have ever lived are living today. These individuals are studying the information gathered in the ever-narrowing fields of specialization in which they work and, in striving to do their jobs as well as they can and to experience approval and praise from members of their in-groups, they put great effort into further increasing knowledge and advancing technology. There are so many scientists in existence today that the innovative contribution of each of them does not have to be very great in order to ensure rapid technological development (although occasional individuals are placed in circumstances such that they succeed in making 'breakthroughs'). The process is essentially a collective one. It is also pertinent to note that, despite the large number of scientists, the vast majority of the population of *Homo sapiens*, even in affluent Western countries, make no contribution whatsoever to the innovative process—beyond the role they play as consumers and as users of novelties, thus providing the economic thrust which lies behind much technological change.

An aspect of the capitalist economic arrangement of importance in this connection is the existence of large industrial and commercial corporations. In their efforts to compete with each other for survival, they put immense effort into technological innovation. This principle is also relevant with respect to another major influence: powerful nations bent on gaining or maintaining military advantage over others devote astronomical amounts of financial and human resources to research and development in the technology of warfare. It is said that 60 per cent of the scientists active in the world today are working on military projects.

Another extremely important factor is the value system of modern society. Innovation and technological change is seen as good. Kudos is bestowed both on the innovator and on the individual who possesses a novelty. Progress is defined in terms of technological change, and a common response to the gloomy warnings of some ecologists and others about the future of humanity is the claim that all predicted difficulties will be overcome by technological advance—the notion of the *technological fix*.

Clearly the nature of the ecological and experiential impacts varies a great deal from one kind of technology to another. It has been claimed, for example, that the new technology which developed in the USA between the years 1948 and 1968 had appreciably greater environmental impact than the technology which it replaced [51]. Machines in general have produced a wide range of different effects both on the biosphere and on human experience. The motor car, for instance, apart from its influences on the quality of the air and on noise levels in cities, is 'widely believed to have changed the shape of our cities, shifting home ownership and retail trade patterns, altered sexual customs and loosened family ties' [52]. In the case of electricity, except for the changes on the biosphere resulting from this conversion of energy stored in fossil fuels, the main impact has been on human experience. Fifty per cent of the people on earth now have electric lighting in the home, and this has had a most significant effect on the way they spend their time and on their social interactions.

An important biosocial aspect of technological change is the tendency, once a new application of technology is introduced into a society, for that society to become progressively more dependent on it ecologically. We have referred to this phenomenon as *technoaddiction* [53], because it is analogous in some respects to the dependence of individuals on addictive drugs. The new technology is usually introduced experimentally, not because it is necessary, but probably because it benefits a small section of the population in one way or another, and partly because some people are attracted by its novelty. As time goes by, however, it spreads in the population, and society begins to rearrange itself to accommodate it, or in response to the facilities which it offers. The society thus becomes increasingly dependent on the technology for normal functioning, and eventually completely so. This has occurred in modern Western society in the case of the technologies which are fuelled by oil and electricity. As Jane Jacobs has written:

Cities today are so dependent upon electricity that their economies would collapse without it. Moreover, if modern cities had no electricity most of their people—if they could not quickly get away—would die of thirst or disease [54].

The same inexorable process is at present at work in the case of computer technology and automation in industry.

Debates about technology

Let us take a brief look at the 'technology debate' of recent times. Many books have appeared over recent years by authors such as Mumford, Roszak, Fromm, and Illich complaining about the impoverishing effect that technological progress has had on the quality of human experience [55]. Many of these works are well known and have been well received by the public, but it is doubtful if any one of them, or even all of them together, have had the slightest influence on the direction and the rate of the march of technological change.

Technological progress has also been under fire from those who believe that new applications of technology in industry will result in increased unemployment. This particular concern goes back to the early 19th century. Between 1811 and 1816 a group of workers in England, who came to be known as Luddites, rioted and resorted to breaking new machinery and to physical violence directed against the importers of such new machinery, which they saw as a threat to their jobs [56]. Today the term 'latter-day Luddites' has been used for those people in present-day society who advocate preventing the adoption of certain new technologies, especially computer-based automation, or who advocate that mechanisms be set up in society to assess the likely effects of proposed new applications of technology on the demand for labour before they are introduced [57]. However, opinions differ about the impacts on employment rates of these new applications of technology, some authors arguing that the overall effect will be to maintain or even increase employment rates [58]. Others take the opposite view. Rada, for instance, in his comprehensive book on *The impact of micro-electronics*, expresses the firm opinion that the new applications of technology involving micro-electronics will displace jobs at a rate greater than they will create them [59].

A third area of concern is represented by those who perceive technology as extremely dangerous, as a serious threat to the viability of the biosphere and to human survival. Their anxiety is deepened by the fear that technological development may be 'out of control'. Once again, expressions of concern on this count are not new. In 1933 an eminent British engineer, Sir Alfred Ewing, suggested a moratorium on invention, so that society might assimilate and integrate the existing mass of inventions and evaluate progress. He was widely denounced as a crank [60].

The reasons why some people today perceive technology as representing a major survival threat to life on earth are obvious enough and were discussed in the last chapter. Apart from the undesirable side-effects of intentional applications of technological devices, there is the threat of harm resulting from human error. In the words of Broomfield:

We have created ... only very recently—a technology intolerant of mistakes; a technology that assumes infallibility but is operated by fallible humans. We are a mistake-making species, and we should never have developed technologies that assume mistakes will not happen. Accidents resulting from human error are a certainty [61].

The view, then, that technology is 'out of control' is both widespread and justified. The overwhelming majority of informed human beings, including, almost certainly, the leaders of the major powers, see the nuclear arms race as worse than insane; and yet it proceeds relentlessly.

Others are more optimistic. Fromm, in his book *The Revolution of hope*, discussing the question how the control of technology might be achieved, believes that technology can eventually be directed toward optimum human development rather than maximum production [62]. Toffler in *Future shock* sees the challenge of controlling technological change as an extremely urgent one: 'In our haste to milk technology for immediate economic advantage, we have turned our environment into a physical and social tinderbox' [63]. His relative optimism is reflected in a number of steps that he suggests might be taken in society in order to bring technology under control.

The various expressions of concern are, as would be anticipated, countered by others—the proponents of 'technology as liberator' [64]. These people see the advantages of technological progress as outweighing the disadvantages to such an extent that they consider any attempt at societal control of technology to be both unnecessary and unwise. Some of them admit that technological change in industry can result in human hardship; but the answer to this problem is seen in terms of human beings adapting more quickly to technological change, rather than in terms of controlling such change. This attitude is reflected in the following statement from the *Report of the Committee of Inquiry into Technological Change in Australia* (1980): 'The preservation of a stable society in Australia depends on the capacity of the workforce and the population generally to cope with and adjust to changes (i.e. technological changes) as they occur' [65].

To conclude these brief comments on technological change, the following words of Barbour are particularly apt:

I believe that we should neither accept the past directions of technological development, nor reject technology *in toto*, but redirect it toward the realization of environmental and human values. In the past, technological decisions have been governed by narrowly economic criteria, to the neglect of environmental and human costs ... . The *redirection of technology* will be no easy task. Contemporary technology is as tightly tied to industry, government, and the structures of economic power that changes in direction will be difficult to achieve. As the critics of technology recognize, the person who tries to work for change within the existing order may be absorbed by the establishment. But the welfare of humankind requires a creative technology that is ecologically sound, socially just, and personally fulfilling. The challenge of our generation is to use technology in the service of these environmental and human values [66].

VALUES

Whether a society responds successfully to changing circumstances or is devastated by them can be influenced to a large degree by its value system, or by the flexibility of that system [67]. The value system of a society may be either consistent with, or inconsistent with, the long-term ecological sustainability of that society, or with the well-being of the various sectors of its human population.

Before discussing the value system of present day high-energy society, a few further observations must be made. First, there is the problem of definition. The term 'value' has been used very loosely in the literature, and many people have spent much effort in trying, without a great deal of success, to move towards a definition which is useful from the scholarly and scientific point of view [68]. For our purposes, it is sufficient to adopt the definitions of Rokeach as follows:

A value is an enduring belief that a specific mode of conduct or end-state of existence is personally or socially preferable to an opposite or converse mode of conduct or end-state of existence. *A value system* is an enduring organization of beliefs concerning preferable modes of conduct or end-states of existence along a continuum of relative importance [69].

Second, it is necessary to be aware of a number of distinctions between different kinds of values. It has been recognized for a long while among students of the subject that some values are dependent on others. For example, a person may consider that healthiness is good (and conversely that illness is bad). This person may believe that vitamin C, which is found especially in fresh fruit, is necessary for health; and the person will then hold the view that 'eating fruit is good'. This is clearly a value statement, but it is dependent both on the *basic value* —that health is good—and on the *understanding* or *belief*, based on available information, that vitamin C is necessary for health and that it is found in fruit [70]. We can thus recognize 'chains' of values, those further down the chain being much influenced by understanding, knowledge and beliefs. Here we shall use the term *motivating values* for the values at the far end of the chain—the values, that is, that directly influence behaviour.

The difficulty of being precise in this area of thought may be illustrated by the fact that the basic value that healthiness is good may be regarded as being derived from an even more basic value—that happiness is good—and from the belief that happiness requires good health. In fact, the basic value that healthiness is good can give rise to many branches of value chains relating not only to diet, but also to many other aspects of the environment and of behaviour, including noise levels, air quality, the habit of smoking, levels of physical exercise, and personal and public hygiene.

The diagram in Fig. 9.1 illustrates the relationship between basic values, knowledge or belief, and motivating values in its simplest form (the value

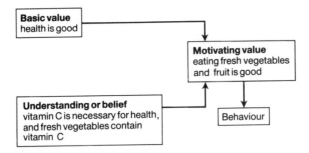

Fig. 9.1. The interrelationships between basic values, understanding, motivating values, and behaviour.

chain, for simplicity, is made to consist of only two links, with a single input from knowledge or belief).

One of the most important implications of this view of the mechanism of value formation is the fact that the motivating values are largely dependent on knowledge, as provided by personal experience or by science. Science, on the other hand, does not influence basic values; it cannot, for example, in the words of Haldane, 'answer why I should be good' [71]. Thus motivating values may sometimes be inconsistent with basic values, if the knowledge and understanding is incomplete or wrong. Countless examples of this phenomenon exist in human history and at the present time.

Two further points must be made about motivating values. First, their special importance lies in the fact that it is these values, rather than the basic values, which directly affect the action, or lack of action, of an individual. Second, although derived from and much influenced by knowledge, they may nevertheless become very fixed. When people holding certain motivating values are exposed to new information which shows that these values are inconsistent with their basic values, they may either change the motivating value in the light of their new understanding or they may deny the new information. Which of these courses of action is taken depends on a number of factors, including the source, nature, and perceived reliability of the new information, and the personality, and possibly the age, of the individual.

Much of the debate which exists concerning the macroecological situation of modern society is the consequence of differences in motivating values. Thus, although there are occasional individuals who claim not to hold the basic values that the survival of the human species or human health and well-being are desirable, these basic values are shared by the great majority of humankind, including people as different in their motivating values as many of the proponents and critics of economic growth. The

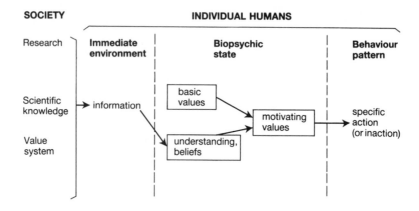

Fig. 9.2. Interrelationships between society and the values, understanding, and behaviour of individuals.

difference between the motivating values of these two groups can be accounted for only in terms of differences in the information available to them, in their personal mental mechanisms for filtering information, or in their interpretation of this information. Thus, returning to the generalization that the value system of a society affects the nature and effectiveness of its response to changing circumstances, the sequence depicted in Fig. 9.2 is important.

In discussions about values, the question naturally arises whether any human values are 'innate'—that is, are they phylogenetic characteristics? This subject is, as we might expect, a controversial one. My own view is that there is probably a phylogenetic basis for certain values. However, innate biologically determined values must be restricted to those which would have been of selective advantage in the evolutionary setting. One can envisage, for example, selective advantage in the evolutionary environment of such values as the following: loyalty toward the in-group is good, disloyalty is bad; protection and care of very young individuals is good. Innate values are unlikely to include, for example, values such as 'it is good to help strangers who are disadvantaged' or 'it is bad to disturb the biosphere in a way such that the human species will cease to exist in 500 years'. While these particular two values are basic, they are probably not innate. For our purposes, discussion on their origins is not profitable: whatever their underlying basis, they are very common, although not universal, and they are held by the author of this book and probably by the majority of its readers.

Another important distinction is that between the values of individual people and those of corporate organizations. Corporations frequently have well-defined values which reflect their original purpose and which determine their behaviour and consequently their societal and environmental impacts. The values of corporations represent an important ecological force in modern high-energy societies.

Nature in the value system and the Western idea of progress

In the context of the theme of this book, probably the most fundamental aspect of the value system of society is the degree of sense of respect that people feel for the processes of nature [72]. It should be pointed out that a sense of respect (or lack of respect) for nature does not fit neatly into the dichotomy between basic values and motivating values. In some individuals, this value seems to be in the basic category, arising entirely from the intuitive or spiritual feelings. In others, it is more the consequence of interaction between other *basic values* (e.g. human health and well-being is good) and *understanding* (e.g. the health of individuals and of populations is dependent on the processes of nature).

Before discussing the value system of modern high-energy society as it relates to nature, some comment is necessary on another related aspect of the value system—namely the Western idea of progress. This topic may be introduced with the following quotation from Pollard:

The world today believes in progress. Indeed, so widespread is this belief among modern nations, that Governments will ignore it at their peril, and the word 'progress' itself has become an unqualified term of praise. As is usual in such cases of general approval an even superficial inquiry will quickly reveal that the term of 'progress' may take in a host of different meanings. It will also reveal that the belief in it is very recent, having become significant only in the past three hundred years or so, and that those who might be expected to have devoted most time to its consideration, the historians and philosophers of history, are the least certain of its validity.

A belief in progress implies that things will in some sense get better in the future, but it has never been limited to this simple idea of melioration. ... . It is, instead, always in the nature of a scientific prediction, based on the reading of history, or at least recent history, and the operation of laws of social development. Thus a belief in progress implies the assumption that a pattern of change exists in the history of mankind, that this pattern is known, that it consists of irreversible changes in one general direction only, and that this direction is towards improvement from 'a less to a more desirable state of affairs' [73].

The notion of continual progress from an inferior to an increasingly superior condition did not fully develop before the 17th century. The ground was prepared for the growth of this idea by the cultural changes associated with the Renaissance—the revival, that is, of the Greek ideal of a rational approach to the study of reality. As discussed earlier, Francis Bacon, in particular, is credited with having contributed, especially in his *Novum Organum* (1620), to the growth of this idea, and to the societal and technological changes which have been associated with it. He emphasized the power of knowledge, and suggested that the accumulation of knowledge meant increasing power for human beings to make the world more comfortable for themselves, and to make human life easier and happier. The contributions of scientists such as Newton, adventurers like Colombus, and philosophers and writers such as Descartes, Locke, Voltaire, and Condorcet all helped to further the general idea; and, in the 19th century, it received further boosts from the findings of geologists and from the theory of Charles Darwin, which together implied apparent 'progress' in the biosphere through evolution [74].

The general *idea of progress* links up with the *attitudes to nature* theme in an important way, because the concept of progress as it developed came to be identified with the view that it was a human prerogative, indeed obligation, to use science and reason to overcome or 'conquer' nature, and that, as this objective was achieved, people would be happier. Thus as Western society and philosophy became increasingly under the influence of city dwellers, merchants, and entrepreneurs, so did the divorce from nature, and even a certain antagonism towards nature, become increasingly evident. People no longer felt either close to nature nor respectful of it; and a sense of reverence for nature was restricted to a very few. So uninteresting and unimportant was nature seen to be, for example, that most schoolboys in Britain in the 1930s and 1940s, although required to learn at least eight subjects, received no instruction whatsoever in biology, and they were not informed of the principles and facts of evolution and ecology. It is no wonder if most of today's political leaders appear ignorant of essential principles relating to the place of humankind in the biosphere.

Another attitude associated with the set of values and assumptions which go with the Western idea of progress is the notion of 'technofix'. This is a product of our societal intoxication with technological change and consists essentially of the assumption that no major problems exist that cannot be solved by further technological effort. While it would be foolish to deny that technology can be very useful—indeed it will have an essential role to play in a transition to an ecologically sustainable society—it is also absurd to imagine that technological advance by itself can overcome the burgeoning threats to the biosphere and to civilization which are inherent in the societal processes of the developed world today.

It is evident, then, that the powerful set of values associated with the Western idea of progress represents a significant ecological force in modern high-energy societies, and it is also clear that it is incompatible with the long-term ecological sustainability of society. It is also questionable whether, even if long-term ecological sustainability were not an issue, progress of this kind leads inevitably to increasing well-being for the mass of the population.

As mentioned above, there has been some spasmodic but serious questioning of the Western idea of progress among a small minority of the population of high-energy societies throughout the present century. Recently, however, there is evidence of a more widespread disillusionment [75], although so far this development appears to have had minimal effect on societal policies.

Let us conclude this chapter by quoting a letter from Marian McLeod which appeared in the *Canberra Times* on the 11th May 1983:

Sir,—I wonder if any newspaper in the world ran as front-page headlines the report by Paul Weitz, the commander of the US space shuttle Challenger, who commented that the whole earth was surrounded by a heavy blue haze of pollution, much worse than he observed during his first flight in 1973 during the Skylab mission and rising much higher? Probably not. But doesn't everything else on our front pages fade into insignificance compared with this horrifying fact? Only by looking at facts like these head-on can we begin to work towards solutions—or have we all given up trying? [76].

NOTES

1. Holdgate, Kassas, and White 1982, p. 387.
2. It has been pointed out, for example, that the cost of a single Trident submarine is equivalent to the average expenditure in underdeveloped countries on a year's schooling for 16 million children (Broomfield 1981).
3. Heilbroner 1970, pp. 222-3.
4. Raskall 1978.
5. *Canberra Times* 31 October 1984, p. 4.
6. Miller 1964, p. 38.
7. Christie 1980, p. 11.
8. Gottschalk, Mackinney, and Pritchard 1969.
9. Galbraith 1958, pp. 52-3; Wilkinson 1973, chapter 6. John Maynard Keynes put down the 'slow rate of progress' from 4000 years before the present to around AD 1700 to the lack of capital accumulation and of technical improvements (see Arndt 1978, p. 5). Wilkinson (1973) puts it down to lack of ecological necessity.
10. Barnet and Muller 1974, p. 48.
11. Leiss 1978, p. 52.
12. O'Riordan 1976.
13. Myers 1980, p. 65.
14. Daly 1973, p. 95.
15. Lewis 1955, p. 420.
16. Quoted by Clarke 1973, p. 281.
17. Nixon 1970, p. 64.
18. Johnson, quoted by Barnet and Muller 1974, p. 34.
19. Burns 1977, p. 117.
20. Meadows, Meadows, Randers, and Behrers 1972.
21. Meadows *et al.* 1972, pp. 23-4.

22. See, for example, Cole, Freeman, Jahoda, and Pavitt 1973.
23. Leiss 1978, pp. 107, 109.
24. For example, Riesman 1950; Marcuse 1972; Mishan 1967; Robinson 1972; Henderson 1981; Zolatas 1981.
25. Robinson 1972, p. 7.
26. Hirsch 1977.
27. Mishan 1967, p. 112. For further criticisms of economic growth from a sociological point of view, see Dowie 1972, p. 172.
28. Coombs 1980, p. 1.
29. Wilkinson 1973, p. 218.
30. Boulding 1966; Daly 1977.
31. See, for example, Goldsmith, Allen and Allaby 1972; Daly 1977; Valaskakis, Sindell, and Smith 1977; Schumacher 1973.
32. Heilbroner 1970, pp. 282-3.
33. Galbraith 1958, 1971; Daly 1975; Valaskakis *et al.* 1977.
34. Roberts 1979, p. 39; see also Wheelwright 1978, p. 145.
35. Passmore 1970, p. 321.
36. See, for example, Maddox 1972; Cole *et al.* 1973; Beckerman 1974; Freeman 1974; Simon 1981.
37. See, for example, Barnett and Morse 1963; and useful discussion in Arndt 1978.
38. Goeller and Weinberg 1976, p. 688.
39. See, for example, Beckerman 1974, p. 241. This author states 'My main argument has tried to show that there is no substance at all in the main fears of the eco-doomsters'. However, he does not discuss, for example, the carbon dioxide problem (although it is mentioned and discussed in a quotation on p. 211), the threats to the ozone layer, the impact of acid rain and other pollutants on forests, or increases in temperature due to the use of extrasomatic energy.
40. Passell and Ross 1972. For example Ehrlich (1968) predicted various major ecocatastrophies in the 1970s, including the extinction of all important animal life in the sea by September 1979. While Ehrlich may well not have intended these prophecies to be taken too literally, the fact that they did not come true has provided ammunition for the anti-environmentalists.
41. Arndt 1978, p. 142.
42. Arndt 1978, p. 144.
43. Arndt 1978, p. 71.
44. Barnett and Morse 1963, p. 250.
45. Tonnies 1955.
46. Galbraith 1971, pp. 59-60.
47. This topic will be treated more fully in a longer version of this chapter, to be published as a CRES paper (Centre for Resource and Environmental Studies, Australian National University).
48. Toffler 1970, p. 25.
49. Wilkinson 1973.
50. Toffler 1970, p. 27.
51. UNCTAD 1978.
52. Toffler 1970, p. 387.
53. Boyden, Millar, Newcombe, and O'Neill 1981, pp. 157-9.

54. Jacobs 1969, p. 48.
55. Fromm 1968; Roszak 1970; Mumford 1971; Illich 1978.
56. Thomis 1972.
57. Coombs 1980.
58. For example, Myers 1980.
59. Rada 1980.
60. Mumford 1971, p. 410.
61. Broomfield 1981, p. 31.
62. Fromm 1968; see also Barbour 1980, p. 51.
63. Toffler 1970, p. 380.
64. Barbour recognizes two types of people. There are those who see technology as 'liberator'. These people, Barbour suggests, belong to the 'domination over nature' school. The other type are those who see technology as 'threat', and this group tends to correspond with the 'unity with nature' school; see Barbour 1980, p. 53.
65. Myers 1980, p. 83.
66. Barbour 1980, pp. 55–6.
67. Davison 1977, chapter 7.
68. Rescher 1969; Rokeach 1973.
69. Rokeach 1973, p. 5.
70. Rokeach (1973) uses the term 'terminal values' for what I call basic values. I am not using his term because it seems to imply (although it is not intended to) that these values are those which are arrived at right at the end of a sequence of interactions or influences. Rokeach also discusses other categories of values (e.g. 'moral values' and 'competence values').
71. Haldane (1928) quoted by Keith 1946, p. 3.
72. We may note that this aspect of the mental state might, some would feel, more appropriately be called an 'attitude' rather than a value. However, in my opinion, the sense of respect, or even of reverence for nature, as opposed to disrespect or indifference, is indeed a value judgement. Consequently, for the purposes of this discussion, I will regard 'attitudes to nature' as a key aspect of the value system.
73. Pollard 1971, p. 9.
74. It is clear that much biological evolution has been 'progressive' in terms, for example, of increasing complexity and of intelligence. However, for individual species such progress has usually come to a sudden halt through their extinction. And sometimes it is retrogressive, as in the loss of visual capacity in cave-dwelling species (see Simpson 1950).

    As pointed out in an earlier chapter, it is in fact questionable whether the pattern of technological change has been in the least affected by the philosophers. Wilkinson (1973) writes: 'Instead of regarding development as a matter of 'progress', towards a better life, 'motivated' by an incurable dissatisfaction with our present lot, we see that it is a process of solving a succession of problems which from time to time threaten the productive system and the sufficiency of our subsistence'.
75. Milbrath 1981, pp. 81–3.
76. McLeod 1983.

10

HUMANS IN THE HIGH-ENERGY PHASE

Introduction

The high-energy phase of human existence has been associated with important changes in the material and psychosocial environments of human beings and in their patterns of behaviour. From the biohistorical viewpoint, it is pertinent to ask: How are human organisms responding to these new life conditions which are so different from those which prevailed in the environment in which the species evolved? This question is value-free: by asking it we do not infer that humans are worse off or better off in the new situation. Nevertheless, the answers to the question, as far as they can be ascertained, make a significant contribution to our understanding of the contemporary human situation and to many of the problems facing humankind in the modern world.

This chapter will be organized in five sections. In this first section we will deal with the patterns and causes of morbidity and mortality in modern high-energy societies, with brief reference also to the situation in other populations elsewhere in the world. The second section will discuss the long-term implications of societal developments for the genetic constitution of human populations. In the third section we will focus on the material or physical components of the life conditions of humans, especially as they affect health and well-being. The fourth section will be concerned with the psychosocial and behavioural aspects of life conditions. The final section will consist of some brief concluding comments about human experience in the modern world.

Causes of mortality

Mortality statistics provide the most definitive information about the health and disease patterns in modern high-energy society. The differences between these patterns and those of early urban societies are due in part to cultural adaptive processes and in part to other changes in life conditions associated with 'modernization'. Two aspects in particular of mortality statistics reveal important differences between modern high-energy societies and other societies, past and present. First, the age-specific death rates are very different and life expectancy is much higher in the high-energy societies. Second, there are important differences in the statistics relating to the *causes* of death in human populations.

236

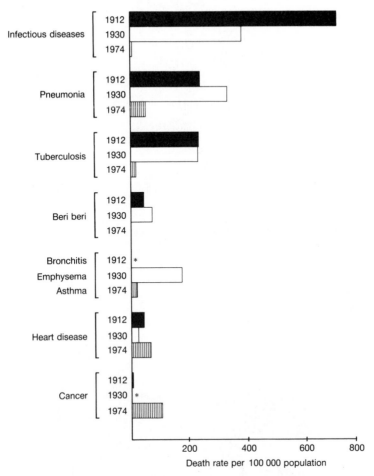

Fig. 10.1 Death rates from main causes of death in Hong Kong in 1912, 1930, and 1974. (*Not listed in official figures.)

The kinds of changes in the causes of death that have been associated with the transition from early urban conditions to high-energy society are illustrated in Fig. 10.1, which summarizes data from Hong Kong for 1912, 1930, and 1974 [1]. Hong Kong is especially interesting in this context because it was, as compared with the Western countries, relatively late in undergoing this transition, and as a result unusually detailed data are available on the causes of death when the society was still in the early urban phase.

By the first decade of the present century the mortality pattern in western Europe and the USA was already very different from that of the early urban situation. Nevertheless, as indicated in Fig. 10.2, it was still very different from that of the second part of the century [2].

1911

1964

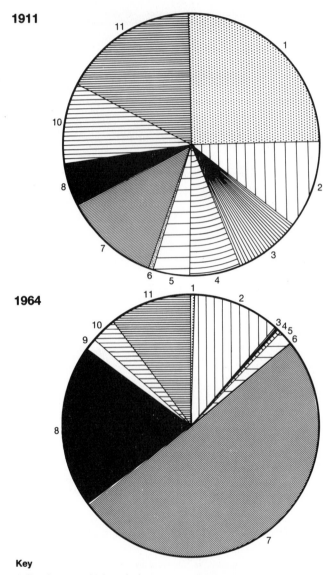

Key

1: Respiratory and tuberculosis
2: Influenza, pneumonia, bronchitis
3: Other infectious diseases and
   parasites
4: Diarrhoea
5: Infant-related
6: Maternal-related

7: Cardiovascular
8: Cancer
9: Motor-vehicle accidents
10: Other accidents and violence
11: Other and unknown
    (including certain degenerative diseases)

Fig. 10.2 Causes of death in England and Wales, 1911 and 1964.

The outstanding feature of this epidemiological transition, shown up clearly in the Hong Kong figures, has been the spectacular drop in the proportion of deaths due to infectious diseases—diseases which had been largely the consequence of particular conditions of life characteristic of early urban societies. The virtual disappearance of infectious disease as an important cause of death is a prime example of humans successfully using their capacity for culture to overcome biological threats brought about by culture itself. A very important aspect of the societal adaptive response has been the introduction of legislation aimed at improving public sanitation (see chapter 7). Other factors contributing to the decline in number of deaths due to infectious disease have included improved nutrition and the use of vaccines, antibiotics, and a range of chemotherapeutic agents. Indeed, through a systematic and highly organized international effort orchestrated by the World Health Organization, one of the most feared of these diseases, smallpox, has recently been eradicated from the face of the earth. This was a massive undertaking, and stands out as an encouraging illustration of the human potential for constructive transnational and transracial co-operation [3].

Although smallpox is the only infectious disease to be completely eliminated by vaccination, artificial immunization has contributed significantly to the decline in incidence of many others, including diphtheria, poliomyelitis, whooping cough, tuberculosis, and cholera.

Other marked changes include a substantial increase in the proportion of deaths from cardiovascular disease, especially coronary heart disease and strokes, and from cancer. Part of this increase can be explained simply by the reduction in deaths from other causes, especially from infectious disease, and the changing age structure of society; but much of the increase is certainly 'real' and represents a response to aspects of life conditions prevailing in high-energy societies.

The causes of mortality vary considerably among different socio-economic groups within the Western developed countries. For example, in Australia in the period 1980–1 the death rate among Aborigines in country areas of New South Wales has been reported to be over four-times that of the general population on an age and sex basis. The life expectancy at birth of Aborigines was calculated to be about 48 years for males and 56 years for females, as compared with 71 years for males and 78 years for females in the total population [4]. Another example is provided by mortality statistics for the city of Melbourne. It has been shown that for the period 1970–6 the death rates in some of the inner suburbs were considerably higher than in outer parts of the city. For instance, the mortality rate from all causes for men of 45 years old in the inner suburb of Fitzroy was nearly three-times that of men of the same age in some of the less central areas [5].

Cardiovascular disease

Of the various disorders which involve pathological changes in blood vessels, the two of greatest importance as contributors to mortality in modern society are coronary heart disease and strokes. The latter are the result of clots in the blood vessels of the brain or of local haemorrhages in the brain tissue. Coronary heart disease is due to infarction—that is, blood clots or thrombi occurring in the coronary arteries—resulting in a loss of blood supply and hence of oxygen to parts of the musculature of the heart. Death from cardiac arrest following fibrillation, or cardiac arrythmia, is also common, and may occur with or without infarction; it is due to irregular discharges in the autonomic nerve supply to the heart muscle from the central nervous system. The current 'epidemic' of coronary heart disease is characteristic of the modern high-energy societies. The percentage of deaths due to this cause in the USA, England and Wales, and Australia in the period 1974–81 was 27 to 29, whereas in Papua New Guinea there is virtually no sign among the indigenous people of coronary heart disease, hypertension, or strokes [6]; and in eastern Africa, cardiovascular disease, associated with hypertension, has become important only in the last 20 or 30 years, as the lifestyle of the population has become more Westernized [7].

The high incidence of cardiovascular disease in the developed countries has generated an immense amount of research ranging from pathological and physiological investigations in the laboratory to large-scale epidemiological studies in human populations. Although it is now generally agreed that environmental and lifestyle factors are largely responsible for the fact that coronary heart disease has become a major public health problem [8], there is still considerable debate and uncertainty about the relative contribution and actual nature of the various influences. Nevertheless, a distinct series of physiological, environmental, and behavioural factors have been identified as being associated with a high risk of cardiovascular disease.

Of the physiological changes associated with a risk of heart disease, one of the most important is atheroma or arteriosclerosis, a condition involving the deposition of fatty material in the walls of blood vessels. Another physiological correlate is hypertension: individuals with significantly raised blood pressure are about three-times as likely to die of heart disease as are those whose blood pressure is relatively low [9].

Another important physiological risk factor is a high level of cholesterol, in the form of *low density lipoprotein* [10], and/or triglycerides in the blood. People with significantly raised blood cholesterol of this type are about three-times more likely to die of coronary heart disease than are the rest of the population [11]. The nature of the relationship between cholesterol and heart attack is not clear; the association between the two does not necessarily indicate that high blood cholestrol is the cause of the pathology.

Another interesting correlation is that between high levels of cholesterol in the form of *high density lipoprotein* and low incidence of coronary heart disease. Indeed, it has been suggested that the high density lipoproteins are in some way protective against coronary heart disease. High levels of these substances are found especially in individuals who take regular physical exercise, who do not smoke, and who do drink alcohol fairly regularly, and they also tend to be higher in women than in men [12].

With respect to life conditions, there are striking associations between heart disease and a number of factors. For instance, heavy smokers are about three-times as likely to die from a heart attack as are non-smokers [13]. Another factor associated with a high risk of heart disease is a sedentary lifestyle. Although there is not complete agreement on this issue, there are now a number of studies which strongly suggest that a relatively high level of physical work, especially involving regular vigorous exercise, is associated with a lower rate of coronary heart disease [14].

There has been a great deal of debate about the possible role of various dietary factors in heart disease. Many authors consider that high levels of intake of saturated fats are conducive to coronary lesions, and in support of this view they cite the fact that the incidence of heart disease is relatively low in populations around the world with a low consumption of saturated fats. Others entirely disagree with this viewpoint [15]. Again, while some authors advocate reducing cholesterol intake in the diet, others point out that there is no evidence at all that people who consume less cholestrol than others are less likely to have heart attacks. Indeed, a retrospective study of dietary changes in the USA during a 70-year period leading up to 1961 shows that there has been only a slight increase in total fat consumption, and that this was due mainly to increase in the consumption of polyunsaturated fats [16]. However, in a large-scale trial recently carried out in the USA, the rate of initial heart disease episodes in middle-aged men with high blood cholesterol levels who were put on cholesterol-lowering diets and medication was half that in a control group [17].

Other authors have suggested that a high intake of sugar is the important dietary factor contributing significantly to the heart disease epidemic [18]. Indeed, there occurred a substantial increase in the per capita consumption of sugar during the first half of this century in the USA, as well as in Britain and Australia [19]. Another suggestion put forward is that a high intake of cereal fibre tends to be protective against heart disease [20].

There are strong indications that stressful life experience is conducive to coronary heart disease, and the death rate from heart disease is higher, for example, among bereaved individuals than among the rest of the population [21]. It is also recognized that the rate of coronary heart disease is considerably higher among individuals who show strongly assertive tendencies, who are hurried, impatient, and competitive, and who seem to be driven by a sense of urgency than it is among more relaxed and easy-going members of the population. Studies have shown that the former 'type A' individuals are more than six-times as likely to suffer a heart attack than are the more relaxed 'type B' individuals [22]. It has also been claimed that deliberate effort by type A individuals who have experienced one heart attack to change their behaviour to the type B pattern may reduce the likelihood of a second attack by as much as 50 per cent [23]. It is noteworthy that stressful experience is known to increase the cholesterol level of blood [24]. Linked to these observations is the view of many workers that catecholamines play an important role in the pathogenesis of myocardial infarction [25].

These, then, are the main environmental and lifestyle factors which have been found to be linked with the incidence of heart disease. However, some authors suspect that there is at least one other important *primary risk factor* which has yet to be revealed.

The situation with respect to coronary heart disease is often described as being one of 'multicausality'—that is to say, cardiovascular disease appears to be promoted by a whole series of different environmental or lifestyle factors. This conclusion does not mean, however, that the *increase* in cardiovascular disease over the past 50 or so years is due to all of these influences. A similar situation occurs with respect to many infectious diseases. In an epidemic of cholera, for example, some individuals will not be affected at all, and some will be only moderately ill, while others will die, and there is little doubt that a search for correlation with environmental and lifestyle factors would indicate links between susceptibility to cholera and such factors as nutritional state, stressful experiences, fatigue, and so on. Nevertheless, the actual *cause* of the cholera epidemic is the introduction into the community of the bacterium responsible for cholera and its spread due to unsatisfactory sanitation. In my own view, the most likely explanation of the great increase in the incidence of heart disease lies in two areas: (1) reduced levels of physical activity, and (2) an increase in chronic mental tension associated with some increase in long-term stressors and, perhaps more important, a decrease in intangible melioric experience (p. 00).

Incidentally, one variable which does not seem to have received much attention in the epidemiological work on heart disease is the frequency of mild infections with viruses. As we discuss elsewhere, the number of different viruses causing relatively mild febrile disease in human beings has now reached several hundred, and it is valid to question whether repeated acute, if mild, febrile illness throughout life, which is so characteristic of urban conditions today, has any cumulative effect on the organs of the body in general, and on the blood vessels in particular.

With respect to hypertension, it is now clear that an increase in blood pressure with age after about 25 years, which is regarded as 'normal' in modern society, is a response to modern urban conditions of life. It does not occur in people living in simpler societies. In the case of the villagers of Papua New Guinea, for instance, blood pressure actually falls progressively after the age of 25, as does body weight. A similar situation has been reported for many other peoples whose lifestyle is essentially that of the primeval and early farming ecological phases [26].

While there is still much more to be learned about the environmental and behavioural causes of cardiovascular disease, the individual who wishes to avoid coronary heart disease or a stroke would be well advised, on the basis of existing information, to act as follows: to take regular (but not excessive) physical exercise, some of it vigorous; to avoid excessive intake of animal fats; to avoid excessive intake of refined carbohydrates; to avoid overeating in general; to avoid unnaturally high levels of salt in the diet; to avoid smoking; to include cereal fibre in the diet; to avoid, as far as possible, situations involving a chronic sense of urgency and of intense competition; to adopt an easy-going relaxed way of life, with warm and supportive human relationships at the family and community levels. Let us note that, while this list is based on epidemiological and medical research on the risk factors in cardiovascular disease, it is also entirely consistent with the principle of evodeviation. All the recommendations are, in fact, recommendations to adopt a lifestyle more akin to that which prevailed in the evolutionary setting of the human species.

It is of interest that the death rates from cardiovascular disease are beginning to drop in some Western countries. Mortality from 'heart disease' in males in the USA dropped from a peak level of 439.5 per 100 000 population in 1960 to 370.4 in 1979. In females the drop was from 300.6 to 297.9 [27]. Similarly, in Australia, deaths in the population as a whole from 'diseases of the circulatory system' dropped from 432.6 per 100 000 in 1976 to 381.5 in 1980, and deaths from 'ischaemic heart disease' dropped from 243.2 to 210.2 per 100 000 [28]. It is not possible at present to ascertain how much of this drop is due to more advanced antidotal adaptive measures, in the form, for example, of intensive care in hospitals, as compared with corrective adaptive responses involving changes in dietary practices, in smoking habits, or in patterns of physical exercise.

Cancer

It is widely accepted now that most cancer in human beings is induced by environmental or lifestyle factors, and is therefore theoretically avoidable [29]. The best evidence for this is provided by data on geographic variation in the incidence of cancer in general and in the incidence of different forms of cancer. For example, it is reported that cancer before the age of 65 is four-times as common in the USA as it is in Ibadan, Nigeria [30]. Further evidence is provided by studies on cancer rates in immigrants in the countries of their adoption. Immigrants to Australia have higher rates of stomach cancer and cancer of the pancreas than native-born Australians, and there is a 25 per cent risk reduction in stomach cancer in immigrants who have been in Australia for a long while. On the other hand, immigrants from continental Europe have half the risk of cancer of the colon as have

Australians; but the longer they are in Australia, the greater their risk of colon cancer [31].

However, the differences in the overall rates of cancer in the developing and developed world are not very great. In general, Westernization and economic development seem to increase the incidence of certain cancers, such as cancer of the lung, colon, rectum, prostate, bladder, pancreas, and breast; but it is associated with a decrease in the incidence of cancer of the mouth, pharynx, oesophagus, stomach, liver, and biliary tract.

The increase in cancer of the lung and bladder associated with modernization is considered to be mainly the consequence of smoking, although air pollution resulting from the combustion of fossil fuels may contribute to a small extent. The increase in cancer of the colon and rectum may be due to increased consumption of meat, although it is possible that the reduction in dietary fibre in the diet is a more important factor. In fact, it has been suggested that dietary fibre is actually protective against colo-rectal cancer, possibly because it causes material to move through the large intestine more rapidly [32].

It is uncertain why cancer of the breast is higher in developed countries, although the lower levels of fertility may contribute. Some authors believe that dietary fat may contribute to cancer of the breast, pancreas, prostate, colon, and corpus uteri [33]; and it has been suggested that 60 per cent of female cancer and 40 per cent of male cancer can be attributed to diet [34].

Despite these various clues, the cause of a great deal of cancer in the high-energy societies is at present unknown, although there is a suspicion that much of it is the consequence of the presence in the environment of a wide range of different biologically novel chemical compounds— products of industrialization. On the other hand, some authors take the view that the cancer risk from environmental chemicals is greatly exaggerated [35].

It is noteworthy that differences occur within societies in the incidence of cancer in different socio-economic groups. This is not simply a matter of occupational hazards, although certain groups, such as asbestos workers and, in the past, chimney-sweeps, are well known to be particularly at risk. In Britain, the mortality from cancer is 24 per cent higher in manual workers than in non-manual workers. It has been estimated that nearly nine-tenths of this difference can be attributed to social variables, such as the habit of cigarette smoking, rather than to occupational factors [36].

Causes of morbidity

Mortality statistics provide information only on causes of death, and they therefore describe but a small part of the total picture of health and disease in a society. With the increasing molly-coddling influence of civilization, illness becomes less and less of a survival disadvantage, and as a result the differences between the causes and rates of death on the one hand, and the causes and rates of ill health on the other, become progressively greater.

Consider the situation which exists, for example, with respect to infectious disease in modern high-energy societies. The great microbial diseases which contributed so much to mortality during the few thousand years which preceded the beginning of the present century are now rare; but the number of relatively mild virus diseases circulating among human populations today and causing inflammation in the respiratory and the gastro-intestinal systems is extraordinarily high, and there is every reason to believe the number is increasing. There are probably at least five hundred such viruses at the present time, but because in general they produce only mild symptoms, they do not figure in the mortality statistics. As discussed in chapter 6, these virus diseases are products of civilization, because they could not have survived in the sparsely populated world of hunter-gatherers and early farmers [37]. The demographic situation in the modern world is such that whenever a new pathogenic virus appears on the scene, whether as a result of a mutation from another previously existing human virus or by transfer from some other animal species, there is nothing to stop it gaining a foothold and becoming established as a new member of the ever-growing band of disease-causing viruses now supported by humankind.

An unusual situation has arisen in recent years, with the appearance in Western countries of a previously unrecognised virus disease known as *acquired immunity deficiency syndrome* (or AIDS). This condition has a high fatality rate. It is spread almost entirely through sexual activity and blood transfusions (and from infected pregnant women to their unborn babies), and it occurs mainly in male homosexuals. Again, the survival success of the virus is assured by the very large numbers of individual hosts at risk in the major population centres of the high-energy societies.

Morbidity studies in a number of Western societies show that upper respiratory infections and mild gastro-intestinal infections account for around 40 to 50 per cent of the visits of patients to general practitioners. In Australia, which has a population of fifteen million, about twenty million cases are reported annually to doctors. The actual incidence of these virus diseases must surely be three- or four-times higher than this, because only a relatively small proportion of them are serious enough to warrant a visit to the general practitioner.

Morbidity studies also show that mental ill health also looms large among the reasons why patients visit their general practitioners in the developed countries. While it is not possible from the kind of data available to determine the precise prevalence of neurosis or other forms of mental disturbance, some indication can be gained from consideration of the level of use of psychotropic drugs taken for medical purposes in modern urban populations (as distinct from drugs like alcohol, marijuana, and heroin which are taken for pleasure). These drugs include the tranquillisers (major tranquillisers and minor tranquillisers), anti-depressants, stimulants, sedatives, and hypnotics. A study in the USA carried out in 1970 and 1971 indicated that about one in three Americans had used psychotropic drugs of this kind during the course of the year. This rate was lower than that in some western European countries [38]. In 1973 it was estimated that about twenty million Americans—that is, about 12 per cent of the adult population—were taking the tranquillisers 'Librium' or 'Valium'. In the same year over one million people in Britain were taking these drugs regularly or intermittently; and valium was being prescribed in Australia at the rate of 139 million tablets per year—that is, about 10 tablets per capita per year [39].

In general, more women than men use psychotropic drugs of this kind. In a study in Sydney it was found that 26 per cent of women and 15 per cent of men regularly take aspirins and pain-relievers, and 13 per cent of women and 6 per cent of men regularly take tranquillisers or anti-depressants [40]. The average per capita consumption of aspirin in 1972 was 295 gm in the USA, 253 gm in Australia, 220 gm in Canada, and 110 gm in western Europe as a whole [41].

In a recent major survey of the health and habits of Australians living in State capital cities, questions were included designed to provide a measure of 'how people perceive themselves to be coping with the ups and downs of daily living'. From the answers given, the investigators classified 11 per cent of men and 17 per cent of women as being 'severely disturbed', and 25 per cent of men and 30 per cent of women as being 'mildly to severely disturbed' [42].

While the rates of use of psychotherapeutic drugs and morbidity surveys provide some indication of the levels of perceived mental ill health in contemporary Western society, they tell us nothing of any changes that may have taken place over the years. While the 'stresses of modern urban living' are suspected of being responsible for the high rate of use of psychotropic drugs, no reliable data exist which show us that neurosis is on the increase, or that it is at a higher level than at other times or in other places. On the other hand, there is no evidence to the contrary, and I share the common view that a great deal of the mental disturbance experienced in modern society is culturally induced and therefore theoretically avoidable. It is pertinent in this context that catecholamine levels in the urine have been found to be higher in individuals who are exposed to certain stressful experiences, such as commuting long distances in crowded trains, and in individuals who experience frustration and general dissatisfaction with life [43]. It can be argued, of course, that since we now have drugs which, in some respects at least, will bring people 'back to normal', there is no cause for concern. On the other hand, this antidotal adaptive process leaves the underlying environmental causes of the problem untouched, allowing them to persist and perhaps intensify, so that more people become adversely affected and the psychotropic drugs have to be used in higher doses.

SECTION 2: THE GENETIC STRUCTURE OF HUMAN POPULATIONS

The process of natural selection

There are broadly three ways by which the changed environmental conditions associated with high-energy societies can affect the genetic structure of human populations. These are: effects on the process of natural selection; effects on the patterns of distribution of existing genes in a population; effects on mutation rates.

With regard to the first of these, it is clear that, theoretically, any important deviation from the 'evolutionary' or primeval conditions of life is likely to influence selection pressures. We noted earlier that some changes in life conditions which followed both the domestic and the urban transitions must have resulted in the relaxation of certain selective pressures which had been important in the hunter-gatherer situation. However, the new lifestyles also introduced new causes of death, especially in the form of various contagious diseases. It is impossible to determine the extent to which natural selection, in the face of this new situation, resulted in changes in the genetic constitution of farming and urban populations associated with a greater resistance to these diseases, although such an effect would seem likely [44]. It is also possible that, especially in early urban societies, specific nutritional diseases may also have influenced the genetic structure of human populations with respect to genes which affect susceptibility to these forms of malnutrition, although no evidence for such an effect has been forthcoming. It is pertinent in this context to recall that, until recently, in all regions of the world only a small minority of human beings actually lived in cities.

In the modern high-energy phase of human existence, however, infectious diseases and nutritional deficiency diseases have become relatively rare as causes of death, and they have to a large extent been replaced in this respect by other disorders, of which cardiovascular disease and cancer are the most important. However, death resulting from these conditions usually occurs relatively late in life, after reproductive age, and consequently the selection pressures which they exert are slight. The main cause of death in younger people, especially adolescents and young adults, in the high-energy societies is motor accidents, and we can speculate whether there may be some negative selection operating against individuals who might be 'genetically' dangerous drivers.

All in all, then, the trend in high-energy societies has been towards a further general relaxation of selection pressures. This is particularly obvious in the case of certain genetically determined disorders which in earlier times would have interfered seriously with survival and reproduction, but which in the high-energy societies, as a result of cultural adaptive measures, are less of a Darwinian disadvantage. Specific examples are provided by such metabolic defects as phenylketonuria, galactosaemia, fructosaemia, all of which are due to recessive genes in the homozygous state and all of which can be cured in the phenotype by appropriate dietary treatment in the infant. The adult form of diabetes mellitus, which appears to be a genetically determined sensitivity to the modern Western diet and lifestyle, can also be cured by lifelong hormone therapy. Indeed, concern about the *dysgenic effects* of modern medicine—that is, the possibility that there is occurring an increase in the number of cases of genetic disease as a consequence of this relaxation of selection pressures resulting from medical treatments—has been a major preoccupation of eugenicists, some of whom have advocated legislation aimed at preventing individuals affected by such diseases from breeding [45]. A more realistic approach is perhaps the improvement of genetic counselling services. While, in the long run, some increase in the proportion of the population manifesting these conditions can be anticipated, several authors have pointed out that such an increase will be very slow, even for disorders in which the relevant gene is dominant [46].

Before leaving the question of selection in the new environment, mention must made of the concept of *positive eugenics*—that is, deliberate breeding programmes (or artificial selection) aimed at increasing the proportion of certain 'desired' genetic characteristics and decreasing the proportion of less desirable traits. Indeed, the notion of 'improving' human populations has been the main basis of the so-called *eugenics movement* which was particularly active in the earlier part of this century, although it still has its adherents today. Needless to say, this idea raises many difficulties, not least of which is the problem of deciding what is 'desirable' and what is not [47]. For example, if intelligence were chosen as the key variable, *what kind* of intelligence should be selected for in the best interests of the population as a whole? (Bearing in mind that the main threats to the survival of humankind today are the consequence almost entirely of the behaviour of individuals with unusually high IQs.) Perhaps the most sensible suggestion is that, *if* positive eugenics is to be practised, it should be aimed at maintaining or increasing genetic diversity—any other approach would 'paint us into an evolutionary corner' [48].

In sum, while some genetic change, as a consequence of changes in selection pressures, is inevitable, the nature of the process is such that the problem is a minor one at present, when compared with the other very urgent biosocial issues which confront humanity today and which concern the very survival of the human species.

Gene distribution

As mentioned in chapter 1, various factors, including especially differing selection pressures in different regions of the world and genetic drift, brought about some genetic differentiation among human populations. Moreover, the earlier geographical patterns of distribution of genes which resulted have been greatly modified by mass migratory movements, such as that of the Indo-European speaking people to the south, east, and west from southern Russia several thousand years ago, and the more recent movement of Europeans and Africans to the American continent and of Europeans to Australia. Another factor has been the selective advantage that certain technologies have conferred on populations which have possessed them.

Within the high-energy societies, two main influences can be recognized affecting gene distribution. The first of these is increased geographical mobility together with a general increase in the number of social contacts experienced by the average individual. This change has greatly reduced the amount of *inbreeding* in populations, and an important effect of this development is a lessening of the likelihood of births of individuals who are homozygous for recessive genes which, in the homozygous state, are responsible for genetic disorders.

The other important influence relates to *assortative mating*. This has been defined as:

... a deviation from random mating in which like individuals preferentially mate with each other (positive assortative mating, also called homogamy) or unlike individuals preferentially mate with each other (negative assortative mating, or dissortative mating) [49].

The analysis of this phenomenon in human populations is difficult, partly because of the difficulty in determining the extent to which various traits which seem to influence mate selection are genetically determined. This is clearly less of a problem in the case of certain anthropometric variables, such as those responsible for the clear observable differences between races. In such instances, assortative mating in mixed populations is very strong. There is also evidence that assortative mating, although relatively weak, applies in the case of such physical characteristics as stature and hair colour in Caucasoid populations [50].

With respect to behavioural and mental characteristics, the situation is even more complicated. Nevertheless, it is clear that in the typical high-energy society, with its emphasis on 'equal opportunity' (even if this is never actually achieved), there is a much increased likelihood that young adults with special abilities (e.g. high scholastic or musical ability) will come in contact with other young adults with the same ability. This fact, together with the tendency for people to be attracted towards others who share their interests and enthusiasms, is likely to lead to a definite deviation from random mating with respect to such characteristics. Thus, to the extent that these abilities have a genetic component, this situation, while not influencing the gene frequencies in the population as a whole, is likely to result in a concentration of the relevant genes in certain occupational groupings within society, and to increase the incidence of homozygosity with respect to the relevant genes.

Mutation rates

The third main way culturally induced environmental change can potentially affect the genetic constitution of human populations is by increasing the rate of mutation in the germ cells in the gonads, and hence increase the number of mutants in the gametes. Concern has been expressed, for example, about the possibility that the chemicalization of the environment in modern industrial societies is resulting in the exposure of human beings to a range of chemical mutagenic agents [51]. It is impossible, however, on the basis of existing data, to assess the importance of this factor. It is known that radiation from cosmic rays and from natural radioactive isotopes is a potential cause of 'spontaneous' mutation [52]. In high-energy societies technological developments have added to these natural background sources of radiation—mainly through the medical use of X-rays and other radiation sources and from radio-active wastes from atomic power plants and explosions of atomic weapons. Taking the population as a whole, it has been estimated that the medical use of X-rays has increased the exposure of the average individual by about 30 per cent. With the exception of the inhabitants of Hiroshima and Nagasaki and certain occupational groups at special risk, the increase in radiation from nuclear power stations and nuclear weapons has been slight. However, it has been estimated that a full-scale nuclear war would increase the mutation rate by more than ten-times [53].

SECTION 3: MATERIAL LIFE CONDITIONS

Introduction

Cultural developments, especially those involving advances in technology, have resulted in a wide range of changes in the immediate environments and behaviour patterns of average human beings living in Western societies. The aspects of life conditions affected by societal developments range from 'hard' variables which are easy to analyse scientifically and to quantify— such as nutritional factors, air quality, and noise levels—through to important psychosocial and behavioural aspects of life experience, many of which are difficult to describe precisely or to measure in a satisfactory way. In this section we will consider some of the most patent of the tangible changes in human life conditions and their implications for health and well-being.

Air quality

One of the most noteworthy changes in human life conditions which has accompanied the transition to the high-energy society has been the contamination of the air that people breathe with various by-products of industrial and other technological processes, especially the combustion of the fossil fuels which are used as sources of extrasomatic energy. The most important of these contaminants are sulphur oxides, nitrogen oxides, carbon monoxide, various hydrocarbon compounds, and lead.

The combustion of fossil fuels results in the release of sulphur in the form of *sulphur dioxide* and, to a lesser extent, *sulphur trioxide*, and these are later transformed in the atmosphere to acid-sulphate aerosols. The sulphur oxides, and more particularly the acid-sulphate aerosols, are believed to aggravate asthma and pre-existing heart disease, and evidence suggests that prolonged exposure over several years will result in increased disease of the lower respiratory tract in both adults and children. Studies in the USA and Britain indicate that mean concentrations of sulphur oxides in excess of 500 micrograms per cubic metre are associated with comparatively high mortality rates [54]. Sulphur dioxide was the chief irritant in the notorious smog episode in London in 1952 which caused the deaths of about 4000 individuals.

The main source of *nitrogen oxides* (NO and NO^2) in the atmosphere is the internal combustion engine, and their main role in air pollution is as components in the reactions which lead to the formation of photochemical smog (see chapter 8). However, nitrogen oxides have distinct biological effects apart from those associated with photochemical pollution. On the basis of existing evidence, it seems likely that they may render people rather more susceptible than they would otherwise be to infections of the respiratory tract, and continuous exposure may contribute to degenerative changes in the lung leading to emphysema [55].

Carbon monoxide is an odourless gas and levels of about 15 micrograms per cubic meter are not uncommon in the air of cities. Its toxicity is due to the fact that it has a greater affinity than oxygen for haemoglobin in the bloodstream, and consequently it interferes with the transportation of oxygen from the lungs to the tissues of the body. It is especially harmful to individuals who are likely to be sensitive to a decreased oxygen supply, such as those with anaemia, chronic lung disease, and cardiovascular disease. Mortality data provided by the National Air Pollution Control Administration of the US Department of Health, Education, and Welfare suggest that exposure to levels of 9 to 16 milligrams per cubic metre is associated with increased mortality from coronary heart disease [56]. It is also suggested that the probability of motor accidents may be higher with greater carbon monoxide exposure. Symptoms of carbon monoxide poisoning include headaches, dizziness, lassitude, flickering before the eyes, ringing in the ears, nausea, vomiting, difficulty in breathing, and apathy.

The combustion of fossil fuels also releases into the atmosphere a range of *hydrocarbon compounds* of variable molecular weight. These include various aromatic polycyclic hydrocarbons which are known to be carcinogenic, and it is likely that these substances contribute, even if only to a small extent, to the incidence of lung cancer in urban environments [57].

The *lead* in the atmosphere in the urban environment is derived mainly from the tetraethyl lead which is added to motor spirit to prevent premature detonation. It is emitted as an aerosol of inorganic salts and oxides, most of the particles being less than one micron in diameter. The lead content of air in modern city streets in commonly between 1 and 10 micrograms per cubic metre and sometimes reaches 25 micrograms per cubic metre; the average is around 2 to 4 micrograms per cubic metre. City dwellers therefore regularly inhale lead. Concentrations of this element in the blood of 10 to 15 micrograms per 100 millilitres are not uncommon. Higher levels are found in people who are frequently exposed at close hand to the exhaust fumes of motor vehicles [58]. Overt symptoms of lead poisoning are associated with considerably higher levels, such as 60 to 80 micrograms per 100 millilitres of blood. However, smaller concentrations are likely to result in some maladjustment. The symptoms and signs of mild lead poisoning include: tiredness; lassitude; constipation; slight abdominal discomfort or pain; anorexia; altered sleep; irritability; anaemia; pallor; and, less frequently, diarrhoea and nausea. If everyone in the city were to suffer from these signs and symptoms, life might well go on much as usual, and perhaps few people would realise that they were anything but normal. Apart from these more immediate effects of mild lead poisoning, epidemiological studies have led some authors to conclude that chronic lead poisoning interferes with mental development in city-dwelling children [59]. The suggestion has even been made that it may contribute to juvenile delinquency [60].

It is important to appreciate that, as in the case of so many potentially harmful environmental factors, there is considerable variability in response to chemical pollutants in the atmosphere. Different individuals react to pollution differently, and the reactions of a single individual may vary over time [61].

Noise

Another obvious change in the human environment associated with machine use is the local increase in noise levels in factories, on busy streets, and near airports. Some typical figures for noise levels in dB(A)* in a variety of locations in the modern world are as follows: just audible, 10; leaves rustling in the breeze, 20; soft whistle, 30; quiet restaurant, 50; freeway traffic (50 metres away), 70; busy city street, 90: heavy traffic in city street,

100; low-flying jet aircraft overhead, 100; discotheque, 120; jet aircraft taking off within 50 metres, 120.

Noise levels in the natural or evolutionary environment are usually about 30 dB(A)—or about 40 dB(A) when people are talking. In a busy city street today the noise level may be up to 100 dB(A)—that is, about one to ten million times louder. It is therefore valid to ask whether this deviation from the natural life conditions of the human species gives rise to any signs of maladjustment. The answer is in the affirmative. It is known, for instance, that 90 dB(A) will cause a temporary reduction in the acuity of hearing, and frequent exposure to such levels over long periods of time produces permanent impairment.

The extent to which noise gives rise to psychological disturbances depends to a considerable extent on how it is perceived by the hearer. Thus some individuals in modern technological society find the extreme levels of noise in discotheques, up to 120 dB(A), to be enjoyable. Others find both the levels and the kind of noise in these places to be a cause of distress [62].

High noise levels can also affect health and well-being indirectly, by competing aggressively with other sensory inputs and thereby interfering, for example, with enjoyable conversation or the contemplation of visually beautiful scenery.

Light

A relatively recent change in the physical life conditions of human beings which is seldom discussed, but which is most significant for the quality of human experience, has been the increase in the hours of exposure to visible light. In a typical city in a temperate zone this increase amounts to about five to seven hours per day in the winter, and perhaps two or three hours in the summer. This change is not known to have resulted in any signs of maladjustment, but it has certainly had a profound affect on the time budgets of human beings. It is worth noting that in some other species the hours of exposure to visible light is an important influence on hormone levels; in many birds, for instance, increased exposure to light promotes sexual activity and egg-laying [63].

Diet

Introduction

What people eat is influenced not only by their biological requirements but also by the customs of the society in which they live, by religious beliefs, by pressures of advertising, by other competing demands on their time, by their knowledge of nutrition, by economic forces, by aesthetic considerations, and by many other factors. Nevertheless, whereas cultural factors are important determinants of the diet of human beings, let it be noted that no cultural measures have produced, nor would be expected to produce, a diet more promotive of health for humans than that to which the species is naturally adapted through evolution.

Quantitative aspects

Before the technological era, by far the most common quantitative deviation in the Western world from the optimal diet was under-consumption. This is still a widespread and serious problem in some parts of the developing world today. In modern high-energy societies the situation is reversed, and the main health problems relating to the quantity of food eaten are the result of excessive consumption beyond the energy requirements of the body. This form of malnutrition has been a risk among the nobility since the first cities came into existence, but in Europe it is said to have become especially important among them from the late-17th century onwards. In the words of Trowell:

The English National Portrait Gallery, wherein portraits date back to the fourteenth century, clearly portrays that gross obesity, previously very rare even in the nobility, suddenly became extremely common in the upper social classes towards the end of the seventeenth century and even more so in the eighteenth century. Portraits even of young men and women usually portrayed protruding tummies and large double chins [64].

Overeating is considered to be an important cause of a multitude of diseases in modern Western societies. It appears to be a major factor contributing to high rates of coronary heart disease, and it is also believed to contribute to certain kinds of malignant disease, including cancer of the breast and of the corpus uteri. It has been estimated that an extra 25 lbs (11.4 kg) above the standard weight on a man 45 years old is associated with a reduction in his life expectancy by about 25 per cent [65]; and it has been pointed out that every extra 10 lbs (4.5 kg) above the optimum weight is equivalent, in terms of its implications for life expectancy, to smoking twenty cigarettes a day [66].

In Western countries obesity has come to be perceived by most people as undesirable, for either health or aesthetic reasons, and this is reflected in the fact that in the late 1960s the annual financial yield of weight-reducing schemes promoted by commercial interests was around a hundred million US dollars [67]. Obesity has recently become a problem among certain populations in developing countries, especially in some parts of Africa [68].

The amount of food which people eat is a function of a range of interacting factors, some tending to promote eating and others to suppress it. In the natural environment of any species, including *Homo sapiens*, these two opposing sets of forces result in a rate of consumption of calories which is sufficient for, but not in excess of, the energy requirements of the body. That is to say, they tend to promote a feeding pattern which is conducive to optimal health and well-being. The various influences affecting the amount of food consumed by individuals in the modern world are summarized in the 'behavioural equation' in Fig. 10.3.

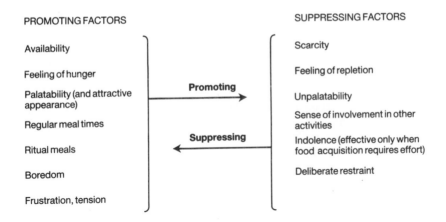

PROMOTING FACTORS

Availability

Feeling of hunger

Palatability (and attractive appearance)

Regular meal times

Ritual meals

Boredom

Frustration, tension

Promoting →

Suppressing ←

SUPPRESSING FACTORS

Scarcity

Feeling of repletion

Unpalatability

Sense of involvement in other activities

Indolence (effective only when food acquisition requires effort)

Deliberate restraint

Fig. 10.3 Behavioural equation for food consumption.

Let us briefly discuss the items on these two lists.

Availability, as opposed to *scarcity*, is so obvious an influence on food consumption that little comment is necessary. For most people in modern high-energy society, food is regularly available, as it was in the typical natural or primeval habitat. However, uneven distribution of material wealth and income still results in undernutrition in a minority of the population in the affluent countries of the West.

A feeling of hunger versus a feeling of repletion is an important aspect of the feeding equation. Several theories have been put forward to explain the mechanism of the hunger drive [69], but these need not concern us here. What is relevant is that a feeling of repletion or fullness in the stomach tends to discourage feeding, whereas an empty stomach is, in healthy individuals, usually associated with the experience of hunger. This mechanism, a product of evolutionary processes, was a satisfactory one for humans on an evolutionary or natural diet. However, the practice of removing the fibre, which is relatively non-digestible, from foodstuffs of plant origin means that much more food energy can be consumed before a feeling of repletion begins to counter the feeling of hunger. This interference with one of the natural regulatory mechanisms controlling food intake in relation to food requirements—pushing the equation in favour of consumption—is one of the more important biological consequences of the production and commercial distribution of refined carbohydrates.

Palatability is another important variable affecting the relationship of food intake to food requirements. In the evolutionary environment, palatability was usually associated with nutritiousness: animals in their natural habitat find palatable those foodstuffs of survival value to them, foodstuffs to which their tastes, their digestive systems, and their metabolism are suited through evolution. In the case of humans in the primeval environment the enjoyment, for example, of the taste of sweetness led to the consumption of fruits and berries, a good source of ascorbic acid and other vitamins, and enjoyment of the taste of salt led to the consumption of meat as a source of essential amino acids and other important nutrients.

The break in this association between palatability and nutritiousness in foodstuffs is one of the most important and interesting consequences of human culture relating to diet. The change began with the culinary art, which sets 'endless traps of sight, taste and smell to break down the restraints of appetite' [70]. More recently the food industry has further contributed to this change, through its use of sophisticated advances in food technology and the addition of a wide and ever-growing range of nutritionally useless chemical compounds to render the products more attractive. One of the consequences of these developments has been the fact that people are more likely than in the past to eat when they are not really hungry. The promoting side in the behavioural equation has thus been further strengthened.

Another important factor tending to promote feeding behaviour in humans is the cultural institution of *regular mealtimes*. In some sections of society there are three or more of these each day; and for the majority of individuals they involve merely sitting at a table—to be served with a meal prepared by someone else. Certainly the whole experience of a mealtime can be a very positive one, consisting of a blend of pleasing tastes, good humour, and friendly conversation, and it is one of the *experiential bonuses* of civilization. On the negative side, this deviation from the evolutionary situation is an additional factor tending to increase the intake of food energy above the requirements of the body. In certain occupations, *ritual meals*, as in the entertaining of business associates, represent an occupational hazard which further contributes to the overconsumption of calories and hence to ill-health.

It is well known that *boredom*, one of the products of modern living— especially for the traditional suburban housewife in her culturally enforced isolation—sometimes promotes the consumption of food beyond levels which are nutritionally desirable. Others take to eating in response to *chronic frustration* and as a means of allaying *tension*, and many worried people find relief in eating.

Indolence is an important factor affecting food consumption in humans under certain circumstances. In the primeval situation this behavioural characteristic of our species must have played an important part in preventing both overeating and overexploitation of the food sources in the environment. The acquisition of food required considerable effort: people did not go to the trouble to go out and gather food until they felt like doing so—that is, until hunger, together with an interest in hunting and gathering and a desire for some excitement, overcame the tendency towards indolence. Because of the ready accessibility of food, this factor plays little part in the control of food consumption in modern high-energy societies.

It is clear, then, that the main effects of culture on the behavioural equation for eating have been to greatly strengthen the various factors which tend to promote feeding behaviour and to weaken those which have the opposite effect. It is therefore not surprising that all the evidence suggests that most people in modern Western society consume calories considerably in excess of their energy requirements and that overeating is a common cause of phylogenetic maladjustment [71]. Needless to say, another factor contributing to this situation is the relatively low energy requirements of the majority of the population due to their low levels of physical work. It is noteworthy that, while a progressive rise in body weight after the age of 25 is usual in individuals living in the developed Western countries, no increase in weight after this age occurs in most people living in relatively simple or 'primitive' societies.

There is, however, one further factor in modern high-energy society which sometimes comes into play and which tends to suppress food intake: it is *deliberate restraint*. Individuals, either because they believe that overeating is bad for their health, or for religious or moral reasons, may make the conscious effort not to consume food in excess of their physiological requirements even when tempted to do so. Certainly, this factor affects food consumption to a greater or lesser degree among a proportion of most modern Western populations. However, deliberate restraint is clearly itself a product of civilization—it was totally unnecessary in the natural environment. The question therefore arises whether the application of self-discipline in this situation, and the anxiety which is sometimes associated with it, may in its own right have some undesirable consequences. Occasionally, especially in young women, the phenomenon of deliberate restraint reaches pathological proportions. In the condition known as *anorexia nervosa*, irrational fear of being 'overweight' associated with various emotional disturbances, results in the individual consuming much less food than is necessary for the maintenance of good health. Loss of weight, general debilitation, and even, in some cases, death may follow [72]. To what extent this condition is a 'disease of civilization' is not clear, but to my knowledge no cases have ever been reported among hunter-gatherer people.

Before leaving the topic of quantitative aspects of food consumption, a few words are necessary on growth rates in children and on the ultimate height of individuals. One of the outstanding changes in human biology that has taken place in the developed regions of the world during the present century has been the progressive increase in stature. A study in Glasgow, for instance, showed that in 1960 boys and girls of thirteen years old were, respectively, four inches and three and a quarter inches taller than their counterparts forty years previously. However, tallness does not seem to be a particular advantage, and some may see it as a disadvantage. Similarly, with respect to infants, 'there is as yet no proof that the biggest baby, vitamin- and protein-stuffed throughout the year, is necessarily the healthiest baby' [73]. Certainly, in times of food shortage, small stature may well be advantageous, because the smaller individual has lower food requirements. Indeed, it has been suggested that small body size is a biological adaptive mechanism which comes into play in populations in which calorie intake is severely restricted [74].

Qualitative aspects

Specific deficiency diseases, so important in some early urban societies, and perhaps in the earlier part of the high-energy phase, are relatively rare in modern Western societies. Nevertheless, nutritional deficiencies of iron, leading to anaemia [75], and of thiamine (vitamin B^1) have not been uncommon in the USA in recent times. However, nutritional deficiency diseases such as beriberi, rickets, and kwashiorkor are still very common among children in the Third World. In fact, the extent of 'protein-energy malnutrition' among children in some areas is considered sufficient seriously to impair their mental development [76]. Another common deficiency disease in particular areas of the world is endemic goitre, associated with iodine deficiency. It possibly affects about one-twentieth of the world's population.

Before considering separately a few of the more important qualitative aspects of diet in modern high-energy society, let us briefly consider the ways by which, or through which, cultural processes can influence the quality of food consumed. Clearly, beliefs, based on information communicated from others, are an important influence on what people eat [77]. Thus the firm conviction in some Eastern countries that rice is a 'good food', not only for adults but also for very young children, results in rice being fed to babies within a few days of birth. Not surprisingly, this unnatural practice often has undesirable consequences for the infants [78]. In Malaya the converse situation exists, in that older children are denied a foodstuff of high nutritional value because the local culture incorporates the view that fish is bad for young people.

Certainly, beliefs about the positive value of certain foodstuffs influence the dietary intake of many families and individuals in the modern world. Nutritional science and educational programmes promoted by health authorities have had a major influence on diet and hence on nutritional state in a high proportion of the population. People have become very conscious of the existence and need for vitamins. Indeed, the consumption of vitamin pills in the high-energy societies has reached extraordinarily high levels, and many people are consuming more vitamins than are necessary or desirable for their health and well-being. The recent conclusion of nutritional science that fibre from foods of plant origin is desirable is also beginning to have impact on the diet of some people.

There has, of course, been a good deal of debate among nutritional scientists over the years about the dietary requirements of the human species, and some of it persists to the present time. The uninitiated may be excused for a certain cynicism and for asking 'What will they be telling us next?'. In Australia, for instance, the consumption of sugar has dropped some 10 per cent over the past ten years, probably because health educators have been emphasizing the importance of unrefined foodstuffs and the perceived disadvantages of refined carbohydrates. Very recently, however, the sugar industry in that country has launched a major advertising campaign aimed at convincing the public that 'sugar is good for you'.

Although information derived from nutritional science and transmitted through health education programmes has influenced the diet of many people, it is also true that a considerable section of the population appear to be oblivious of this information.

Religious beliefs are also an important influence on the quality of diet with respect to specific components. Jews and Moslems must not consume the meat of the pig, while many sects associated with Eastern belief systems are not permitted to eat meat of any kind. Although the latter is a significant deviation from the evolutionary diet of humankind, it has been shown that, with careful selection of appropriate plant foodstuffs, it is possible to achieve adequate nutrition on a purely vegetarian diet [79].

To follow habit is a common behavioural tendency of humankind and a major determinant of a great deal of human behaviour, and it is a very important influence on what people eat. Food preferences are usually determined early in life [80]. Dog and snake meat is relished by some Chinese people, but the idea of eating either is abhorrent to many people with an Anglo-Saxon background. The power of habit can be very strong: a European explorer in the 19th century in the eastern part of Australia died of malnutrition because he refused to eat the foods of the local Aborigines with whom he was in contact [81].

It is a truism that there is much variability with respect to food likes and dislikes among individuals in the high-energy societies, where there is a broad range of different foodstuffs available. While this does not usually result in specific deficiency diseases, certain predilections resulting from conditioning in early life are potentially detrimental to health. For example, the majority of adults in Western society today, having been conditioned to unnaturally high levels of sodium chloride in their diets in early life, are unable to enjoy their food unless it contains substantial quantities of added salt.

Another aspect of culture which has had a profound effect on diet in ecological phase four society has been *technological change*. One of the best examples of this in the early part of the modern phase was the change that took place when new fine-silk bolting cloths (sieves) and steel roller-mills displaced the ancient process of grinding grain between stones. This change began late in the 19th century and was almost complete by 1890. It resulted in a drop of about two-thirds in the average per capita intake of vitamin B^1 [82]. More recently, of course, societal adaptation to this undesirable culturally induced change has involved legislation in some countries (e.g. Britain and the USA) requiring that vitamin B^1 and other vitamins be put back into bread.

We have only to look around at the extraordinary range of foods offered in the typical modern Western supermarket to see how the quality of what people eat—in terms, for instance, of consistency, pappiness, fibre content, colour, and taste—has been modified by technological developments in the food industry.

It is well appreciated that industrial and commercial forces, especially some of the larger corporate organizations, are an important influence on the nature of the food that people consume. In their own struggle for survival in the world of commerce, these organizations put much effort into producing products which people will perceive as desirable; and this process, together with subtle advertising, almost certainly modifies the tastes of the consumers. The ethics of this behaviour on the part of the food industry has been the subject of lively debate at many a nutritional science conference.

There are many other cultural influences affecting the quality of the diet of human beings in indirect ways, including those which put additional demands, associated with work, travel, or television, on the time budgets of individuals. As the 'pace of life' accelerates, so is there less time for preparing and consuming food. The food industry has successfully exploited this situation with the production of *convenience foods*, which are often nutritionally inferior to, and certainly very different experientially from, the family meal prepared in the home and consumed with a certain amount of ritual and relaxed interaction at the dining table.

We will now turn to consider in a little more detail some of the more outstanding ways in which the typical diets of people in the high-energy technological societies differ qualitatively from those of human beings in the evolutionary environment.

Refined carbohydrates and fibre   A most significant change in the diet of humans associated with ecological phase four has been the introduction of refined carbohydrates. The two most important of these are: *sugar* (a soluble disaccharide), about 60 per cent of which comes from sugar cane and the rest from sugar beet; and *white flour* (digestible polysaccharides), which is produced by refinement of wheaten flour.

There has been an enormous increase in the consumption of sugar over the past five or six generations in both Britain and America. In Britain about twenty five times as much sugar is now consumed per capita as was the case in the mid-18th century, and it now accounts for about one-fifth of the total calorie intake of the population [83]. Before 1850, sugar consumption in Britain had been minimal. After that it rose steeply, and had reached 80 grams per day per capita in 1870, and it continued to rise to 100 grams per day in 1900; after that date the amount consumed continued to rise, except during the periods of the two World Wars, until 1958 when it became stationary at about 164 grams per day *per capita* [84].

This change in sugar consumption is shown in Fig. 10.4 [85]. The figure indicates the relative contributions of fat, sugar, starch, and protein, as well as the salt and fibre content, in the diets of typical individuals in primeval, early farming, and modern high-energy societies, respectively.

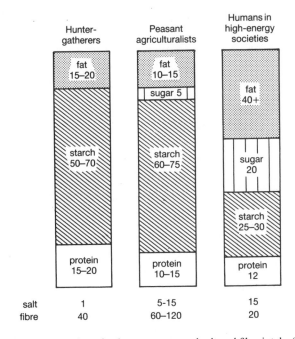

Fig. 10.4 Percentage of energy from food components, and salt and fibre intake (gm/day).

The interesting history of the milling of wheat has been well told in a book by McCance and Widdowson entitled *Breads, white and brown* [86]. The grinding of grain, including wheat, goes back to Palaeolithic times, but the first sieves designed to separate the coarse bran (the outer coating of the grain, once the husk has been removed) were introduced in Egypt at least three thousand years ago. Some Egyptians were skilled in the art of bread-making, and although flour from barley and millet was mainly used, white bread made from fine wheaten flour was available for those who could afford it [87]. It is probable that the whiter bread was preferred because it was likely to contain less grit than coarser products. In Greece and Rome the upper class consumed off-white bread made from flour from which some of the bran had been extracted, and from that time onwards whiteness in bread became associated with high social status. As a result, people in general came to aspire to white bread, and by the mid-19th century millers and bakers were adding alum to the flour in order to make the bread whiter and thus to increase their sales. Around 1850, a chemist by the name of Accum living in London analysed some of the bread sold in that city and showed that it was adulterated with alum, and he published his findings in a pamphlet. However, the defensive backlash from the vested interests—the millers and bakers—was severe, and Accum was eventually forced to leave the country. Many years later his work was confirmed and legislation was introduced aimed at preventing adulteration of bread with any substance not approved by the authorities [88]. During the present century various bleaches have been added to bread made from white flour in order to make the bread appear even whiter. Concern was voiced in some quarters when it was demonstrated that when agene, one of the bleaching agents commonly used for this purpose, was fed to dogs it caused them to suffer fits [89].

In what ways, we may ask, is the human organism responding to deviations from the evolutionary diet of the species resulting from the refining of carbohydrates? Is the change giving rise to signs of phylogenetic maladjustment? Although a considerable debate still exists in this area, an increasing number of authors consider that the removal of fibre from the diet is responsible for many forms of ill health in modern society. The least controversial case is that of dental caries, which is generally accepted to be the consequence of the consumption of sugar and white flour; the teeth of people in populations which are not on a Western diet show very little sign of this condition, although it becomes common when refined carbohydrates are introduced into the diets of such populations [90].

In the case of dental caries, the evidence strongly suggests that the purified carbohydrate itself is the cause of the pathological condition, through its effect on the bacterial flora on the teeth. In other pathologies associated with the consumption of refined carbohydrates it is less clear whether the concentrated carbohydrate is the cause of the abnormality, or whether the pathology is rather the consequence of the lack of fibre in the diet [91]. As mentioned earlier in this chapter, the refinement of carbohydrate, involving the removal of fibre from plant foodstuffs, often results in people consuming more food energy than would otherwise be the case. Moreover, the lack of fibre in a diet rich in refined carbohydrate has a marked influence on the physical consistency of the contents of the intestine, and this affects the ease and rate at which material travels through the alimentary canal. Thus pathological conditions associated with the consumption of refined carbohydrates may in some cases be the consequence of a generally high intake of energy and in others the result of modifications in the consistency of the gut contents.

The maturity-onset form of diabetes mellitus is a well documented example of a pathological condition associated with diets rich in energy. It is well known that in populations on this kind of diet, diabetes mellitus in adults is an inheritable condition, thus providing example of *genetic vulnerability* to a particular deviation from the evolutionary conditions of life (see page 42). The question arises why the gene or genes associated with this genetic susceptibility to diets rich in refined carbohydrates should be so high; presumably this is a case of balanced polymorphism, and in some other circumstances this particular genotype may confer some advantage on the individual [92].

Other conditions for which refined carbohydrates are blamed by some authorities include diverticular disease of the colon, cancer of the colon and rectum, ulcerative colitis, varicose veins and haemorrhoids, coronary heart disease, duodenal ulcer, appendicitis, and gall stones [93].

Fats   The unresolved controversy about the possibility that a high rate of consumption of animal lipids—saturated fats and cholesterol—might be conducive to coronary heart disease has been discussed above. While cross-culturally there is a strong correlation between the mortality rates from heart disease (as well as cancer of the colon) and the consumption of animal fats and cholesterol [94], there is still debate on whether reduction in the intake of cholesterol and of saturated fats will decrease the risk.

Nevertheless, it is clear that the average daily intake of fatty substances of animal origin is considerably higher than it was in most populations in the primeval stage of human history. This is partly because the amount of meat consumed on average in the modern affluent societies is above that likely in most, but not all, hunter-gatherer situations. Another important consideration is the fact that the fat content of meat from modern domestic food animals differs both quantitatively and qualitatively from that of free-living or wild animals. It has been shown that muscle samples from domestic bovids (after the attached fatty tissue has been removed) contain around twice as much lipid as do samples from wild buffalo and that the fat content of intensively reared cattle is considerably higher than that of animals left to graze naturally on grassland. Moreover, a higher proportion of the fat in domestic animals is saturated. The ratio of polyunsaturated to saturated fats in domestic bovids has been reported to be 1:50, as compared with 1:2.3 in free-living animals [95].

The average meat-eating person in modern technological society is therefore likely to consume much more saturated fat than the average hunter-gatherer in primeval society.

Additives   Under this heading we include any substance deliberately added to food for a specific purpose—for example, to improve its flavour, to make it more visually attractive, to act as a preservative, or for other reasons. The distinction between 'ingredients' and 'additives' is not a sharp one. Thus herbs and spices may be added to a stew, but whether they should be regarded as additives or as ingredients is a somewhat academic question. Here we are especially interested in distinct chemical compounds, many of them synthetic, which are added by the commercial component of the food-processing chain rather than by the cook or the consumer. These agents include preservatives, anti-oxidants, colouring agents, flavouring agents, sweeteners, sequestrants, filling agents, stabilizers, emulsifiers, and other 'improving agents' [96]. The battery of flavouring agents alone which are incorporated in commercial food products today includes well over one thousand different chemical compounds.

First, however, let us consider a chemical additive which has been used by humankind since ancient times—sodium chloride, or 'salt'. It may be added to food at various stages in its preparation or at the time of its consumption. A taste for salt appears to be a basic human characteristic. The hunter-gatherers of the Kalahari Desert, who certainly have not been conditioned to eating salt by the consumption of salted baby foods in infancy, have a strong liking for salt [97]. However, in the hunter-gatherer setting free salt is uncommon, and the biological basis for this characteristic presumably lies in the fact that enjoyment of the salty taste contributes to the appetite for meat, thus promoting the consumption of animal protein. In the modern setting, however, cultural processes have resulted in salt being available in unlimited quantities. Because no such excess of salt existed in the evolutionary environment, no innate natural regulatory mechanism exists in human beings to prevent its consumption beyond the needs of the body. In the case of salt, the majority of people in modern technological society consume ten to fifteen times more than is necessary to satisfy their metabolic requirements. Salt is added to vegetables during cooking and to soups and sauces, and it is often shaken liberally onto the plate; and most canned and processed foods available in supermarkets also contain added salt. In fact, one slice of a conventional loaf of bread provides sufficient salt to satisfy an individual's physiological needs for one day.

The consumption of such large quantities of sodium chloride is clearly a deviation from the evolutionary life conditions of the human species, and the question therefore arises whether this change in diet is responsible for any form of phylogenetic maladjustment. The answer appears to be in the affirmative, at least for about one in five of the adult population, and perhaps to some extent for everyone. As early as 1904 two physicians in France showed that when some patients with hypertension were put on a salt-free diet, their blood pressure dropped considerably [98]. Since that time, evidence has been slowly accumulating to the effect that the consumption of salt at levels above those found naturally in foodstuffs leads to high blood pressure in at least 20 per cent of people in Western populations [99]. It is also known that the blood pressure of people living in certain developing regions of the world, such as Papua New Guinea and some islands in the South Pacific, who do not add salt to their food *decreases* slightly as the individuals grow older after the age of about twenty five [100]. This is in marked contrast to people living in Western technological societies where a progressively increasing blood pressure after the age of twenty five is considered normal. It has also been reported that blood pressure patterns in East Africans have recently come to resemble those typical of Westerners, and this change has been associated with the introduction of added salt to the diet [101]. Moreover, a strong correlation has been found to exist between salt intake and blood pressure within a single (Western) population [102].

As indicated above, apart from sodium compounds, a fantastic array of chemical substances are deliberately added, mainly by the food industry, to human foodstuffs. Around 3500 chemical additives are in use today. Although they are usually present in food in extremely small quantities, as many as thirty may be added to a single product [103]. Again it is pertinent, from the biosocial standpoint, to ask: What are the consequences for the human organism of these deviations from the evolutionary diet? We have already discussed the role of flavouring agents in encouraging the consumption of food in the absence of hunger; but another important question is whether these unnatural substances exert any direct harmful effect on the tissues. Certainly, some artificial flavouring and colouring agents which were used in the past, especially some of the coal-tar dyes, have been found to be carcinogenic. The societal adaptive response to threats of this nature has involved legislation. Thus the Food and Drug Act in the USA requires that no chemical additive can be used in commercial products that has not been tested for toxic effects in at least two species of experimental animals [104]. The 'burden of the proof' now rests with the food industry, to demonstrate that a new additive does *not* produce observable lesions in the experimental animals even in concentrations one hundred times greater than those in the food products on the market. This legislation is consistent with the point of view arising out of the principle of phylogenetic maladjustment—that is, that any deviation from the evolutionary conditions of life should be under suspicion until proven harmless.

Unfortunately the approach has a number of important defects. First, it does not involve testing for possible synergistic effects of different compounds which may appear in combination in foods or in meals. Second, the possibility always exists, even if slight, that an agent may be harmful for one species, such as *Homo sapiens*, but not harmful for certain others, such as the two species used in the tests. Third, and especially important, the tests are unlikely to detect slight changes in mental processes—changes of a kind, for example, which might affect mood or rationality. The relevance of this point is brought home by the claims of Feingold, who reported that a high proportion (48 per cent) of children exhibiting 'the hyperkinetic syndrome' can be effectively treated by removing artificial colouring agents from their diets. He believed that a certain proportion of children are 'genetically predisposed' to suffer adverse reactions triggered by one or more chemicals contained in synthetic flavouring or colouring agents. If the Feingold hypothesis is correct, this syndrome provides us with another example of the principle of genetic vulnerability. While there is considerable controversy about Feingold's claims, some authors have repeated his treatment of hyperkinetic children and report similar, or even more convincing, results [105].

It is possible that in the near future there may be more public debate about the costs and benefits of food additives. Clearly, preservatives often play an important positive role in protecting food from decomposition, and it is hard to imagine a food system in the modern world that does not make use of them (as it is, some 20 per cent of the world's harvest is said to be lost through the activities of microbial, fungal, and animal pests). It is more difficult to put forward a convincing case for the use of artificial flavouring or colouring agents, or even emulsifiers and texture agents. In the past, and in some parts of the world today, people have thoroughly enjoyed eating without using them at all. It is not impossible that cultural adaptation will eventually take the form of legislation banning the use of all such agents.

Contaminants   Undesirable contaminants of foodstuffs are, broadly speaking, of two kinds: biotic and chemical. The biotic contaminants, especially bacteria, may hasten decomposition of the foodstuff and hence affect its taste; or they may produce toxic products, as in the case of *Clostridium botulinum*, which produces the toxin which causes botulism. Other contaminating organisms may multiply and cause disease in the consumer of the food, as in the case of food poisoning due to salmonellae. However, ill health due to the contamination of foodstuffs with biotic agents is not common in modern technological society.

Incidental contamination of food with various chemical products of the industrial society is a more serious problem. Especially important is contamination with various pesticides used in the treatment of crops or of harvested plant food. An extensive literature now exists on the contamination of human food with DDT and other biocides and on the consequences for human health and well-being of the consumption of these substances.

Within the developed countries there has been growing awareness of the risks associated with chemical pollution of foodstuffs, and much progress has been made towards the regulation of the use of potentially noxious substances. Nevertheless, the situation is still far from satisfactory. At the time of writing, for example, concern is being expressed about the contamination of various grains with ethylene dibromide, a substance with pesticidal properties which is still widely used in the USA for spraying crops, and was used in Britain for directly fumigating grain until 1981. It is also added to leaded petrol as a 'scavenger' to prevent lead accumulating in the moving parts of the engine. Ethylene dibromide has been shown to be one of the most potent carcinogenic agents known in experimental animals [106].

In sum, the additives and contaminants present in the food of the majority of people living in modern technological society are not, either singly or in combination, causing immediate and obvious ill health. Nevertheless, there are two causes for concern. The first is the possibility that some of these may be agents promoting cancer in a small proportion of the population. The second relates to the concept of *subliminal toxicity* [107] and the possibility that some of these substances, singly or in combination with others, are giving rise to low-grade chronic conditions such as lethargy, fatigue, irritability, and general malaise—states which can easily come to be accepted as normal. Would many people feel more full of life if the contaminants and additives were not there? Scientifically, this is a fair question, but it cannot be answered on the basis of existing knowledge. This being the case, there is some justification for avoiding, as far as possible, foods which are likely to contain unnecessary alien compounds of any kind.

Infant feeding

The natural diet of infants is human milk, sucked directly from the maternal breast. While deviations from this diet involving the artificial feeding of milk from cows or other ungulates sucked by the infant from a bottle or similar container date back at least to 2000 BC [108], it was not until the modern high-energy phase of human history that it became common practice. In the early part of the present century, 60 per cent of babies in England were not receiving breast milk by the second week of life. From the standpoint of the life conditions and biopsychic state of the infant, there are three aspects of this change in diet and behaviour which need to be considered.

First, it is reasonable to ask what the implications are for the infant of the qualitative differences between human and bovine milk [109]. In general, it seems that the small differences in nutrients which exist are not important for the majority of individuals. Indeed most adults of middle-age in modern Western societies were raised, apparently successfully, on cows' milk. However, a small number of infants in these populations are intolerant of bovine lactose, although the proportion is much higher in populations in other parts of the world. Rather more important is the fact that human milk (and especially colostrum) provides better protection than does cows' milk against various bacterial infections which are likely to be encountered by the human infant [110].

Second, artificial bottle-feeding of infants greatly increases the chances of contamination of milk with potentially pathogenic micro-organisms in the environment. In order to avoid problems due to this cause, strict precautions are necessary in the preparation of the bottle. In some developing countries—in many African cities for example—artificial feeding of infants with reconstituted cows' milk made from milk powder has been promoted by large commercial companies based in the developed countries. Indeed, as a consequence of promotion campaigns launched by these companies, artificial feeding of infants has become a status factor in these populations. However, because the artificial feeding has been carried out without the necessary precautions of hygiene, which in any case are usually more difficult to apply in these regions, the practice has resulted in high mortality rates from infantile diarrhoea [111].

The third cause for concern relates to psychology. It is now widely accepted that the interaction and physical contact between mother and infant which occurs in breast-feeding contributes importantly to the establishment of satisfactory bonds between them. Artificial feeding clearly deprives both the infant and the mother of the benefit of this experience [112].

Diet—summary and comment

The history of dietary change in human populations over the millennia, and especially in the modern era, is interesting because it provides many clear examples of a number of important biosocial principles. One of the most important of these, commented upon many times in this book, is the principle of evodeviation and phylogenetic maladjustment. Slowly and painstakingly, nutritional science is demonstrating that there is no diet more conducive to human health and well-being than the diet to which our species is genetically adapted through evolution—that is, the varied diet of the typical hunter-gatherer, made up of a broad range of fresh vegetables, fruits, nuts, and roots, and some cooked lean meat. Deviations from this evolutionary norm have frequently been found to give rise to signs of maladjustment of one kind or another. If this simple principle had been widely understood in historical times, all the malnutrition over the ages could have been avoided, without any knowledge of the specific essential nutrients. Similarly, if it were thoroughly appreciated in modern Western society today and if dietary habits were based upon it, then the population would be a good deal healthier than it is.

Iatrogenic disease

While medical science has clearly contributed to the reduction in death rates and to the remarkable increase in life expectancy, it is also the cause of a good deal of maladjustment in contemporary populations. It has been reported, for instance, that such *iatrogenic disease* was responsible for 20 per cent of the admissions to a university teaching hospital in the USA, and that half of these patients were suffering from adverse reaction to drugs [113]. Other causes of iatrogenic disease include infection (said to be responsible for 10 per cent of iatrogenic deaths), diagnostic procedures, and surgery.

The problem of iatrogenic disease is a difficult one. In many cases, while a physician may appreciate that there is a risk of an adverse reaction to a certain treatment, he or she has to weigh this risk against the risk of the patient's condition worsening through lack of treatment. Nevertheless, there is much to be said for the view, with respect to medical science as a whole, that the cultural adaptive process has overshot the mark, and that a serious effort should be made to bring it properly under control [114].

SECTION 4: PSYCHOSOCIAL, BEHAVIOURAL, AND INTANGIBLE ASPECTS OF LIFE CONDITIONS

Introduction

It is self-evident that modern high-energy society, with its many kinds of machines and electronic devices, has produced far-reaching changes in the psychosocial environments of people and in their lifestyles. The first problem, when discussing or writing about this topic, is one of organization: the changes are so numerous, so different one from the other, and yet so interwoven and interrelated that any form of classification is bound to be unsatisfactory from one standpoint or another. While I am fully aware of some of the disadvantages of the arrangement which I have selected for the remainder of this chapter, it is the one that seems best to suit the theme and purposes of this book. Let us begin by briefly recalling a number of biosocial concepts of relevance to this theme.

Intangible aspects of life experience and phylogenetic maladjustment

The hypothesis was put forward in chapter 3 that the principle of evodeviation applies not only to material or physical environmental influences, but also to psychosocial, behavioural, and intangible aspects of life conditions. It was later suggested (chapter 4), on the basis of this hypothesis, that the following factors might be expected to be conducive to health and well-being: effective psychological support networks; co-operative small-group interaction; some variety in daily experience; life conditions which offer incentives and opportunity for creative behaviour and the exercise of learned manual skills; short goal-achievement cycles; an aesthetically pleasing environment; life conditions which promote a sense of personal involvement, a sense of purpose, a sense of belonging, a sense of responsibility, a sense of comradeship and love, a sense of challenge, a sense of interest, and a sense of confidence [115].

There is nothing particularly surprising about this list. All the items on it, although often expressed in different words, have been claimed by numerous authors, on the basis of intuition and personal experience, to be important for human well-being.

The melior–stressor concept

This concept suggests that the health and well-being of an individual is in part a function of his or her position, at a given time, on a hypothetical continuum between a state of distress at one extreme and a complete absence of distress at the other [116]. The position of the individual on this continuum is the function of two kinds of experience. On the one hand, there are psychosocial, behavioural, and intangible experiences which tend to push the individual towards a state of distress; they are referred to as *stressors*. On the other hand, there are experiences which have the opposite effect and which we call *meliors* [117]. Consequently, societal changes which result in the erosion of meliors may be regarded as being as undesirable as those which result in an increase in stressors.

A few words are necessary on the question of terminology. When Selye introduced the concept of *stress* in 1936 (see chapter 3), he deliberately chose the word 'stress' rather than 'distress' mainly for the sake of simplicity, but this term was not actually used in print until later (see Selye 1936 and 1950). In the present volume, however, the term *distress* is used in preference to stress, because it has a slightly different connotation which more accurately reflects the meaning intended here [118].

Although distress is certainly not peculiar to modern high-energy society, there is reason to suspect that the *nature* of distressing experience is different for many people than it has usually been in the past, in that it may go unrelieved for long periods and the reactions to it are often passive rather than active. It is well established that unrelieved chronic distress often leads to undesirable physiological conditions. It is considered, for example, to be an important contributor to coronary heart disease, and it probably contributes to a large number of other forms of ill health, including high blood pressure, strokes, and ulcers. Moreover, the state of mind associated with chronic distress is itself undesirable.

Much less attention has been paid in the biomedical literature to the role of experiences of the kind which are here referred to as *meliors*. Nevertheless I suggest that they are of immense importance both as a protection against the unwanted effects of stressors and as positive health-promoting experiences in their own right. In our biosocial study of Hong Kong, my colleagues and I were impressed by the fact that, although the level of environmental stressors appeared to be very high, most people seemed to be moderately healthy, both physically and mentally [119]. We believe this to be largely due to effective meliors associated with the lifestyle of the people in that city—involving experiences which tend to promote, for instance, a sense of personal involvement, a sense of belonging, and a reasonable degree of excitement and of variety in daily experience.

Clearly, both stressor and melior levels of individuals or of groups of individuals are much influenced by developments at the societal level. However, they are seldom taken into account in policy formulation and decision-making by governments; nor do they feature prominently on political platforms.

Variability

Much of the discussion in this section of this book will necessarily be somewhat generalized. We will note, for example, that the tendency in the high-energy phase of human society has been for people to lead increasingly sedentary lives. However, *some* people are as active physically, or even more active, than were any hunter-gatherers in the primeval phase. Thus a range of factors, such as occupation, socio-economic group and status, religion, and individual enthusiasms result in much experiential diversity.

One cause of variability with respect to behaviour is society's perception of the roles of the two sexes. We have noted that in hunter-gatherer societies adult males and females spend their time rather differently. This general statement also applies to the great majority of societies in ecological phases two and three, in which the economic arrangements at the family level resulted in a definite tendency for females to be more associated with the home and with the raising of young children, and for males to be more involved in economic activities away from the home, such as farming, manufacturing, and travelling, even though at times and in some societies females have certainly, on occasion, worked in the fields and in factories. During most of the high-energy phase this pattern has continued, and has possibly become more exaggerated: females have played the role of housewives, males the role of 'bread-winners'. It is apparent that attitudes towards the sex role are determined very early in life [120].

In recent years, however, a notable change has been taking place, with an increasing proportion of females joining the work-force and earning money. Nevertheless, differences in the sex roles still account for a considerable degree of the variability in life conditions among human populations in technological society. There is less variability among children who, in most Western countries today, are required to attend formal schooling at least until they are fourteen years old, and many of them continue at school for several years beyond this age.

It is important also to bear in mind the fact that there is considerable variability in the response of humans to any given set or aspect of life conditions. We have already noted that individuals may differ, for example, in their reactions to air pollutants in their environment. Such variability may be due to genetic differences or to differences in previous experience. The fact that an occasional heavy smoker lives to a ripe old age does not mean that smoking is a harmless pursuit for all the rest of the population; or the fact that a hermit survives many decades living alone in a remote cave does not mean that all other people could tolerate long spells of isolation without adverse effects.

Let us now consider a number of areas of life experience in which the emphasis is on behaviour, that is, on what people do and on how they spend their time. The discussion on these topics will be followed by comments on: consumer behaviour; learning experience; values and aspirations; other intangible aspects of life experience.

Physical exercise and resting and sleeping patterns

The lifestyle in the primeval or evolutionary setting involved a considerable degree of physical exercise for everyone, although there must certainly have been a good deal of variability from day to day, from place to place, and in some cases from season to season. But, in general, most people would have walked long distances quite frequently, and bouts of vigorous physical activity would have been common; on the other hand, there would also have been periods of complete relaxation. Broadly, deviations from this pattern in civilization have included 'quantitative' changes—resulting in people performing much more or much less physical work than that which was usual in the primeval habitat—and 'qualitative' changes, resulting in the use of muscles and limbs in different ways (e.g. blacksmiths and operators of word processors).

The most obvious change associated with the modern Western lifestyle and affecting the great majority of the adult population is the relatively sedentary existence and generally low levels of physical work. Evidence from biomedical science strongly suggests that this deviation in behaviour from the evolutionary pattern contributes importantly to cardiovascular disease [121]; and it clearly contributes to obesity with its attendant problems. There is also a growing body of evidence suggesting that physical fitness resulting from regular, and sometimes vigorous, exercise is associated with a general sense of well-being and enjoyment of life.

Societal adaptation to the undesirable consequences of low levels of physical exercise has included a growing appreciation among medical scientists of the health value of exercise and the communication of this appreciation to the public. In recent years this societal phase in the process has led to widespread *individual adaptation*, with large numbers of people engaging in regular physical activity for the sake of their health. On biomedical grounds, we can expect this new development to have a positive effect on the patterns of health and well-being in the high-energy societies. However, the principle of optimum range is relevant in this context (see page 37), and it is not unlikely that physical exercise at levels *greatly in excess* of those characteristic of the primeval lifestyle would tend to be detrimental to health and well-being. Another point to be borne in mind is the question whether vigorous physical activity devoid of an immediate purpose is physiologically the same as physical activity which is directed towards an immediate goal, such as catching prey, escaping predators, or, perhaps, competing in sporting events [122].

Resting and sleeping patterns are another area of human experience where definite deviations are observable from the situation that existed in the evolutionary setting. In hunter-gatherer society, individuals usually rest or sleep as a spontaneous response to the urge to do so, whenever that urge occurs. Consequently, the primeval sleeping pattern is usually polyphasic rather than monophasic—most sleep being taken in the hours of darkness, but with one or more short periods of sleep occurring during the daytime. In modern high-energy populations cultural norms interfere with this spontaneity in resting and sleeping behaviour, although it is noteworthy that adults on holiday, or after retirement, often revert to the hunter-gatherer pattern. Moreover, the literature is full of anecdotes about famous individuals, from Pliny the Elder to Napoleon and Winston Churchill, who have insisted on having periods of sleep during the course of the day, and who, because of their position in society, have been able to achieve this aim without difficulty. Indeed, sleep periods during the daytime may well have been common among peasants in pre-industrial times—the 16th century Flemish painter Pieter Breughel's sleeping harvester was probably not an exceptional being. Today, however, the great majority of adults, other than those in retirement or on holiday, do not sleep during the daylight hours, either because this behaviour is not permitted by socio-environmental constraints or because they have been conditioned to feel guilty if they do so.

Population density

One of the most conspicuous changes in the human condition that has accompanied civilization is the great increase in population densities, as observed especially in cities. As a consequence of this change, there has been a great increase in the number of social interactions that a single individual is likely to experience in a single day. From the biohistorical viewpoint, we may ask: How does the human organism react to life conditions which are so different in this way from those which prevailed in the evolutionary environment? This question, which has given rise to an extensive literature over the past couple of decades, is pertinent both to local planning in contemporary cities and to serious discussion about the patterns of societal organization and the economic arrangements which might characterize an ecologically sustainable and peacable society of the future. Broadly, some authors in affluent Western societies recognize the existence of two opposing schools of thought relating to population density. On the one hand, there are individuals who, perhaps partly on the basis of intuition, take the view that high population density is in some way undesirable and who, if their economic situation allows it, move out of cities into rural areas. On the other hand, there are others, particularly town planners, who see economic advantages in high population densities; and they express doubts about the assumption that high population densities have undesirable effects on humans [123].

Population densities are usually expressed in terms of the number of people per hectare (or per acre). The estimates presented in Table 10.1 give some idea of the degree of difference between overall population densities in primeval situations as compared with a range of different urban communities at different times in human history [124].

Table 10.1. *Population densities (number of people per hectare) in different situations*

Situation	Population density
Typical hunter-gatherer communities	0.5
Babylon, 430 BC	300
Rome, AD 100	500
Genoa, late Middle Ages	600
Typical European cities today	100–200
Calcutta today	300
Hong Kong, 1971	380
Wanchai (census district in Hong Kong), 1971	2288

The number of people per hectare is an important indicator of the overall situation, but it is not a satisfactory measure for all purposes. For instance, the population density within a hunter-gatherer camp when everybody is there at the same time, if extrapolated to density per hectare, would presumably be more like 100 than 0.5; and, within Hong Kong, one hectare of a housing estate, with its high-rise buildings, may hold a population of 12 200 [125], although the density at any particular floor level may be only a few hundred per hectare.

Let us return to the question: How does the human organism respond to these unnatural conditions? Interest in this subject has been expressed in several different academic disciplines including medical science, epidemiology, ethology, sociology, and social psychology, and the nature of the analysis of the problem has differed accordingly [126]. Here we can attempt no more than a brief summary of some of the most essential points.

First, there is the effect on host–parasite relationships—the host in this case being the human organism and the parasites being a wide range of potentially pathogenic organisms, especially microbial agents. This has been discussed earlier in this chapter.

Then there is the question whether high population densities have behavioural and psychological impacts. Considerable interest was sparked off some years ago by experiments which showed that animals of a number of different species, when exposed to unnaturally high population densities, exhibited various behavioural abnormalities. In a population of crowded rats, for instance, there was an unusually high incidence of aggression, rape, and homosexuality. Various physiological responses have also been observed in experimental animals living at unusually high densities, and these include enlarged adrenal glands, increased rates of abortion, and a weakened resistence to infectious disease [127]. These studies raise the important question: Do human beings respond in a similar way to high population density?

The concept of *personal space* is relevant to this question. According to this notion, there exists around each adult individual an area in which other individuals are not welcome. The personal space hypothesis has been based on observations of human behaviour in many different countries and situations and on the responses of individuals when they become close to each other. While the concept appears to be universally applicable, the actual dimension of this personal space differs very markedly from place to place and from time to time. Thus just how close an individual is likely to permit another individual to come depends partly on the nature of the relationship between them [128]. Indeed, the personal space may become non-existent in, for example, interactions between parents and children and between lovers. It is also very much affected by cultural norms and expectations. Certainly, there is no reason why the personal space requirements of individuals cannot, at least with appropriate cultural norms, be accommodated under the conditions of population density which prevail in typical modern cities.

Another concept of importance in relation to population density is that of *cognitive overload* (or sensory or psychic overload) [129]. There appears to be a preferred level of environmental stimulation in human beings, and either understimulation or overstimulation may produce adverse effects. It has been suggested that in urban conditions involving high population density individuals may be exposed to so many human interactions that the level of stimulation exceeds that which the central nervous system is capable of handling. Thus, through this mechanism, crowding may be seen as a potential source of distress. While distress from this cause might itself be seen as undesirable, some of the psychological mechanisms which can come into play to protect individuals from such distress may, for society as a whole, be deemed even more undesirable. Thus individual adaptation to cognitive overload may involve the 'screening' of stimuli, so that only a fraction of the stimuli received is actually registered by the central nervous system. This effect, together with the deliberate adoption of a lifestyle which reduces involvement with other people, can result, for example, in a reluctance to give assistance to others when they are in distress. The end result is a 'dehumanization' of society [130].

Many studies have been carried out aimed at showing whether there is a relationship between population density and various undesirable social phenomena, like crime and mental or physical illness. Such relationships have indeed been found to exist, but the situation is complicated by the fact that high population densities are usually associated with certain other aspects of life conditions, such as low socio-economic status and poor housing. In sum, there is at present no clear evidence that high population density itself gives rise either to aggressive or criminal behaviour or to physical or mental ill health [131].

Hong Kong has the highest population density of any city in the world; and yet to the visitor from overseas the people appear, on the surface at least, to be no less healthy, mentally and physically, than the residents of other far less crowded towns and cities. This impression was supported by the results of a large survey—the Biosocial Survey—which was carried out as a part of our study on the ecology of Hong Kong in 1974 by Sheila Millar and colleagues in the Chinese University of Hong Kong, and which provided information on the state of mental and physical health of the population [132]. Following this work Millar found herself in agreement with other authors who have emphasized the importance of *perception* as a mediator or determinant of the response of people to high population densities [133]. Indeed, it seems that whether or not individuals *feel* that there are too many people around them—that is, whether they *feel crowded*—is likely to be by far the most important determinant of their psychological response to high population density. In other words, maladjustment is more likely to be associated with what has been termed *affective density* than with actual physical density [134]. Affective density is influenced by a variety of factors. These range from *immediate influences*, such as what individuals wish to do in their environment, whether or not they are tired, and whether the other people around interfere with their access to certain facilities or amenities, to influences related to their *previous experience*, such as the degree of physical density to which they became accustomed in childhood and local cultural attitudes to crowding. It is not impossible that genetic factors may also affect perception of the situation.

We do not suggest that a state of high affective density necessarily, and under all circumstances, gives rise to a state of maladjustment in the individual. In fact, we postulate the existence of a series of possible intermediate states between high affective density experience and any consequent maladjustment. In certain circumstances, as for instance in the primeval environment, a feeling of being crowded may lead to a simple behavioural response which will immediately resolve the problem: it may prompt the individual to move away from other people, thereby bringing to an end the state of affective density, which will then have served its behavioural purpose and will have led to no harm. If, however, the inclination to move to a lower physical density situation is thwarted, as must usually be the case in Hong Kong, then the high affective density state exists and a state of frustration may develop. If no effective form of coping behaviour comes into play and the state of frustration persists, it is likely to lead to a state of chronic distress which, in turn, may give rise to undesirable psychological symptoms.

Accordingly, we come to the view that cultural factors have an important influence on the response of human beings to high population density. Certainly, the results of Millar's work in Hong Kong are entirely consistent with the suggestion that the traditional Chinese way of life is in some way conducive to the development of tolerance to high density conditions. She found that people who had lived all of their lives in Hong Kong were less likely to be tolerant than those whose childhood had been spent in the adjoining province of Kwang Tung, while people educated in the Western tradition were likely to be even less tolerant. A strong positive relationship was found between density tolerance and age, and this fact is also consistent with the idea that the main explanation of differences in response to high population density lies in socialization and learning experience, and that the people who have had the greatest exposure to traditional Chinese culture are the more likely to be tolerant to high density living.

If it is correct that a major determinant of the human response to high density is the individual's perception of the situation, and if this perception is to a large extent a function of the individual's socialization experience, then possibilities can be envisaged for deliberate societal adaptation to high density living, in the form of programmes of socialization especially designed to render people tolerant of such conditions. This possibility, however, raises some difficult problems. It is true that, from the point of view of individuals living in Hong Kong, a kind of socialization which leads to an attitude of tolerance is likely to be beneficial, allowing them to endure the persisting conditions of high population density with a minimal degree of distress. In addition, such tolerance might be beneficial to society as a whole, by virtue of its tendency to allay the development of discontent which might foster antisocial and criminal behaviour. On the other hand, the capacity to tolerate conditions of extraordinarily high population density can be seen as an example of the kind of adaptation which we refer to as *habituation*, and which Dubos had in mind when he wrote: 'The frightful threat posed by [human] adaptability ... is that it implies so often a passive acceptance of conditions which really are not desirable for mankind' [135]. Thus the fact that people are, through habituation, tolerating the situation in Hong Kong with remarkable equanimity does not mean that they would not enjoy life a great deal more if they had more room.

There is one other important form of adaptation which comes into play if all others fail: medicines may be used to counter the effects of environmentally induced distress. In the Biosocial Survey conducted in Hong Kong, 20 per cent of the respondents reported that they regularly took medicine, and estimates suggest that the rate of consumption of various psychotropic drugs, such as tranquillisers, antidepressants, and sedatives, is increasing. Needless to say, this coping procedure has all the usual disadvantages of antidotal adaptation, including the fact that the widespread use of drugs to offset the undesirable effects of unsatisfactory conditions effectively blocks the initiation of societal moves aimed at improving the conditions, which are thus allowed to persist or to deteriorate further.

Finally, before leaving the topic of crowding, mention must be made again of the role of meliors in protecting individuals from the undesirable consequences of environmental stressors. My colleagues and I are personally convinced that melioric experiences—resulting, for example, in a sense of personal involvement, of responsibility, of belonging, and of comradeship—make an important contribution to the ability of people in Hong Kong to cope with the environmental stressors associated with high population density. Consequently, any societal or economic trends in that city which tend to erode these intangible meliors are likely to be detrimental to the health and well-being of the population.

The use of psychotropic drugs for pleasure

The availability to human beings of drugs which produce a pleasurable alteration of mood is not new in history. We can only speculate, of course, about the possible use of 'natural drugs' in the primeval or evolutionary setting; but it seems reasonable to assume that the psychological effects of some of the drug-containing plants would have been discovered and that they would have been used on occasion for much the same purposes as they are today. However, since the natural drugs are found in different ecological niches scattered around the world, only a very few, if any, would have been available in the natural environment of any particular hunter-gatherer society. Alcohol is an exception to this statement, since it could have been produced, through the fermentation of plant material, in any place. In primeval society, however, its consumption is likely to have been limited to the occasional meal of fermented fruit. It is known that some animals in their natural habitats, such as elephants and monkeys, occasionally come under the influence of alcohol by this means.

A further point relating to the primeval situation is the fact that socially undesirable behaviour resulting from the consumption of psychotropically active substances would have been unlikely because of the close-knit nature of hunter-gatherer groups and the powerful mechanisms of social control which tended to prevent behaviour perceived to be inappropriate.

The modern situation with respect to psychotropic drugs differs from the primeval one in several basic ways. First, the range of different agents which are available, legally or illegally, is much greater. In typical high-energy societies today, for example, the following drugs are all used to a greater or lesser degree: alcohol, tobacco, marijuana, amphetamines, hallucinogens such as LSD, cocaine, heroin, and other derivatives of opium (to say nothing of coffee and tea). Another basic difference is the ready accessibility of at least some of these agents—alcohol, tobacco, and marijuana in particular [136].

Alcohol

Once our ancestors had adopted the farming way of life, they lost little time in learning to brew alcoholic beverages. There is evidence, for instance, that beer was in use by 6400 BC [137]; and by the third or fourth millennia BC people were cultivating the vine—'the finest of all sources of alcoholic enthusiasm' [138]. Possibly, as Hawkes suggests, they were attempting to compensate themselves for the monotony of the peasant life. She writes:

... since the time he became a cultivator man has been ready to lavish time, labour, patience, ingenuity, and sensibility on the business of making himself more or less drunk, showing a divine discontent in his determination to alter the state of consciousness to which evolution has brought him. He at once longs to change it, just as he changes his face with paint and strange ornaments. From the Rhine to the Mediterranean, from the Caucasus to Palestine untold lifetimes of work, untold acres of soil have been willingly devoted to the vine. Now the patchwork of vineyards has spread throughout the world. In the Orient when populations are near starvation rice is still fermented and distilled, and in the Americas the spikey plants of the Mezcal cactus stand in vast vegetable armies, their straight lines rising and falling over hills where soil is precious, waiting for the Indians to cut out their hearts and collect the gathering liquid for pulque and tequila. ... . Grapes, mezcal, potatoes, barley, rice, sugar, apples, plums, juniper—it seems half the vegetable kingdom has been forced to pay tribute in alcohol (see note 138).

Alcohol is usually consumed simply for its enjoyment value, although it is also used in many religious and other rituals. It may also be consumed for practical purposes. In the fifth century BC Herodotus reported that the Persians of his day made deliberate use of alcohol as an aid to decision-making [139]. If an important decision had to be made, the matter was discussed when the members of the decision-making group were drunk and a decision was reached; and the following day the decision was submitted for reconsideration when they were sober. If they still approved it, it was adopted; if not, it was abandoned. Conversely, any decision they made when they were sober was reconsidered afterwards when they were drunk. The *Holy Bible* contains numerous references to wine and to drunkenness.

Alcohol in modern high-energy societies comes in a range of different drinks prepared from the fermentation of various fruits and grains. The alcohol content of some of these drinks is as follows: beer, 4-6 per cent; ale, 6-8 per cent; wine, 11-14 per cent; fortified wine, 19-21 per cent; whiskey, brandy, rum, vodka, and gin, 45-50 per cent; liqueurs, 50-85 per cent.

There is considerable variation among Western countries in the per capita consumption of alcohol, with a four-fold difference between the highest and the lowest rates [140]. In Britain, the USA, and Australia the per capita alcohol consumption has been increasing over recent decades. In the USA it rose by 25 per cent in the 1960s and by 8 per cent in the 1970s [141]. This change has been mainly due to an overall increase in the proportion of the population who imbibe alcohol, but especially to an increase in the number of young people who drink. In 1979, 61 per cent of the population of the USA over the age of twelve used alcohol [142]. In Australia in 1980 the proportion was 70-80 per cent of women and 80-90 per cent of men [143].

The effects of alcohol consumption on the individual human organism are well known. On the positive side, it gives rise to an enjoyable state of mind and is a facilitator of social interaction. Indeed, in many situations the consumption of alcohol can be viewed as an adaptive response of individuals to the requirement by modern society that they interact a great deal with relative strangers. Under the influence of alcohol, natural reserve and even suspicion often give way quickly to an atmosphere of relaxed conviviality. Also on the positive side, if the statistics can be believed, is the fact that people who fairly regularly drink a small amount of alcohol each day have slightly less chance of a heart attack than do non-drinkers, and they have slightly longer life expectancies.

On the negative side, short-term ill-effects of larger quantities of alcohol include lack of co-ordination, blurred vision, interference with judgement, and sometimes, but not inevitably, aggressive behaviour. In the modern setting these changes are especially undesirable when the drinker is in the driving seat of a motor vehicle. In the USA between 35 and 64 per cent of drivers in fatal accidents have been drinking alcohol before the accident [144]. It is estimated that in Australia about 50 per cent of serious road accidents can be blamed on alcohol. It is also estimated that 50 to 68 per cent of drowning victims in the USA had been drinking alcohol and that about 26 per cent of deaths of adults from fire involve alcohol.

Alcoholism—that is, dependency on alcohol—is associated with a range of undesirable consequences. These include social factors, such as loss of jobs and disruption of families, mental changes, such as loss of short-term memory, and various physiological disorders, one of the most important being cirrhosis of the liver.

In Australia, alcohol-related deaths in 1980 accounted for about 3.3 per cent of the mortality in that year, representing an increase of 20 per cent since 1969 [145].

Tobacco

Tobacco smoking does not have as long a history in the Western world as does the drinking of alcohol. It was introduced into Europe around 1550, and by the early sixteenth century its distribution was worldwide, with only a few societies resisting its introduction. Although tobacco could be taken in a number of ways—for example, it could be chewed or swallowed as a tincture—it was smoking that eventually became the main means of administration. In the USA today around 36 per cent of the population over the age of twelve regularly smoke. In Australia, about 40 per cent of men and 31 per cent of women are smokers; and in 1977 2800 million cigarettes were smoked every month, which is equivalent to about six or seven cigarettes per capita of the whole population each day [146].

With regard to the effects of this behaviour, on the positive side smokers experience a sense of well-being and some relaxation of tensions, associated with the absorption of nicotine into the blood. Also important, at least in some individuals, is a sense of gratification resulting from manipulating the cigarette with the hands, holding it in the mouth, and sucking it. On the negative side, tobacco smoking easily becomes addictive; severe withdrawal symptoms occur when a heavy smoker tries to, or is forced to, give up. The constant inhalation of smoke from the burning tobacco leaf into the lungs also frequently gives rise to phylogenetic maladjustment in the form of lung cancer and emphysema, as well as disease in other parts of the body, such as heart attack, cancer of the bladder, and gangrene of the limbs. The contribution that cigarette smoking makes to the death rate is greater than that of alcohol. In Australia, for example, it is estimated that 80 per cent of all drug-related deaths in 1980 were due to tobacco; this is 14.9 per cent of all deaths [147].

Societal adaptation has mainly taken the form of educational campaigns about the undesirable consequences of smoking tobacco, and in some countries there have been limited restrictions on the advertising of cigarettes. As would be anticipated, the situation has been characterized by powerful counter-attacks by vested interests—in this case the tobacco companies. Advertising is geared to equate smoking not only with desirable status situations and sexual attractiveness, but also with a healthy outdoors lifestyle. Another aspect of the counter-attack has involved the distribution of pamphlets intended to throw doubt on the scientific evidence linking smoking with cancer.

Unlike the situation with regard to alcohol and marijuana, there has been no attempt to prohibit the sale of tobacco by law, although in many countries there is heavy tax on tobacco products and this may be seen as an attempt to discourage smoking. Basically, the choice to smoke or not to smoke lies with the individual.

Marijuana

Marijuana is one of the names applied to the plant *Cannabis sativa*. It is a native of southern Africa and has been used medicinally in China and India since at least several centuries BC. It did not come to be used widely in most developed countries until the last few decades. It has been estimated that at present somewhere between 2.4 and 9.6 per cent of the population in Australia use marijuana [148]. The proportion of people over the age of twelve in the USA who smoke marijuana is estimated to be around 13 per cent, and about 6 per cent of high-school students smoke it on a daily basis [149].

The effects of smoking marijuana range from dizziness, light-headedness, and a dry mouth to subjective feelings of euphoria, time and space dilation, extreme hunger (especially for sweet things), and a general feeling of well-being. Marijuana is almost invariably smoked in social situations, and the group of smokers tends to become close-knit, with bonds forming between the group members. The symbolic role of the drug is said to be often more important than its physiological and mental effects.

The undesirable consequences of smoking marijuana in the short term seem to be few. There is some interference with co-ordination, but this is no real problem unless the individual attempts to drive a motor vehicle. There is some uncertainty about long-term consequences, but as yet there is no definite evidence of harmful effects. Nevertheless, from the biological viewpoint, it would seem unlikely that the frequent inhalation of the smoke of this plant would have no effect on the lungs, nor that repeated modification, through the action of the drug, of the activity of cerebral neurons would be without long-term consequences for mental processes.

Opium derivatives

Opium is the milky exudate of the incised, unripe seed-pods of a poppy, *Papaver somniferum*. The exudate is dried in air to form a brownish, gummy substance. Opium was probably used medicinally and in religious ceremonies in western Asia and the eastern Mediterannean area before 2000 BC, and it is possible that it was in use in the Mesopotamian cities by 3000 BC [150]. Theophrastus (1961, pp. 269–81), writing in the third century BC, made a definite reference to the drug. It was carried by Arab traders to India and China, where it eventually became well established, initially mainly for pharmacological use but later as a source of pleasure.

The three main pharmacologically active derivatives of opium are morphine, heroin, and codeine. Morphine has been used since ancient times as an analgesic for relieving pain and as a sedative. It is used less by drug addicts than is heroin, which is two to three times more potent as an analgesic, although it produces a similar euphoria. Heroin, however, is about half as bulky as is morphine and is therefore easier to smuggle and handle. It is used in solution and is injected intramuscularly or intravenously. It has become the classic 'hard' drug of addiction in the Western world.

It is estimated that in the USA about 1.4 per cent of the members of the population over the age of twelve have used heroin [151].

Heroin produces a state of euphoria and a relaxed detached attitude towards the environment. Other effects include analgesia, drowsiness, some impairment of mental and physical performance (especially of complex learning behaviour), reduced sex and hunger drives, mental clouding, sometimes some excitation, inability to concentrate, and apathy. It subdues all urges, and does not promote aggressiveness. The opium or heroin addict ceases to become a participating member of society, loses weight, and becomes generally feeble and lives in a dream world. From the standpoint of the rest of society, heroin addiction has two main undesirable consequences. First, individuals cease to contribute economically, in money or in action, to their families, and this is frequently a cause of great distress. Second, affected individuals are usually not earning a regular income and, because the cost of heroin is very high, they often resort to crime as the only means of acquiring the money with which to purchase the drug.

Societal adaptation aimed at overcoming problems associated with addiction to heroin and other opium derivatives involves legislation prohibiting the possession and sale of these substances.

Comment

It is apparent that considerable inconsistencies exist in the societal responses to the undesirable consequences of the use of different psychotropic drugs. In biosocial perspective, it is clear that the ready availability of these substances is a deviation from the conditions of life under which the species evolved. If the drugs were not there, then the various problems for individuals and for society associated with them would not exist (on the other hand, it can be argued, nor would people be able to enjoy the perceived benefits of their use). Thus legislation aimed at removing the drugs from the environment is an attempt at corrective societal adaptation, in that it aims to restore the natural conditions; but experience has shown that this corrective approach, however rational it may seem to be, does not work. Nor, however, do antidotal measures involving punishment for drunkenness or for using marijuana or heroin. Ultimately it is individuals who decide whether they drink alcohol, smoke tobacco or marijuana, or inject themselves with heroin. This decision itself depends to some extent, no doubt, on the inborn genetic personality characteristics, although environmental factors certainly play a major role as well.

The complexity of the drug problem became particularly apparent to my colleagues and myself during our work in Hong Kong. Our interpretation of the situation with respect to opium and heroin use in that city is summarized diagrammatically in Fig. 10.5.

The pattern of drug use in a society is thus the function of a complex set of interacting variables, and societal adaptive responses to the drug problem is unlikely to be successful unless a truly comprehensive approach is taken to the analysis of the problem.

Psychosocial aspects of life experience

This section is concerned with the ways in which culturally induced societal changes have modified those aspects of the immediate environments and behaviour patterns of individuals that involve interaction with other people, and the implications of these changes for health and well-being.

We will begin by considering a number of concepts which are particularly relevant to this theme under the following headings: *social support networks; alienation; anomie; biosocial perspectives.*

Social support networks

Much has been written to the effect that close supportive relationships with one or more other individuals are conducive to health and well-being [152]. This fact is generally well accepted among social psychologists. However, two points must be made. First, some recent work in Australia has suggested the possibility that the critical factor may not be so much the actual amount of interaction with potentially supportive friends and relatives, but rather whether individuals at risk *perceive* that their support network is adequate under conditions of adversity [153]. If these findings are confirmed, they may mean that the actual determinant of health and well-being in this context is basically whether or not people feel themselves to be deprived of support. Second, the emphasis in the literature is on the health-promoting value of *receiving* psychological support in difficult times. It is suggested here that at least as important is the opportunity to *give* support to other individuals in trouble. The giving of support is an important source of such feelings as a sense of personal involvement, a sense of self-respect, a sense of responsibility, and a sense of belonging. As such, and if not in excess, support-giving can be expected to contribute to an individual's health and well-being.

Alienation

An extensive literature has developed in the social sciences, especially during the past 20 years, dealing with the concepts of *alienation* and of

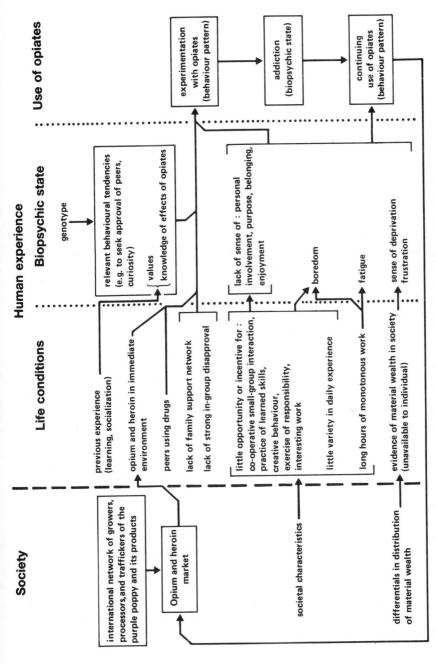

Fig. 10.5 Behavioural aspects of human experience with respect to the use of opium and its derivatives.

anomie, and much of it is both confused and confusing [154]. Each of these terms has been used in a variety of different ways, and when some authors write about what they call 'alienation', they are in fact writing about what others call 'anomie', and vice versa. With respect to the differences between alienation and anomie, some sociologists have suggested that the concepts are so similar that attempts to distinguish between them should be abandoned [155]. Indeed some *have* abandoned these distinctions and give the two terms the same definition [156]. Others regard them as being quite distinct, even as almost contradictory concepts [157].

In my view, alienation and anomie, at least as discussed respectively by the young Karl Marx and by Emile Durkheim, are sufficiently different to be considered as two different aspects of human experience. Marx's initial interest in alienation derived from his deep concern about the impacts of industrialization on the life experience of ordinary people (see Etzioni, (ed.) 1971). He recognized different kinds of alienation, such as alienation from the products of labour, from the work process, and from nature, and alienation of the individual from himself or from other people. With time, the word *alienation* has come to be used more and more loosely, and yet usually as if the concept were a well-defined, discrete, specific aspect of human experience. In fact, 'alienation of man from his true nature' is surely very different from 'alienation from the products of labour', which in turn is different from 'alienation from the political process'. Not only are these different sorts of alienation different kinds of experience, possibly with differnt kinds of consequences, but they are likely to be the result of different societal conditions.

The confusion arising from these diverse concepts of alienation has, some believe, robbed the original concept of a good deal of its usefulness. A recent author has written: '... the extension of the term to explain almost everything ... leaves us with a concept that is dangerously rhetorical and emotive, one that is synonymous with a general and often vague feeling that "something is wrong somewhere" ' [158].

We even find confusion and disagreement with regard to whether alienation is a good or a bad thing. It is true that the great majority of authors have assumed it to be, even if inevitable, undesirable—interfering with the enjoyment of life of individuals and also contributing to unwanted societal phenomena such as crime and vandalism. But others have suggested that we should view it as desirable, or at least appreciate that it may sometimes have positive consequences and that, rather than being destructive, it may be creative, and therefore good. It has been pointed out that many well known philosophers, poets, artists, and religious leaders appear to have been alienated in one way or another. Referring to the *Holy Bible*, Kaufmann writes:

What we do find is a succession of imposing figures who not only keep telling people that they should be different, but who themselves *are* different—and thoroughly alienated from their own society. Moses, Elijah, Amos, Hosea and Jeremiah are outstanding examples [159].

Kaufmann argues 'that alienation has always been the price of autonomy' (1973, p. 165). He also points out that many famous and 'success' characters in fiction, Oedipus and Hamlet to mention but two, are deeply alienated, indicating a particular fascination among people in the alienated individual. However, the suggestion that these supposedly positive aspects of alienation should be seen as a vindication for life conditions which tend to induce feelings of alienation in the majority of the population is surely debatable.

The confusion which reigns in the literature on the subject of alienation and anomie is due, in large part, to the lack of a conceptual framework in sociology which distinguishes between the different levels and dimensions of human situations. Thus both 'alienation' and 'anomie' have been variously described as forms of behaviour in individuals, or as a feeling or state of mind in individuals—that is, as aspects of the *behaviour pattern* or of the *biopsychic state* of individuals, and, on the other hand, as a condition of *society*. The biohistorical conceptual framework used in this book has something to offer towards the clarification of this problem. In this discussion alienation and anomie will be treated as distinct, although certainly related, concepts, and, in keeping with the original notions of Hegel, Marx, and Durkheim, both terms will be defined as intangible aspects of the biopsychic state of individuals—that is, as states of mind.

For our purposes, *alienation*, is defined as an aspect of the biopsychic state of an individual characterized by a sense of separation or isolation from something in the environment, involving also a sense of estrangement or helplessness. It is associated with a lack of sense of personal involvement and of purpose, a lack of sense of responsibility, of comradeship, of challenge, and of interest; and it is associated with a sense of apathy and of boredom. As a consequence, the individual may develop psychosomatic symptoms or neurosis.

It is clear from this definition that the term always requires further qualification: *alienation from what?* For the purposes of the present discussion, let us restrict our attention to alienation from the products of labour and the processes of production. The behavioural adaptive responses to such alienation may involve: satisfying leisure activities to compensate for an unsatisfactory work situation; the consumption of psychotropic drugs; withdrawal; vandalism; or participation in various protest activities, including strikes. These behavioural responses in turn affect society in various ways. Thus legislation may be introduced to control vandalism and the use of psychotropic drugs, and economic palliatives may be offered in an attempt to compensate for the unsatisfactory aspects of work situations. These various interrelationships are summarized in Fig. 10.6.

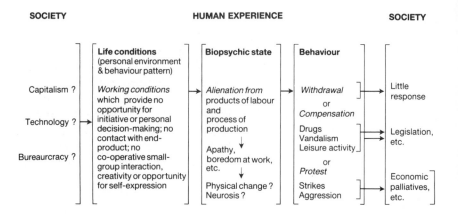

Fig. 10.6 Alienation from products of labour and the process of production—some important interrelationships.

With respect to the causes of alienation at the level of life conditions of individuals, the situation seems fairly clear. Jobs which require no exercise of responsibility or initiative, which are monotonous, which do not involve effective creative interaction with members of a small group, which offer no opportunity for personal decision-making, and which offer no contact with the end-product of the process lead naturally to a sense of separation or isolation both from the end-product and from the whole process of its manufacture.

The 'cause' at the level of society is more debatable. Is it, as Marx believed, all a product of capitalism? Or is it, as others have suggested, rather a product of technological advance or of the growth of bureaucracy? In this connection, it is pertinent to ask whether alienation of this sort is less a feature of human existence in socialist and communist nations.

While alienation of the kind we are discussing may well be a particular feature of modern high-energy society, it would be unreasonable to suggest that feelings of being isolated and separated from important happenings is a new phenomenon in human experience. It has occurred to a greater or lesser extent, at least in some members of communities, throughout recorded history. Nevertheless, if we go back to the primeval or natural habitat, alienation from, for example, the products of labour or the political process would have been inconceivable, and such alienation is also unlikely to have been common in early farming societies. Thus, alienation of the kinds which have been seen as undesirable by so many authors is clearly a product of civilization.

Anomie

The notion of *anomie* was introduced by Emile Durkheim who, like Karl Marx, was concerned about the undesirable effects, as he perceived them, that industrialization and urbanization were having on human beings. Durkheim's concept was based on the observation that traditional norms were breaking down and that a new situation was developing, characterized by variability of value systems among different societal groupings. This change was leading to a sense of confusion and rootlessness in individuals, and in turn to behaviour reflecting the tensions and stresses arising from the efforts of individuals to meet the obligations of two or more irreconcilable norms [160]. Merton further developed this concept, especially in relation to moral issues connected with making money and was concerned about the 'demoralization' of the means of achieving success [161].

The word 'anomie' (or anomy or anomia) has been variously used to describe: (1) the *societal state* characterized by loss of an overriding tradition and consequent variability in norms, (2) the undesirable *state of mind* (i.e. an aspect of the biopsychic state) of individuals affected by such a situation, and (3) the *behaviour* of such individuals.

Here our definition of anomie is restricted to the biopsychic state: it is the sense of confusion, rootlessness, and consequent distress felt by an individual who experiences personal conflict in a social environment in which relevant norms are changing, are unclear, or are different in the various in-groups or reference groups to which he or she belongs. While anomie as so defined is obviously not an entirely new phenomenon in human experience, it is reasonable to accept the view of Durkheim and others that it is especially prevalent in modern high-energy society, and as such it is a biopsychic consequence of cultural developments and societal change arising out of industrialization and technological progress.

Although the behaviour of individuals who experience anomie is often deviant and perceived by most people to be societally undesirable, it has been claimed that in some cases anomie may give rise to 'extraordinary accomplishments in the arts, sciences and other areas of culture' [162].

Family and community experience

Homo sapiens is a social animal. That is to say, people tend to organize themselves into groups of interacting and mutually supportive individuals. The advent of modern high-energy society has greatly modified the nature of these groups—a fact which, in turn, has had significant repercussions on the quality of the social experience of individuals.

Focusing first on the quality of family experience, it is still true that the reproductive unit, consisting of an adult male and an adult female and their offspring, is the most fundamental and all-pervasive social and economic unit in society. This statement is true despite the fact that in Western society fewer of the male–female unions are permanent than was the case fifty years ago, and despite the increasing number of single-parent families. An important trend has been a significant decline in the latter part of the high-energy phase of human history in the number of young born to the average adult female. From the standpoint of human experience, this development means that children have fewer siblings around them and that adults have fewer children to rear.

Another important change affecting human experience is a pronounced weakening, as compared with earlier situations in the Western world and with most non-technological societies today, of the bonds between members of families beyond the unit of the nuclear family. This change is not easy to quantify and is not particularly well documented, but few would doubt its reality. There is certainly much less interaction and interdependence among members of the extended family than was the case in previous generations. Many factors contribute to this change, one of the most important being an increase in geographic mobility, due primarily to economic forces, so that many more people than previously take up residence far away from their place of birth. Such direct face-to-face interaction as does occur between members of extended families (i.e. members of the family beyond the nuclear unit) is usually costly in terms of energy, involving travel, often over long distances, in vehicles powered by fossil fuels.

With respect to community experience, we must appreciate that the concept of 'community' is one which has interested social scientists a great deal. By 1955 at least 94 definitions of community had been proposed [163]. For our purposes, it is sufficient to say that we see a community as an aggregate of human beings characterized by relatively informal relationships between individuals who know each other moderately well and who have certain shared interests, concerns, and responsibilities. A family, of course, can be regarded as a community, but the term will be reserved here for interactions and interrelationships beyond the immediate family, as seen, for example, in the typical village-type situation [164].

In the present chapter we are interested in attempting to identify trends in community experience particularly characteristic of modern high-energy society, especially as they affect the quality of life of individual human beings. We noted in chapter 9 how concern about the impact of industrialization on community structure goes back at least to the last part of the nineteenth century when Tönnies emphasized the distinction between what he called *Gemeinschaft* and *Gesellschaft*. Gemeinschaft is seen as a societal arrangement characterized by 'strongly personal, communal, traditional and stationary social ties' [165]. Gesellschaft refers to social situations in which the majority of relationships are much less personal, and where the 'bonds' between people are based on economic, political, or professional transactions rather than on kinship and comradeship, and where there are strong tendencies toward individualism and secularism.

The modern Gesellschaft situation is, therefore, very different in terms of human experience from the Gemeinschaft situations of the past, in that the majority of social contacts are between individuals who are not known well to each other. As George Simmel and others have pointed out, an emotionally 'meaningful' relationship with each of these contacts would lead to a state of total psychic exhaustion [166].

Many social scientists after Tönnies, and notably Durkheim, have noted the general changes in community experience, and while many of them have considered the changes to be undesirable, they have also seen them as inevitable. More positive in their approach have been those authors who are interested in town planning who have taken the view that modifications in the physical design of the urban environment can help to restore this loss of 'community spirit' and lack of neighbourliness which are so characteristic of affluent societies today [167]. Others, however, have suggested that the physical environment is relatively unimportant, and that the main determinants of the degree and kind of community interaction are cultural [168].

Certainly, values and attitudes in society can affect the degree and kind of interaction at the level of both the extended family and the community. In some residential areas in Britain, 'good neighbours' are those who 'keep themselves to themselves', while in other parts they are neighbours with whom one has an easy-going, informal, reciprocally supportive relationship involving a good deal of spontaneous interaction.

The demise of the extended family and the local community has contributed to another phenomenon characteristic of modern affluent society—the suburban housewife. Trapped in an environment in which very little happens of interest value for hours on end, she experiences levels of environmental stimulation which are unnaturally low and where the receiving and giving of psychological support are absent. Her experience is characterized by sheer boredom and lack of sense of personal involvement and of belonging. These states may in turn, lead to mental depression or to various signs of behavioural maladjustment including, for example, over-eating and heavy smoking.

In summary, it is apparent that the changes in family structure and in community experience associated with the high-energy phase of human history have interfered with some of the important psychosocial health needs of human beings. Thus the shrinking of the extended family and the disintegration of the community have resulted in: less effective psychological support networks and less likelihood of giving or receiving emotional support; less responsibility for local affairs; fewer opportunities for co-operative small-group interaction; and fewer activities in general which give rise to a sense of personal involvement, belonging, comradeship, and responsibility. These changes may then contribute to a sense of isolation or alienation. Moreover, individuals, in searching for group membership— that is, in seeking a sense of belonging and responsibility—are often faced with conflicts of values and loyalties. This in turn frequently leads to a state of anomie and other undesirable consequences.

If we accept that cultural developments have brought about some disadvantageous changes in family and community experience, it is pertinent to enquire into the nature of the underlying causes of these changes at the societal level. Basically, there seem to be several interlocking factors. First, there is the rise of bureaucracy, the process of centralization, and the increasing scale of human organizations and activities. This set of factors has resulted, for example, in further spatial separation of the work-place from the home. Moreover, the economic system, the occupational structure of society, and the employment situation are such that many people, in order to have paid employment, are forced to move every so often from one place of residence to another. An additional factor, very much a feature of the high-energy societies, is the emphasis on *socio-economic mobility*. A promotion, a step up the conventional ladder of success, associated with increased status and more money, often means that the whole family must tear up its roots and migrate half way across the country, if not across the world—away from relatives and from the local community. Thus while there may be a gain in status and income there is a loss of sense of belonging and of personal involvement in family or community networks.

Work experience

The very concept of *work* is a product of human culture. It did not exist in primeval society, where no sharp division existed between behaviour which contributed to subsistence and that which did not. Nowadays, work involves the setting aside of seven or more hours each day, for five or six days a week, during which individuals engage in specific activities which contribute in some way to the satisfaction of the needs and wants of others in the community. In return for this behaviour they receive remuneration with which they try to satisfy their own needs and wants.

The amount of time devoted to 'work' in western countries possibly reached its peak at the beginning of the technological era. In Britain in the mid-nineteenth century at the height of the industrial transition, the average worker's week was around 72 hours, compared with 35 to 40 hours at the present time. The trend, however, has been in the opposite direction in some kinds of occupation. For example, in western Europe in 1750, merchants worked for 39 to 51 hours a week, while in 1850 the average was 50 or more hours a week. In 1972, top executives in large corporations in West Germany worked for almost 60 hours a week [169].

When considering the implications for human health and well-being of changes in work patterns associated with modernization, it is necessary to keep in mind the extraordinary variability that exists with respect to work experience. Once again, we shall have to be content with very broad statements, appreciating that exceptions exist to all generalizations.

The continuing trend towards greater occupational specialization has meant not only that different individuals experience very different working conditions, but also that many of them have monotonous tasks, often representing only an infinitesimal part of the whole process of production, marketing, or administration. For over two hundred years, authors have expressed concern about this trend. In 1776, Adam Smith wrote:

The man whose whole life is spent in performing a few simple operations ... has no occasion to exert his understanding, or to exercise his invention in finding out expedients for removing difficulties which never occur. He naturally loses, therefore, the habit of such exertion, and generally becomes as stupid and ignorant as it is possible for a human creature to become [170].

Most authors, while agreeing that the trend towards specialism often has undesirable consequences for the worker, see the change as inevitable and as a necessary cost of increased efficiency, which often seems to be judged more important than worker well-being. Marx saw the trend as undesirable, but in his earlier writings he expressed the belief that communism would solve the problem. Later, however, in the words of Holland:

... he distinguished the sphere of production which would remain in 'the realm of necessity' and the leisure time which would be 'the true realm of freedom' in which 'the development of human potentiality for its own sake' could take place. Nevertheless he never abandoned the idea that necessary work could itself become to some degree a liberating and educative experience [171].

Another individual who expressed strong views on this subject in the second part of the last century was the author, poet, designer, and political thinker William Morris, who was concerned about the impacts of industrialization and machines on the work experience of the individual worker. Unlike some others, however, he believed that steps should be taken to halt and reverse the undesirable trends in industry, restoring to the worker opportunity for initiative and creative behaviour [172]. If subsequent literature is any guide, few people have disagreed with Morris and many have supported his call for the 'rehumanization' of the processes of industrial productivity. Nevertheless, his views and those of his supporters seem not to have had any impact on the apparently inexorable processes of society.

More recently, a vast literature has grown up on aspects of the changing conditions of work in the high-energy societies. Broadly, we can recognize two distinct, although often confused, areas of concern. First, there are those authors who take the view that production should be based on *industrial democracy*—that is, on a system in which workers, through their elected representatives, participate in management and in the decision-making process. It is assumed that this would result in the work experience being more meaningful and satisfying for the individual worker, perhaps resulting in a greater sense of personal involvement. Whether it would, in fact, have this effect for the majority of workers is questionable. However, even if industrial democracy does not greatly affect the actual working experience of individual workers, it may nevertheless be desirable for other reasons.

The other approach focuses on what workers actually *do*—on how they spend their time. It is concerned with whether the conditions in the work-place are conducive to physical health and the extent to which the work represents an enjoyable and self-fulfilling experience for the individual. Considerable progress has been made in the developed nations with respect to the physical aspects of occupational health, although some problems still exist in this area; asbestosis and tenosynovitis are examples. The situation with respect to the psychological implications of new patterns of work experience is more complicated, and it is clear that tremendous variability in this regard exists in modern society among different occupations. Nevertheless, it is self-evident that technological advance and machine use, as well as the increasing size of industrial, commercial, and governmental organizations, have had important impacts.

Many studies have been carried out by social psychologists on the question of *human relations* in the work situation [173]. As a consequence of some of this research, a number of organizations have introduced changes aimed at improving the quality of work experience. For example, in the USA the large company, General Foods, introduced a series of changes aimed at integrating the technological and social systems, at reducing status differences between employees, and at creating 'a social climate that enhanced self-esteem, feelings of responsibility and identification with the organization's goals'. After eighteen months of operation the outcome of the reorganization was 'overwhelmingly positive'. Absenteeism declined by 80 per cent, productivity increased, and morale was said to have improved markedly [174].

Concern has been expressed recently about the effects of 'high technology' (i.e. automation, use of computers, etc.) on both work experience and employment levels (see chapter 9). With respect to the quality of human experience these developments, like other technological advances, may have either desirable or undesirable consequences, according to circumstances. If the new sophisticated devices perform work which people would find monotonous or otherwise unpleasant, and if this work *has* to be done, then surely they are performing a service to humankind. If, on the other hand, the devices take away work which can be done just as well by humans, who, in fact, find it enjoyable and rewarding, then, from the human point of view, there is nothing to be said in their favour.

Another related cause for concern, at least in the eyes of some authors, is the notion that the growth of the economy will in future depend more and more on the intellectual work of a few [175]. If this trend persists, then lack of sense of personal involvement will increasingly become a characteristic of the life experience of the majority of the population.

It is worth emphasizing that boredom and lack of sense of self-fulfillment in the work-place are important not only in terms of the health and well-being of the individual worker, but also for society. Deficiencies in human experience in the work situation can be expected to give rise to *compensatory behaviour* outside the work-place—an unconscious striving for a sense of personal involvement, sense of belonging to a group, and sense of challenge and excitement. While such behaviours may well be adaptive as far as the individuals performing them are concerned, they may sometimes be undesirable for society as a whole—as, for example, when they involve violence, vandalism, and other destructive activities .

Let us conclude this brief section with a list of some of the more important 'desirable' attributes of work experience expressed in terms of the tentative list of *universal health needs* discussed in chapter 4. Work should provide: opportunity for co-operative small-group interaction; variety in daily experience, and also some change from day to day; incentives and opportunity for creative behaviour; incentives and opportunity for the exercise of learned skills; conditions which promote a sense of personal involvement, of purpose, of belonging, of challenge, of responsibility, of comradeship and self-esteem.

Moreover, work conditions should allow the expression of common behavioural tendencies, such as the seeking of approval of peers, the seeking of status, and the expression of loyalty in ways that are enjoyable and self-fulfilling.

Leisure experience

Patterns of organization of time imposed on human beings by culture in the twentieth century have resulted in the recognition of a category in the time budget known as *leisure* or *leisure time*. Broadly, this is any period of time when the individual is not engaged in 'work', and it includes the time when people are sleeping. In the present context, however, we shall be concerned only with the wakeful hours of leisure time. Like work itself, there is much variability in the populations of technological societies in the manner in which people spend their leisure hours.

Two sets of variables are relevant to the impact of recent societal changes on leisure time, and the implications of this impact for health and well-being. First, there is the question of *how much* leisure time people have. This, of course, is affected by economic and societal conditions in general and by the fact that at present an increasing number of people are unemployed. That is to say, more and more adults are experiencing leisure time on a full-time basis. The second consideration, and the one on which we shall focus here, is the question of *what people do* in their leisure time.

Culture, of course, has enormous influence on the ways people spend their leisure time. It affects their leisure behaviour directly, for example, by providing the learning experience which determines what they most enjoy doing—whether it be active involvement in music-making, opera, football, tennis, gardening, bush-walking, racing motor cars, religious activities, or painting—or the passive and vicarious enjoyment of these various pursuits. Culture can also influence leisure behaviour indirectly through technological change—by making available, for example, motor vehicles, power-boats, television sets, and electronic games.

The leisure-time activities of children are usually somewhat different from those of their parents, and also different from those of children in earlier periods in human history. Once again, the cultural environment is a major influence. Organized activities, such as team ball games, are especially important in many affluent societies, with parents often playing the role of chauffeurs, driving their off-spring to and from their various recreational pursuits. One has the strong impression that there is considerably less spontaneous active leisure-time behaviour among children in modern Western society than there was in the village-type situations which existed in the past or which exist elsewhere in the world today. The other striking change, a direct consequence of technological advances, is the amount of time that children spend in passive recreation with their eyes fixed on the television screen.

The leisure-time behaviour patterns of adolescents and young adults in the modern affluent society are also rather different from those of earlier times or other places. They involve a good deal of time spent watching television programmes or driving motor vehicles, and a marked interest in and enthusiasm for certain kinds of music- and song-making. A considerable proportion of young people participate actively in the music scene, a few of them playing instruments or singing, but many more dancing in discotheques. Many young people in the large cities have an interest in competitive sports, but for most of them the interest is a passive one.

Returning to adults, the most significant development in recent years affecting their leisure time has been, as in the case of children and adolescents, the introduction of television. In Australia in 1981 the average adult city dweller spent 2.5 hours per day watching television. Although this form of recreation is entirely passive, it has obvious entertainment value, and many people would take the view that it thereby contributes positively to the quality of life, perhaps providing meliors which tend to counteract some of the stressors of daily life and the experiential deficiencies of work situations. Apart from its entertainment value, television provides a most important input into the learning process, and there is no doubt that the messages which it conveys influence people's understanding of the world and their attitudes and aspirations.

While there is plenty of room for debate about the advantages and disadvantages of television for young people and working adults, there is little doubt that it has contributed positively to the enjoyment of life of old people who have been left on their own and for whom modern society, with its present organization and value system, no longer has any use. It provides them with a substitute for human company and helps to while away the hours of solitude.

Another ecologically important development associated with leisure activities has been the growth of the cult of tourism. Apart from weekend travel (few people go through a weekend entirely in, or even within walking distance of, their home), an increasing number of people are travelling long distances for their holidays. In the countries of Europe, for instance, between 1970 and 1980 the number of international arrivals of tourists coming from other European countries of Europe grew, by over 100 million, to more than 280 million [176].

It is noteworthy that much leisure activity in the modern Western society is very costly in terms of energy and resources. It is therefore influencing technometabolism and other aspects of the interrelationships between human society and the biosphere.

Finally, from the biosocial point of view, we may ask the question: To what extent does leisure experience compensate for deficiencies that we may perceive in the work component of life conditions? Does it provide people with any of the following aspects of life experience which may be lacking in the work situation: physical exercise, co-operative small-group interaction, creative activity, the practice of learned manual skills, variety in daily experience, short goal-achievement cycles, a sense of personal involvement, of purpose, of belonging, of challenge, of interest, of comradeship, of responsibility, of personal fulfillment?

Some writers take the view that leisure behaviour never can, nor should, compensate for deficiencies in work experience. De Jouvenel writes:

I recognize that the interest taken in 'the problem of leisure' is well meant, but I think it is a false track. It is treating men as children to seek means of entertaining them harmlessly. Indeed, children themselves sense it as a promotion when they are invited to do something apparently significant. You cannot make men contented through entertainments but only through achievements. It is therefore 'the problem of work' which is essential. You cannot have a good society if you do not offer to each man a man-sized job, which he can take joy and pride in. This is an immensely difficult problem but quite an essential matter [177].

Perhaps for most people a real sense of personal involvement and of self-fulfillment can only be achieved through subsistence activity, whether direct or indirect, through defence of family and community against perceived enemies, or through religious or artistic enthusiasm.

Learning experience

The pattern of learning in the modern high-energy societies is quite different from that of earlier ecological phases of human existence. It differs with respect both to *how* people learn and to *what* they learn.

Broadly speaking, learning has two kinds of outcome. On the one hand, it largely determines an individual's beliefs and general understanding of the nature of things, and consequently his or her motivating values. On the other, it shapes a person's skills, both mental and motor. Both these consequences of learning in turn influence the individual's behaviour pattern.

There are various ways of learning which human beings share with many other animals. These include learning to associate certain sensations with pleasure or with pain, trial-and-error behaviour, and mimicry. In humans, however, there is another dimension to learning of overwhelming importance—that is, cultural learning. It is this kind of learning with which we are concerned here. In modern technological society we can recognize three important and distinct pathways of cultural learning. We will consider these one by one under the headings: *informal learning; formal learning*; and *learning through the media*.

Informal learning

During the primeval phase, and, indeed, probably throughout the early farming and early urban phases, learning from members of one's family and other acquaintances in an informal way was by far the most important form of learning. It is clearly less so in modern technological society: sons no longer learn their occupational skills from their fathers, and the attitudes and beliefs of parents are now often in competition with those communicated through formal education and the media. In the informal learning process most of the information is communicated incidentally and without deliberate educational purpose. The individual thus learns spontaneously, simply by listening to casual conversation and by observing everyday behaviour. Informal learning in high-energy societies differs from that in other periods of human history, not only in the fact that it contributes a smaller part to the learning process as a whole, but also in its content. The nature of the information and the attitudes which are imparted even at the informal level reflect overall cultural trends and the influence of technological developments on human thinking.

Formal learning

The supplementation of spontaneous informal learning with schooling goes back to the time of the early Mesopotamian cities. Already in those early days a small number of boys were required to attend educational institutions for formal instruction, and since that time, and until quite recently, this kind of learning experience was restricted to a minority of young males in the population. Today, however, the great majority of children, both male and female, from around the age of five or six years are required to attend school at least until they are about fourteen years old.

Schooling differs from informal learning mainly in the fact that spontaneity is replaced by regimentation. Interest and enthusiasm as a motive for learning is sometimes, but not always, replaced by fear, if only fear of disapproval. It is probably true to say that for the majority of students the formal learning process is not an enjoyable experience most of the time.

A small proportion of adults of all ages choose to attend formal learning programmes which are arranged by universities or other educational authorities.

With respect to the content of educational programmes, we can recognize at a glance one outstanding characteristic. Even at the primary and secondary school levels, as well as at the tertiary level, the great majority of courses are on specialized subjects, reflecting the structure of the academic world. Certainly, specialization in intellectual pursuits is absolutely necessary in the complex society in which we live, with its vastly expanding body of knowledge. However, extraordinarily little effort is devoted in education to the intellectual process of integration and to attempts to improve understanding of the dynamic interrelationships between the different aspects of reality—aspects which are studied independently in the different areas of specialism and which are covered separately in different courses.

One of the consequences of the extreme emphasis on specialization in formal education is the fact that the human products of such education from different areas of specialism have extraordinarily different world views and pictures of 'reality'—so much so that effective communication between them on serious societal topics is often very difficult, sometimes almost impossible. For example, the majority of graduates in economics see the problems and options facing modern society in a very different way from most graduates in the biological sciences.

Learning through the media

Under this heading we include newspapers, popular magazines, radio, television, and, of less importance nowadays, the cinema. The exposure of individuals on a big scale to these sources of information and of attitudes and opinions is a new phenomenon in the history of our species. There can be no doubt that the media play an immense part in the learning process. Consequently, as in the case of formal education, they have great potential for influencing the level and nature of understanding of human situations—global, local, and individual—and thereby for affecting motivating values, if not basic values, of people. The media thus have tremendous potential for contributing to either desirable or undesirable societal change. The important questions are: (a) Who determines the nature of the information communicated? (b) What is the nature of this information?

An important part of the information communicated, especially in television, is that which is incorporated in the messages designed to persuade individuals to purchase commodities, and the view is often expressed that advertising of this kind strongly influences people's priorities in life and affects their attitudes and behaviour. Whether or not it is desirable that the vested interests, in the form especially of large corporations, should have this influence, and indeed should have the right to exert it, is a matter of controversy.

There has also been much debate on the question of whether presentation of scenes depicting violent behaviour on television, either in documentaries or in fiction programmes, tends to induce similar behaviour in viewers. Does the sensation of vicarious participation result in a 'draining off' of aggressive urges, or does it actually stimulate such urges? And does it contribute to the general acceptance by society of such violence? [178]. In my own view, the observing of violence on television probably does not promote violent behaviour in the majority of viewers, but a minority may well be affected in this way; and it takes only a minority manifesting violence in society to create a highly undesirable situation. It is noteworthy, although not surprising, that many programmes on television, both documentary and fiction, reflect the prevailing fascination and faith in machines and technological progress.

Consumer behaviour

The term *consumer behaviour* as used here includes all behaviour of individuals which is associated with, and which often results in, the purchase of commodities, especially manufactured goods, but also various services, such as transport and tourism. As discussed in chapter 9, such consumer behaviour—as a significant aspect of the lifestyle and aspirations of the average individual—is a characteristic of contemporary high-energy society, and it is a key component of the set of processes which are responsible for the continuing intensification of technometabolism. Here we will briefly consider the causes and consequences of consumer behaviour as an aspect of human experience.

The biopsychic factors tending to promote consumer behaviour can be seen as expressions of a number of the common behavioural tendencies of humankind. The following are especially important: the tendency to seek approval and status; the tendency to avoid ridicule; and the tendency to seek novelty. Another important common behavioural tendency in this context is the tendency to seek the attention of others—that is, to try to be the centre of attention. As discussed earlier, a common way of achieving this is the communication of new (and occasionally false) information. Another way, characteristic of present-day society, is to display newly purchased goods. Yet another influential biopsychic factor is the tendency to strive to avoid feelings of deprivation. Thus a major force promoting consumerism is the observation by some individuals that other individuals possess goods which they do not.

Boredom, loneliness, and other causes of mental depression and the general absence of arousal are also important factors contributing to consumer behaviour. A common response in the affluent societies to attacks of depression is an expedition to a shopping centre. Window-shopping alone may alleviate the undesirable condition, but the biggest relief is experienced through the actual acquisition and taking home of new goods. Consumer behaviour thus plays a most important role by providing a source of meliors which may help to protect individuals from the undesirable effects of environmental stressors. In fact, it is a reasonable hypothesis that consumer behaviour in modern Western society compensates in a most important way for the decline of various intangible meliors—meliors associated, for example, with creative behaviour, a sense of responsibility, and a sense of personal involvement.

In the present cultural setting, consumer behaviour is therefore *necessary* in order that the individual gains in-group approval, achieves self-esteem and status, and does not feel deprived. These are basic universal health needs, and culture has determined that in high-energy societies the criteria for the satisfaction of these health needs are mainly in terms of material possessions. Any successful transition to an alternative arrangement which is less costly in resources and energy and which involves lower levels of consumption of material goods by individuals would demand a fundamental change in the value system of society and, in particular, in the criteria for status, approval, and success.

Common behavioural tendencies

For those who hold the view that there is such a thing as human nature, in the sense that members of the human species share some *common behavioural tendencies* (page 48), certain rather important questions come to mind. For example, we ask: In what specific ways do these innate behavioural characteristics find expression in the new high-energy habitat? This question alone warrants a chapter to itself, if not a whole book. Here we must be content with a few basic generalizations.

First, it is abundantly clear that, unlike in the primeval situation, such powerful basic behaviours as the tendencies to seek the approval of the in-group and to seek to maintain or improve status result in totally different specific behaviours in different sections of the population. That is to say, there is enormous variability in the criteria of approval and disapproval in the vast array of different subcultural settings in which people find themselves in present-day society. Thus specific manifestations of basic behavioural tendencies in work situations include such very different activities as, for example, shearing sheep, sitting at desks and arranging economic transactions over the telephone, selling commodities in department stores, preparing academic publications, filming current events for television, and driving buses.

We are thus led to the somewhat unoriginal conclusion that culture determines to a very large extent what people actually do, and how they spend their time. We must then ask: What are the implications of these different culture-induced behaviours both for the health and well-being of the individuals who perform them and for society as a whole? The first part of this question has received some attention earlier in this chapter.

As far as the implications for society are concerned, it is necessary to introduce another variable—the end-product of the specific behaviour. Individuals in different work subcultures may carry out very similar specific behaviours, but through doing so promote quite different societal processes. This occurs, for instance, when people behind desks in different kinds of corporate organizations, in seeking approval do similar things with similar immediate objectives and similar immediate rewards. An individual, for instance, who is an executive in a munitions factory and a senior official in a nature conservation organization, although carrying out similar specific behaviours, are contributing to very different societal outcomes. In both cases, the individuals have identified with the organization of which they have become a part, and are seeking approval by doing their jobs as well as they can, and in this way they are promoting the aims and objectives of the organization. This divorce between work experience and its eventual societal outcome has serious implications for humankind and the biosphere far beyond those to which Karl Marx drew attention.

This is not to say, of course, that all humans are necessarily completely hopeless victims of the cultural system as far as their behaviour and its societal consequences are concerned. Personal choice can have some influence at any stage, especially in early adulthood when individuals are selecting their occupations. Change is still possible later: individuals may elect to leave an occupation, either because they are dissatisfied with the particular kind of specific behaviour required of them or because they disagree with the ultimate societal consequences of this behaviour. Options in this regard tend, however, to diminish as people grow older; and, with high levels of unemployment, the range of choice even at the early stage is becoming increasingly limited. Furthermore, the occupations which young people enter are often not those which they would choose for themselves; but once they have accepted employment in a given area, they become committed to the occupation in question and all that it entails.

The above comments are made mainly in relation to work experience. However, culture clearly is a major determinant of the specific expressions of common behavioural tendencies in other spheres of human activity. We have already discussed consumer behaviour in these terms; and the same principles apply to the behaviour of human beings in relation to art, music, crafts, drinking, smoking, television viewing, bush-walking, and sport. In all these instances, and in many others, the specific behaviour of individuals is influenced to a high degree by prevailing cultural or subcultural norms. The same principles apply to criminal behaviour and even to casual delinquency.

One further problem related to human nature must be mentioned. It is the human tendency to identify with an in-group (and its set of values and belief system), to show loyalty to the in-group, and to act with suspicion towards out-groups—behavioural tendencies which lie behind the terrible and persisting conflicts which exist in the modern world between various religious and racial groups. In such situations, cultural processes of the past have determined the values and attitudes of the members of in-groups today, and individuals find themselves born into one or other such groups— born, that is, into a state of perpetual conflict and hatred. In such cultural settings the common behavioural tendencies of humankind result in an immense amount of unnecessary human suffering and distress, providing a clear example of the 'flaming moth principle' (see page 52).

Additional factors

In a single chapter we have not been able to attempt more than a general survey of some of the ways that societal conditions in modern technological society impinge on the immediate environments, patterns of behaviour, and biopsychic state of individual human beings, and we have not been able to deal in depth with any of the influences and interrelationships described.

Moreover, there are as many other subtle ways, some of them no less important than those discussed above, by which culturally induced conditions of modern society affect the life conditions and biopsychic state of human beings. We could usefully have discussed in this context the implications, for example, of the rates of change in the social environment, the lengths of goal-achievement cycles, the design of the built environment, and the keeping of pets.

SECTION 5: CONCLUSIONS AND COMMENTS

It is clear that, biologically speaking, the human organism is on the whole coping well in the very artificial environment of modern high-energy society. The credit for this desirable situation must largely go to cultural adaptive processes which have effectively countered the nutritional deficiency diseases and the pestilences which bedevilled earlier civilizations. Nevertheless, most of the forms of ill health which cause distress and which contribute to mortality in contemporary society are the direct consequence of environmental conditions and lifestyle, and are in principle preventable. The main emphasis in medical institutions today, however, is not on preventive medicine, but rather on 'miracle cures' and dramatic high-technology surgery. While concern is frequently expressed about this imbalance, it nevertheless persists and deepens [179].

It is worth emphasizing again the role of intangible meliors in human experience, and also the importance of our taking them into account in the consideration of societal and individual options for the future. The risk is very real that, because of the difficulties we experience in coming to grips with intangibles, societal and economic changes may be allowed insidiously to erode away important sources of enjoyment and of self-fulfillment; indeed, such erosion may well have been under way for some time. In our efforts to comprehend the real significance for human beings of societal changes, it is imperative that we take deliberate steps to consider these intangible and unquantifiable aspects of human experience in a systematic and thorough way. It is reasonable to suspect, for example, that in the present situation the lack of sense of personal involvement, of self-fulfillment, and of belonging, as well as the presence of a sense of deprivation, are common causes of chronic distress, interfering with the enjoyment of life and sometimes leading to behaviours which are against the interests not only of the affected individual but also of other members of society.

I will conclude this chapter on a personal note by referring to an hypothesis of relevance to human well-being and to deliberations about societal change. As discussed earlier in this book, when we look at some of the tools made by people in the late Palaeolithic era we see that a great deal more effort went into shaping them than was necessary in order to make them useful. Furthermore, the people of the late Palaeolithic cultures devoted time and effort to making various kinds of ornaments out of bone and wood—objects which had no practical value whatsoever. Similarly, recent hunter-gatherers spend time creating patterns of one kind or another, apparently simply for the sake of creating patterns. I suggest that the capacity to derive enjoyment from creative behaviour of this kind is a fundamental and innate characteristic of *Homo sapiens*, and that artistic creativity has an important potential role to play in promoting health and well-being in human populations. In my view any society that fails to provide appropriate incentives and opportunities for creative behaviour and consequently deprives the majority of its members of this kind of experience is performing a serious disservice to humanity. Most contemporary high-energy societies fall into this category.

NOTES

1. Boyden, Millar, Newcombe, and O'Neill 1981, chapter 10.
2. Sources: United Nations 1967; Preston, Keyfitz, and Schoen 1972; Tuve 1976; United Nations 1980.
3. Fenner, Henderson, Arita, Jezek, and Layny, 1987 (in press). For an interesting account of the role of smallpox in human history, see Hopkins 1983.
4. Julienne, Smith, Thomson, and Gray 1983, p. 9.
5. Adena, Montesin, and Gibson, 1983.
6. Sinnett 1975.
7. Trowell 1981.
8. Epstein 1965, 1979.
9. Kannel, Dawber, Kagan, Revotskie, and Stokes 1961; Kannel 1976.
10. Lipids, including cholesterol, are carried in the blood linked to protein. These lipoproteins in the blood consist of two fractions—the low density lipoproteins and the high density lipoproteins; together they carry more than 90 per cent of the cholesterol in the blood (see Vines 1984).
11. Kannel *et al.* 1961.
12. Kannel 1976; *New Scientist* (1981), **89** (1234), p. 19.
13. Sackett, Gibson, Bross, and Pickren 1968; Kannel 1976.
14. Morris, Adam, Chave, Sirey, and Epstein 1973; Naughton and Hellerstein 1973.
15. For different viewpoints, see Woodhill, Palmer, and Blacket 1969; Keys 1975; McMichael 1979; Gould 1979; Shekelle, Shyrock, Paul, Lepper, and Stamler 1981.
16. Antar, Ohlson, and Hodges, 1964.
17. See Lipid Research Clinics Program (1984*a*, *b*) *Journal of American Medical Association*.
18. Yudkin 1967.
19. Antar *et al.* 1964.
20. Morris, Marr, and Clayton 1977.
21. Raab 1966; Lancet 1970; Benjamin 1971.
22. Caffrey 1968; Rosenman and Friedman 1974; Wood 1981.
23. Thoresen, Fredman, Gill, and Ulmer 1982; Wood 1981.
24. Insel and Moos 1974, pp. 5, 6.
25. Raab 1970.
26. See for example, Kaminer and Lutz 1960; Maddocks 1961; Barnicot, Bennett, Woodburn, Pilkington, and Antonis 1972; Sinnett and Whyte 1981.
27. US Department of Commerce 1983.
28. Australian Department of Health 1982.
29. Cairns 1975; Higginson 1976; Doll 1977; Hall 1981.
30. Doll and Armstrong 1981; Pearce 1983.
31. McMichael, McCall, Hartshorne, and Woodings 1980.
32. McMichael *et al.* 1980; Hunt and Sali 1979.

33. Carroll 1975; Carroll and Hopkins 1979.
34. This is the view of the National Cancer Institute in Bethesda, Maryland; see Hall 1981.
35. Baklien 1981.
36. Doll and Armstrong 1981, p. 98.
37. Fenner 1970; Cockburn 1971.
38. Hordern 1976, p. 254.
39. Hordern 1976, p. 254.
40. Gibson 1977.
41. Healy 1975.
42. Australian National Heart Foundation 1983.
43. Harrison, Palmer, Jenner, and Reynolds 1981.
44. The best documented example of the selection in human beings for resistance to an infectious disease is the case of the influence of malaria (due to the more malignant forms of *Plasmodium falciparum*) on the distribution of genes for sickle-cell anaemia. Individuals who carry a single sickle-cell gene manifest a mild form of anaemia. In homozygotes the anaemia is severe and they have relatively low Darwinian fitness. The anaemia is associated with an abnormal form of haemoglobin known as haemoglobin S. The sickle-cell gene has been found to be very common in populations in parts of Africa where *Plasmodium falciparum* is also common, and it has been shown that heterozygotes for the sickle cell are considerably more resistant to malaria than are individuals who are homozygous for the 'normal' (non-sickle-cell) gene. This is thus a case of *balanced polymorphism* in which the selective advantage, in the malaria-infected environment, of the sickle-cell gene in heterozygotes outweighs its selective disadvantage as a cause of both the mild form (in heterozygotes) and the severe form (in homozygotes) of sickle-cell anaemia (see the discussion in Cavalli-Sforza and Bodmer 1971, chapter 4).
45. Eugenicists are individuals who study methods of improving the genetic quality of human populations, especially through selective breeding (or are the proponents of such action). For useful discussions on eugenics, see Dobzhansky 1962; Cavalli-Sforza and Bodmer 1971, chapter 12.
46. Rendel 1970; Cavalli-Sforza and Bodmer 1971, chapter 12.
47. In 1966, H. J. Muller advocated the storage of semen of 'excellent men' with a view to ensuring eugenic advance toward 'excellence' (see Cavalli-Sforza and Bodmer 1971, p. 768).
48. Wills 1970, p. 107.
49. Cavalli-Sforza and Bodmer 1971, p. 537.
50. Spuhler 1968. For general discussion on assortative mating in humans and its implications, see Cavalli-Sforza and Bodmer 1971, chapter 9; Bennett 1979, chapter 7.
51. See, for example, Cavalli-Sforza and Bodmer 1971, p. 788.
52. The natural background radiation reaching the gonads is estimated, on average, to be about 100 millirads per year (there is considerable variation from place to place). Cosmic rays are thought to contribute about 28 per cent, and gamma rays, from disintegration of radio-active uranium, about 50 per cent.

53. The total radiation from the testing of atomic bombs between 1956 and 1965 reaching the gonads by the year 2000 is estimated to be about 76 millirads. Radio-active fall-out up to December 1965 has added about 2 per cent to the genetically significant dose due to the natural background radiation (and about 8 per cent to the dose significant for leukaemia). See Cavalli-Sforza and Bodmer 1971, pp. 786-7.
54. Coffin and Knelson 1976.
55. WHO 1972, pp. 199-201.
56. National Air Pollution Control Administration 1970.
57. Stokinger and Coffin 1968.
58. Bullock and Lewis 1968; Christopher 1980.
59. Needleman, Gunnolleviton, Reed, Peresie, and Maher 1975. There is considerable disagreement about this conclusion. For opposing viewpoints, see Rutter and Jones 1983.
60. Bryce-Smith and Waldron 1974.
61. Moos 1976, p. 201.
62. The effects of high noise levels on humans (other than noise-induced deafness) are well discussed in chapter 6 of Moos 1976.
63. Hammond 1953; Murton and Kear 1975.
64. Trowell 1981, p. 18.
65. Brock 1963.
66. Cowhig 1972.
67. Mayer 1968, p. 4.
68. Trowell 1981, pp. 14, 20-1.
69. See Cohn and Joseph 1962.
70. Brock 1963, p. 48.
71. Mention must also be made here of one further factor which tends to promote the overconsumption of food energy in modern society, and that is the consumption of alcoholic beverages, not to relieve hunger, but for the effect they have on the mental state of the individual. This topic will be discussed in a separate section.
72. Bruch 1978.
73. Dubos 1965, p. 79.
74. Brooke-Thomas 1979.
75. Ohlson 1969; Butterworth 1972.
76. Monckeberg 1969; Latham 1975.
77. For an interesting discussion on this topic, see Pyke 1968.
78. Harrison, Weiner, Tanner, Barnicot, and Reynolds 1977, p. 421.
79. Albanese and Oslo 1968. Special care is necessary on a purely vegetarian diet to ensure that sufficient iron, calcium, vitamin B12, and lysine are consumed. Breast-fed infants of vegetarian mothers sometimes show signs of B12 deficiency.
80. Brosin 1968.
81. Gollan 1978, p. 18.
82. Drummond and Wilbraham 1958, p. 298; Trowell 1975, p. 44.
83. Yudkin 1964, p. 19.
84. These figures are based on estimates of 'apparent consumption'. While they give a good indication of overall trends, some caution is necessary in the interpretation of apparent consumption estimates; see Cashel 1981*a*, 1981*b*.

85. Adapted from a paper by Burkitt 1982.
86. McCance and Widdowson 1956.
87. Chick 1958.
88. Drummond and Wilbraham 1958, pp. 289-93.
89. Pearson 1970, p. 180.
90. Dubos 1965, p. 160. In recent years the addition of sodium fluoride to drinking water has, some claim, markedly reduced dental caries in some Western populations (although there is not universal agreement about this, since a similar decline has been observed in populations not exposed to fluoride). This is an example of cultural adaptation which, if it is effective, is both *preventive* and *antidotal* (rather than corrective). In fact, it represents an evodeviation, and it would not be surprising if continual intake of sodium fluoride were found eventually to give rise to some form of phylogenetic maladjustment.
91. Cleave and Campbell 1966; Finlay-Jones and McComish 1972; Tudge 1977.
92. For discussion on the 'thrifty gene', see Neel 1962; Neel, Fajan, Conn, and Davidson 1965. See also Steinberg 1965; Keen, Thomas, Jarrett, and Fuller 1978.
93. See various papers in Burkitt and Trowell 1975. Also Cleave and Campbell 1966.
94. Keys 1975.
95. Crawford, 1968. Harries, Hubbard, Alder, Kay, and Williams 1968.
96. Selinger 1976.
97. Thomas 1959, p. 149.
98. Ambard and Beaujard 1904.
99. See Beard, Gray, Cooke, and Barge 1982.
100. Maddocks 1961; Sinnett 1975; Sinnett and Whyte 1981.
101. Trowell and Burkitt 1981.
102. Finn, McConnochie and Green 1982. Not all authorities agree about this relationship between salt consumption and blood pressure.
103. Millstone 1984.
104. Kermode 1972.
105. Feingold 1974; Joyce 1980; Cook and Woodhill 1976.
106. Joyce 1984.
107. Cook 1977, p. 90.
108. May 1965; Banks 1969; Jelliffe and Jelliffe 1978.
109. Gyorgy 1971; Jelliffe and Jelliffe 1978, chapter 3.
110. Gerrard 1974, 1975a, 1975b.
111. Banks 1969.
112. Newton 1971.
113. *Iatrogenesis* is derived from the Greek words *iatros*, meaning physician, and *genesis*, meaning origins. For a full discussion of the problem of iatrogenic disease, see Illich 1975; Taylor 1979.
114. Taylor 1979, p. 64.
115. See also Boyden and Millar 1978; Boyden 1980.
116. It can be argued that complete absence of stress is also undesirable. See Levine 1971.
117. Boyden *et al.* 1981, pp. 343-7.

118. Strictly speaking the word 'distressors' should be used for those experiences which promote a state of stress—but for reasons of convenience 'stressor' will be used here for this purpose.
119. Boyden *et al.* 1981.
120. Lipman-Blumen 1972.
121. Morris *et al.* 1973; Naughton and Hellerstein 1973.
122. Minc 1960, 1967.
123. Newman and Hogan 1981.
124. Sources: Chandler and Fox 1974; United Nations 1980; Boyden *et al.* 1981.
125. Millar 1979; Boyden *et al.* 1981, chapters 9 and 10.
126. For a discussion on some of these different approaches, see Newman and Hogan 1981.
127. See, for example, Calhoun 1962; Christian 1961.
128. Hall 1966; Worchel 1978.
129. Simmel 1969; Milgram 1970; Wohlwill 1974; Baldassare and Fischer 1977.
130. Baum, Harpin, and Valins 1975.
131. Galle, Gove, and McPherson 1972; Baum and Epstein 1978.
132. Millar 1979.
133. Rapoport 1975. Choi, Mirjafari, and Weaver 1976.
134. Millar 1979; Boyden *et al.* 1981, chapter 14. See also Rapoport 1975.
135. Dubos 1965, p. 279.
136. This discussion does not include reference to psychotropic drugs prescribed by doctors, such as tranquillisers and anti-depressants, although some of these may be to some extent addictive (e.g. valium).
137. Blum 1969*a*, 1969*b*, 1969*c*, chapter 2.
138. Hawkes 1954, pp. 155-6.
139. Herodotus 1921, p. 134.
140. This figure is adapted from Bexenius 1981.
141. DeLuca 1981.
142. US Department of Health and Human Services 1981.
143. Williams 1980.
144. National Clearinghouse for Alcohol Information 1981.
145. National Information Service 1982.
146. Williams 1980; Hill and Gray, 1982. The proportion of women in the population smoking tobacco is at present increasing; see Jacobson 1981.
147. National Information Service 1982.
148. National Information Service 1982.
149. Johnston, Bachman, and O'Malley 1982.
150. Blum 1969*c*.
151. Williams 1980.
152. See Cobb 1976; Cassel 1976.
153. Henderson 1981, p. 396: 'The evidence is that, as a causal factor in neurosis, the crucial property of social relationships is not their availability, but how adequate they are perceived to be when the individual is under adversity'.
154. See Burrows and Lapides 1970.
155. See, for example, Nisbet 1970, p. 274.
156. Susser and Watson 1971, p. 87. 'Anomie is a state of confusion, inchoate values, and loss of cohesion in social groups. The impact of such situations

on the mental state of individuals, it is held, gives rise to uneasiness, a sense of separation and isolation, frustration, powerlessness and apathy. This subjective state is variously referred to as *anomia* or alienation'.

157. Holland (1975, p. 258), for example, referring to Marx's views on alienation and Durkheim's concept of anomie writes: 'Marx and Durkheim express contradictory judgements on the basis of human needs. Durkheim stresses security and stability, and Marx creativity, autonomy and responsibility'. Horton (1964, p. 289) writes: 'Anomie is basically a utopian concept of the political right ... . Alienation is a utopian concept of the radical left'.

158. Greenberg 1974, p. 108; see also Lee 1972.

159. Kaufmann 1973, p. 165.

160. Nisbet 1970, p. 274.

161. Merton 1957, p. 138.

162. Nisbet 1970, p. 278.

163. See Young 1974 referring to Hillery 1955.

164. The strength of the sense of community in the traditional English village is well described by Laslett 1971. We should note, however, that the typical village-type situation is itself different in important basic ways from that of primeval society. Indeed, in some respects, the city environment is more like that of hunter-gatherers in that, for instance, it often involves a greater sense of challenge and of excitement, and events are, perhaps, somewhat less predictable than in the village. Village life can be extraordinarily monotonous, although cultural influences may ensure that this is not the case.

165. Tönnies 1955; see also Nisbet 1970, p. 32.

166. Simmel 1969; see Wolff 1950.

167. See, for example, Keller 1972.

168. Gans 1968, 1976.

169. Scitovsky 1976, p. 100.

170. Smith 1910; see also Moos 1976, p. 264.

171. Holland 1975, p. 257.

172. Briggs, 1962.

173. See, for example, Dunphy 1972, pp. 21, 22; Jenkins 1974.

174. Moos 1976, pp. 268-9.

175. Barnet and Muller 1974, p. 328.

176. Holdgate 1983.

177. De Jouvenel 1972, p. 170.

178. See Shulman 1975.

179. Feelings of concern about this situation are well summed up in an article 'On the limitations of modern medicine' by Powles 1973, and in another, 'Promoting health', by Beard 1979. There can be no doubt, however, that vaccination programmes in particular have had a major impact on the incidence of a number of important contagious diseases. Smallpox and poliomyelitis are good examples. Nevertheless, as Anderson and May (1982 p. 411) have written: 'There comes a time in every successful vaccination immunization campaign when the risks of vaccination seem to out-weigh the danger from the disease. But if vaccination is not continued, the epidemics began again'. This dilemma is frequently the cause of vigorous societal debate.

11

POSSIBILITIES FOR THE FUTURE IN BIOHISTORICAL PERSPECTIVE

A SUMMARY OF THE PRESENT ECOLOGICAL SITUATION

Looking back over the history of the interplay between human culture and the processes of nature, we see that there have recently occurred three developments of outstanding ecological significance. The first has been the explosive growth of the human population during the past fifty to one hundred years. The second has been the invention, manufacture, and stockpiling of weapons of mass destruction which threaten the survival of humanity and of the biosphere as we know it. The third has been the introduction, on a massive scale, of machines powered by extrasomatic energy derived mainly from fossil fuels, and the associated discovery and use of electricity. This change has had major impact on the technometabolism of human populations, involving a rapidly increasing use both of renewable and non-renewable resources and of extrasomatic energy, and a concomitant increasing discharge of waste products into surrounding ecosystems. The scale of the growing technometabolic load imposed by human society on the biosphere can be grasped by appreciation of the fact that, if recent trends were to continue for a further fifty years, the amount of energy, somatic and extrasomatic, flowing through human populations would be equal to that flowing through all other animal and plant populations put together.

In considering our present state of knowledge about the impacts of cultural processes on the biosphere, it is important to appreciate the distinction between the *certainties* and the *uncertainties*. Taking the three major issues one-by-one—growth of the human population, weapons of mass destruction, and industrial growth—the important certainties and uncertainties may be summarized as follows:

With respect to the growth of the human population the certainties are: the population is now about 1000 times greater than it was before the advent of agriculture; due mainly to population growth in developing countries, the world population is now increasing at a higher rate than ever before; and continued growth of human population will eventually outstrip food supply, giving rise to starvation and human suffering on a vast scale. The uncertainties are: how many people the biosphere is capable of supporting in a state of health and well-being; and whether, or how soon, deliberate control of population growth will become effective on a world-wide basis.

311

In the case of weapons of mass destruction, the certainty is that if the massive stockpile of existing weapons, or even one thousandth of it, were exploded, there would be damage to the biosphere and human carnage and suffering the world over on a scale never seen before in the history of humankind. The uncertainties are: whether, after a nuclear war, the biosphere would continue to function as a bioproductive system capable of supporting human life; and what proportion, if any, of the human population would survive a full-scale nuclear war (that is, if the life-supporting processes of the biosphere remain intact).

With respect to industrial growth and the biosphere, the certainties are: there are limits to the absorptive capacity (of waste products and toxic substances), resilience, and adaptability of biological systems—a principle which applies as much to the biosphere as a whole as it does to local ecosystems and to individual organisms; and the present pattern of increasing intensity of technometabolism in high-energy societies is unsustainable in the long run: if it is not brought under control through deliberate societal action, it will come to an end either as a result of resource depletion or, more seriously, as a consequence of irreversible damage to the biosphere caused by technological waste products. The uncertainties are: how much longer the biosphere can tolerate increasing industrial productivity in its present form; and which particular culturally induced environmental changes represent the greatest threat to the integrity of the biosphere.

CULTURAL ADAPTATION—THE PRESENT SITUATION

The future of humanity depends on whether or not society works out and successfully implements effective cultural adaptive measures to overcome the threats inherent in the present situation. We can recognize two fundamental and essential aims of this cultural adaptive process. First, it must aim to satisfy the 'health requirements' of the world's ecosystems, so that bioproductivity is maintained and the biogeochemical cycles remain intact. This means that human society must strike a state of ecological balance with the biosphere; that is, the size of the population and the per capita rate of resource and energy use must be maintained at more or less constant levels. It also means that all societal activities which threaten the integrity of the biosphere, including the manufacture, storage, and deployment of nuclear weapons, must cease. Second, the cultural adaptive measures must aim to satisfy the universal health and survival needs of human beings. Society must be organized so that the life conditions of people are health-promoting, with regard both to material aspects of experience, such as diet, air quality, and housing, and to behavioural, psychosocial, and intangible aspects which affect, for example, physical fitness, sense of personal involvement, sense of belonging, and sense of self-fulfilment.

What, we may ask, is the likelihood that such cultural adaptation will come about and be successful? The study of cultural adaptative processes in human history does not help us to answer this question: all we can say is that some groups in the past have successfully adapted through cultural processes to undesirable culturally induced changes in the biosphere or in human life conditions, while other groups have not been successful and have perished. Now, for the first time in human history, the nature and magnitude of the threats to humankind are such that the whole human population can be regarded as a single group. This group is an extremely large one, and so, indeed, are the problems which it faces. On the other hand, modern humanity has the benefit of a great deal of scientific knowledge and accumulated experience which was not available to earlier peoples facing ecological threats. The key question is whether or not humans will, collectively, make wise use of this knowledge. In the words of Dimbleby:

... however ingenious we are at dodging the ecological showdown, in the long run we have to accept the limitations imposed by our environment. We see that by failing to do so man has brought troubles upon himself in the distant and more recent past ... . In the past men acted in ignorance; we do not have that excuse today [1].

Let us attempt to summarize the state of the cultural adaptive process with respect to each of the three major ecological threats facing humankind. In doing so, we will bear in mind the four basic prerequisites for successful cultural adaptation on the societal level (see chapter 2). These are: *recognition* that an undesirable state exists; *knowledge* of the cause of the threat or of ways and means of overcoming it; the *means* to deal with the undesirable state (in terms of human and other resources); and *motivation* to take appropriate action on the part of those in society who make the relevant decisions.

Considering first the population issue, most governments the world over now recognize the seriousness of the problem and the importance of limiting population growth. Moreover, knowledge is available on ways and means of achieving this; and control of population size is not, in general, prevented by lack of resources. The chief impediment to effective cultural adaptation at this time seems to be lack of 'motivation to take appropriate action on the part of those in society who make the relevant decisions' (see p. 24). In this particular case, those who make the relevant decisions are not governments, but the individual couples who decide whether or not to have children. Nevertheless, governments can, of course, play a major role through educational programmes and by providing incentives for restriction of family size.

In the case of nuclear weapons, the existence and seriousness of the threat is widely recognized at all levels in society. The main impediment to effective cultural adaptation appears again to be lack of sufficient motivation on the part of those in society who make the relevant decisions; but in this instance those who make the relevant decisions are not individuals, but powerful corporate systems—mainly governmental, but these are much influenced by military and industrial organizations. The process is also hampered by lack of knowledge and understanding of the true nature of the underlying causes of the problem in human behavioural terms. It is true that numerous international meetings and conferences have been held between representatives of the superpowers, ostensibly aimed at reducing and eventually eliminating the nuclear arsenal; but so far they have achieved nothing. Clearly some new element has to be brought into the adaptive process—an element, perhaps, which has something to do with a sense of perspective and which transcends national, political, religious, occupational, and racial boundaries.

The cultural adaptive responses relating to the undesirable impacts of technological processes on the ecosystems of the biosphere need to be considered at two levels, the local and the global. At the local level, there has been some encouraging progress in the high-energy societies. However, it should be noted that, although about a century has passed since George Perkins Marsh in the USA and Alexander Ivanovich Woekof in Russia expressed concern about the abuse of the biosphere by humankind and emphasized the need for society to achieve a state of balance with nature [2], it is only within the last twenty-five years that protection of the natural environment has become a major public issue. Progress has taken the form, for example, of societal adaptive measures in Western cities aimed at reducing air and water pollution. Outstanding examples of relative success include the partial clearing of the air in London and Pittsburgh, the ecological recovery of Lake Washington, and the return of fish to the River Thames. The societal adaptive process at the local level has been characterized by the vigorous activities of local groups of 'environmentalists' and has led to the establishment of numerous authorities at different levels of government with responsibility for environmental quality. Indeed, lively societal debate between environmentalists and various kinds of vested interests on local environmental and land-use issues has become a permanent feature of most Western populations. The environmental movement in general has gained a great deal of support from certain United Nations' activities, especially the Stockholm Conference of 1972 and Unesco's Man and the Biosphere Programme. However, progress in developing regions has been slight.

At the global level, the societal assessment process in relation to the ecological situation presents a different picture. In this case, the cultural adaptive process appears to be blocked at the level of the first prerequisite for successful cultural adaptation. Certainly, there exist some individuals who recognize a potential threat to the biosphere in the continued intensification of the technometabolism of society, and who hold values which would be consistent with a transition to a fifth ecological phase of human society characterized by ecological balance and sustainability. The view of this group is well expressed in the following quotation from a small book written by Xenophon Zolotas, a Governor of the Bank of Greece:

It seems that mankind's only hope lies in the universal acceptance of a philosophy founded on a respect for life in the broadest possible sense, including respect for nature in all its aspects. For there can be no doubt that affluent societies are to a large extent societies of destruction. Having been brought up in effect on anti-humanistic and purely materialistic principles they have fostered the creation of 'consumer-man'. ... he is oppressed by various factors such as the manufactured wants of affluent societies and the dominant technology of the 'megamachine'. The only hope of weakening the oppressive influence exerted on people by consumer society lies in the awakening of the conscience of the ordinary man in the street [3].

Individuals who hold views such as these are, however, very much in the minority, and so far have had little, if any, influence on societal policies. The majority of the population and of decision-makers are unaware that any such threat exists. It is true that the International Union for Conservation of Nature and Natural Resources, in association with some other international agencies, has published the *World conservation strategy*. This document emphasizes some fundamental ecological truths, as, for example, in the following statement:

Earth is the only place in the universe known to sustain life. Yet human activities are progressively reducing the planet's life-supporting capacity at a time when rising human numbers and consumption are making increasingly heavy demands on it... . The combined destructive impacts of a poor majority struggling to stay alive and an affluent minority consuming most of the world's resources are under-mining the very means by which all people can survive and flourish.

Humanity's relationship with the biosphere ... will continue to deteriorate until a new international economic order is achieved, a new environmental ethic adopted, human populations stabilize, and sustainable modes of development become the rule rather than the exception. Among the prerequisites for sustainable development is the conservation of living resources.

Living resource conservation has three specific objectives: *to maintain essential ecological processes and life-support systems* (such as soil regeneration and protection, the recycling of nutrients, and the cleansing of waters), on which human survival and development depend: *to preserve genetic diversity* (the range of genetic material found in the world's organisms), on which depend the breeding programmes necessary for the protection and improvement of cultivated plants and domesticated animals, as well as much scientific advance, technical innovation, and the security of the many industries that use living resources: *to ensure the sustainable utilization of species and ecosystems* (notably fish and other wildlife, forests and grazing lands), which support millions of rural communities as well as major industries [4].

The publication of this document is itself an important step in the cultural adaptive process at the global level. However, it falls short of discussing the underlying societal causes of perceived threats to the biosphere as a whole, and of discussing the implications of its statements for economic and industrial policies. In any case, despite the fact that in many countries *National conservation strategies* have now been published, evidence is so far lacking of any governments taking the matter seriously.

We have noted elsewhere that the debate about the ecological consequences of continuing industrial growth at the global level has two quite distinct, but often confused, facets—the one concerned with the question of the exhaustion of non-renewable resources, and the other with the possibility of irreversible damage to the bioproductive systems of the biosphere. In this book, the emphasis has been on the second aspect which, from the biological standpoint, is by far the most crucial. With respect to this aspect of the debate we can recognize three broad categories of participants. I refer to them, respectively, as: *the supreme optimists*; *the moderate optimists*; and *the pessimists*.

The pessimists are those who are convinced that the present pattern of increase in resource and energy use and in technological waste production can lead only to the collapse of the global ecosystem, and who are of the opinion that the societal processes behind this pattern are out of control and will remain so. They see global ecocatastrophe for the human species as inevitable. Although individuals holding this view are fairly frequently encountered, their opinions are not well represented in the literature, possibly because they are so convinced of the inevitability of the ultimate destruction of the biosphere that they see little point in spending effort in communicating so gloomy a message.

The supreme optimists are those people who have complete faith in the existing system, and who believe that the best hope for humanity lies in further accelerating technological innovation and further boosting economic growth based on increasing industrial production. They include the authors of such books as *The year 2000*, *In defence of economic growth*, and *The ultimate resource*, as well, it seems, as most politicians and decision-makers in the governments of the high-energy societies [5]. The supreme optimists receive strong support from the powerful commercial and industrial corporations whose economic survival depends on increasing industrial productivity and consumer spending.

The *moderate optimists* share with the pessimists the view that the present pattern of resource and energy use is ecologically unsustainable in the long run. They are nevertheless hopeful that if human society wakes up in time to the nature of this predicament, it may be able to rearrange itself in such a way that the health needs of human beings can be satisfied without increasing indefinitely the per capita consumption of resources and use of energy. Moderate optimists include, among many others, the authors of the following books: *A blueprint for survival*; *The limits to growth*; *The sane alternative*; *The selective conserver society*; *Economics, ecology, ethics*; *Building a sustainable society*; and *The Gaia atlas of planet management* [6].

Reference must be made again here to another area of concern of biohistorical significance. It is the debate about the impacts of societal conditions in the modern technological era on the quality of human experience. On the one hand, there is general appreciation that scientific progress has brought many benefits to humankind. Some of these benefits have taken the form of the removal of important sources of human suffering characteristic of the earlier stages of civilization, and others can be seen as *experiential bonuses* of high-energy society—such as are provided, for example, by motor vehicles, hi-fi stereophonic equipment, hot-water systems, and television. On the other hand, there are many people who are worried about the effects of recent developments on important aspects of human experience. Two of the most persistent and outspoken of these critics are Mumford and Illich. The former sees technology, the power complex, and 'the relentless pursuit of pecuniary gain' to be at the root of undesirable changes in the quality of life. He writes of 'the collective obsessions and compulsions that have misdirected our energies, and undermined our capacity to live full and spiritually satisfying lives' [7].

Some of the authors who have expressed concern about the effects of modern high-energy society on intangible aspects of life experience have pointed out that important interrelationships exist between the technometabolism of society, on the one hand, and the quality of human experience, on the other. Illich, for example, writes:

The energy policies adopted during the current decade will determine the range of social relationships a society will be able to enjoy by the year 2000. A low energy policy allows for a wide choice of lifestyles and cultures. If, on the other hand, a society opts for high energy consumption, its social relations must be dictated by technocracy and will be equally distasteful whether labelled capitalist or socialist [8].

Other writers have similarly drawn attention to the fact that the choice of extrasomatic energy source in the future will have societal consequences far beyond the merely technological ones. It has been suggested, for instance, that while the nuclear power option would necessarily result in a very centralized and elitist society, the solar power option would tend to lead toward decentralization [9].

One encouraging outcome of a biohistorical analysis of the contemporary human predicament is the fact that it indicates that the kind of societal changes which would be necessary to restore balance with the biosphere, involving a reduction in the rate of use of resources and extrasomatic energy, would be entirely consistent with the kinds of changes which are called for by those people who complain about the excessive materialism of our society, the erosion of 'human values', and the decline in the quality of human relationships.

CULTURAL ADAPTATION—SOME FURTHER POINTS

Some further comment is appropriate here on some of the characteristics of cultural adaptation mentioned elsewhere in this book, as they relate particularly to the major ecological and biosocial predicaments of humankind in the modern world. We observed, for instance, that one kind of cultural adaptive response which has come into play when the productivity of ecosystems has become insufficient to sustain a population has been for people to *give up and move on*, and that this has often been successful when there have existed other hospitable places to receive them. Clearly this adaptive response has no future in relation to the three main threats on a global level mentioned at the beginning of this chapter, except in the eyes of those individuals who imagine that technological advances will make it possible for the whole human population to migrate to some other planet where conditions are favourable for humankind.

Another principle relevant to the present predicament is the fact that cultural adaptation is in general more likely to be successful in the long run, and less likely to lead to further undesirable disturbances in natural systems, if it is *corrective* rather than *antidotal*: that is, it is more likely to succeed if it is aimed at correcting the root cause of the unsatisfactory situation rather than at treating the symptoms or consequences of the problem. Corrective cultural adaptation against undesirable changes in biological systems requires some understanding both of the systems themselves as well as of the societal processes which lie behind the undesirable changes.

In earlier chapters we noted a number of factors which sometimes interfere with successful cultural adaptation. The most important of these impediments to effective adaptation is ignorance—a point to which we will return later. Let us first summarize some of the other interfering factors relevant to the present situation.

Any mooted societal change is likely to be, or to be seen to be, a disadvantage to some people or, more importantly in the modern situation, to some autonomous corporate systems. That is to say, there are almost always vested interests which will automatically resist any proposed societal change and which will mount counter attacks of one kind or another against the advocates of such change. In the case of corporate organizations the counter-attack may take the form of propaganda aimed at promoting their own image as benefactors of society and at discrediting the information on which the arguments for change are based; or the organizations may simply use their financial or military power to block moves at the governmental level to introduce the proposed change. These responses are predictable; and it must be remembered that the individuals who are responsible for them are simply behaving in a typically human manner; that is, within the subcultures in which they find themselves they are merely seeking approval, avoiding disapproval, seeking status, showing in-group loyalty, and attempting to ensure their continued employment (indirect subsistence behaviour). They are the relatively innocent pawns of the corporate system to which they have become subservient.

Another set of factors which tend to slow down cultural adaptive processes are to do with 'human nature'. We have noted earlier in this book that the innate behavioural characteristics of the human species are the product of natural selection operating in an environment very different from that in which culture has placed us today. Are the behavioural propensities of humankind, which were entirely appropriate in terms of survival and successful reproduction in the primeval setting, adequate in the biologically novel environments of modern high-energy societies? I suggest that the answer to this question is in the negative, and that humans are disadvantaged in this new situation by a number of *phylogenetic ineptitudes*.

Phylogenetic ineptitudes are broadly of two kinds. First, in some situations certain common behavioural tendencies of the human species, such as the tendencies to seek approval of members of in-groups and to compete with peers, may lead in the new environments to specific behaviours which are against the interests of the individual or of the population. In chapter 3 we referred to this possibility in terms of the 'flaming moth principle'. Similarly, despite laws and religious teachings, the tendencies to show loyalty to in-groups and to be suspicious of out-groups is still a major influence on human behaviour and it can be exploited by vested interests, leading sometimes to societal activities of a totally unreasonable and often dangerous kind. This factor is relevant both to political processes within nations as well as to the great international problems of the present era.

The second kind of phylogenetic ineptitude relates to the concept of *common behavioural deficiencies*. These deficiencies include the relative inability to deal easily and effectively with intangible aspects of human situations, the lack of spontaneous interest in long-term issues, the difficulty experienced in thinking integratively and interrelationally, and the lack of innate concern for the well-being of members of out-groups. These behavioural characteristics are, of course, only 'deficiencies' in the artificial culturally induced environment of civilization; they posed no problem in the evolutionary habitat.

I suggest that these phylogenetic ineptitudes interfere in various ways with the processes of cultural adaptation in the modern world. In my view, the cultural adaptive process is not likely to be successful in the long run unless it incorporates, as it has on occasion in the past, deliberate measures which compensate for these 'weaknesses' (in the modern setting) of the phylogenetic behavioural constitution of the human species [10].

A few more factors which sometimes interfere with the cultural adaptive process are worth brief mention, although they are on the whole less important than those already discussed. We can call them *obstructive attitudes*, and they are:

Belief in infinite human adaptability. This is the common assumption that the adaptive capacities of the human species are almost infinite. 'We' have adapted before and will do so again. This attitude is linked to the technofix syndrome—that is, the culturally induced faith in the human capacity for technological innovation as a sure means of overcoming any difficulties that the future might hold.

The inevitability syndrome. This is the idea that 'you cannot stop progress', even if you happen strongly to disapprove of it. Such things are beyond human control; we are too far along the track, and there is no turning back. This view ignores the fact that cultural adaptation in the past has often involved 'going backwards'. Examples are the reintroduction of fresh fruit and vegetables to overcome scurvy and the present wave of enthusiasm for physical exercise as a health-promoting activity.

The positive proof syndrome. This common state of mind is reflected, for example, in the argument that there is no positive proof that certain industrial by-products will affect the ozone layer, and that therefore no action is necessary. It is a comforting response to the uncertainties of our time. Its weakness lies in the fact that, once positive proof exists, it may be too late to take useful action.

Number fixation. This is the unscientific tendency, prevalent in technological society, to treat as unimportant those aspects of human situations which cannot easily be measured or expressed in quantitative terms.

The main lesson to be learned from these considerations is that, if cultural adaptation is to be successful in the highly complex and dangerous situation that exists today, it must involve deliberate and systematic action aimed not only at promoting understanding of the human situation in ecological and biosocial perspective, but also at overcoming the various impediments that are liable to stand in the way of effective societal adaptative responses.

THE TRANSITION TO AN ECOLOGICAL PHASE FIVE OF HUMAN EXISTENCE— SOME CONCLUDING COMMENTS

Earlier in this book we discussed a biological evolutionary phenomenon which we referred to as *dead-end adaptation*. This term applies to the situation which occurs when natural selection gives rise to specific new evolutionary adaptations or specializations which are of selective advantage in the conditions in which they evolve, but which turn out to be a disadvantage if the life conditions of the species change. As a consequence, the species may become extinct. It is pertinent to ask whether the human capacity for culture may be an example of such dead-end adaptation. Clearly, this characteristic was of selective advantage in the primeval situation; but will it continue to be an advantage in the new and changing habitat that culture itself has produced? Or will the human capacity for culture ultimately lead to the extinction of the species?

The future will provide the answer to this question. Certainly, culture is at present, in some spheres of human activity, dangerously out of control, and its consequences threaten the survival not just of local communities, but of the whole human species. As pointed out by White, it seems as if we do not possess culture but are possessed by it [11]. And yet, of course, the only possible means of bringing culture properly under control is through the application of the human capacity for culture—that is, through cultural adaptation. Some important progress has, of course, been made, in that we have recently left behind the absurd notion that humankind's task is to conquer nature. But society as a whole has yet fully to wake up to the fact that the real threats to humanity are posed, not by nature, but by culture. In other words, it is culture, not nature, that has to be subdued.

The greatest single impediment to successful cultural adaptation has always been ignorance. First and foremost, therefore, effective adaptation to the major culturally induced threats to the world's ecosystems and to humanity will depend on a greatly improved understanding in society as a whole of the human situation in the biosphere. Such improvement in understanding would lead to a shift in the dominant outlook in society [12], and this in turn would lead to changes in societal policies and activities—changes compatible with long-term ecological sustainability and human health and well-being.

NOTES

1. Dimbleby 1976, p. 12.
2. Thomas 1956, pp. xxvii–xi.
3. Zolotas 1981, p. 191.
4. IUCNNR, UNEP, and WWF 1980.
5. Kahn and Wiener 1967; Beckerman 1974; Simon 1981.
6. Goldsmith, Allen, and Allaby 1972; Meadows, Meadows, Randers, and Behrens 1972; Robertson 1979; Valaskakis, Sindell, and Smith 1977; Daly 1980; Brown 1981; Myers 1985.
7. Mumford 1971, p. 167.
8. Illich 1974, p. 16.

9. Hayes 1977; Lovins 1977; Lonnroth, Johansson, and Steen 1980, p. 156.
10. A good example of such compensatory cultural adaptation is the introduction of laws in the early cities to compensate for the lack of innate concern in humans for members of out-groups. Religious teachings (e.g. the story of the Good Samaritan) provide another example.
11. White 1949, p. 126.
12. The word *outlook* is used here as a comprehensive term for the complex of mental states, including values, opinions, aspirations, intentions, and attitudes, which determine both the behaviour of individuals and, in democratic countries, the aims of societal policies.

APPENDIX: CONCEPTUAL SCHEMES

INTRODUCTION

The conceptual framework which my colleagues and I have been using reflects the sequence of events in the history of life on earth mentioned above. That is, we recognize three fundamental clusters of variables—those relating to the *biosphere* (minus the human and societal components), those relating to *humans* (as biological organisms), and those relating to *society* (and culture), and it is concerned with the interrelationships between these different sets of variables.

Various models can be constructed for use in different situations (or for different purposes) incorporating these three fundamental sets of variables and various subsets. Here I will mention just two models or *orientations* which we have used most frequently—concerned respectively with two different and essential sets of relationships. The first is referred to as the *total environment* and the second as *human experience*. In both cases the emphasis is on the interrelationships between *society* and *biological systems*:

Biological system(s) ←————————→ Society

THE TOTAL ENVIRONMENT ORIENTATION

This orientation is concerned with 'the system as a whole' and particularly with the ecosystems of the biosphere and their interrelationships with society:

Biosphere (and its ecosystems) ←————————→ Society

The main subsets of variables in the system are depicted in Fig. A.1.

With respect to the *interrelationships* between the biosphere and society we are concerned especially with the flow of energy, of renewable and non-renewable resources, and of organic and inorganic wastes, and with impacts of societal activities on soil, the atmosphere, the oceans, and populations of plants and animals. We are also concerned with the responses of society when societal activities give rise to changes in the biosphere which are perceived to be undesirable. The important interrelationships between society and the biosphere are listed in Fig. A.2.

If desired, this framework can be used as a starting point for exploring the societal factors which lie behind a given societal activity seen as responsible for

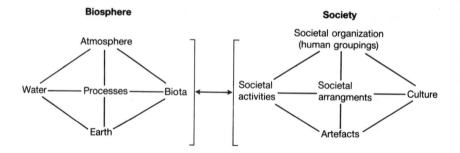

Fig. A.1 The total environment.

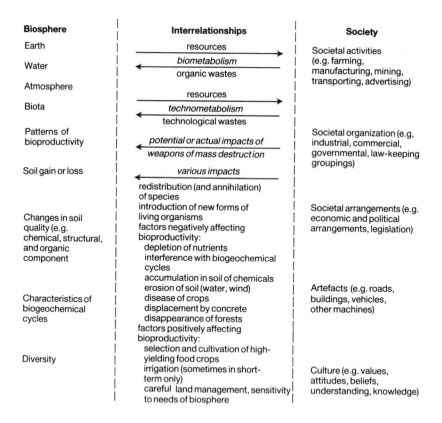

Fig. A.2 The total environment, depicting important interrelationships between society and the biosphere.

changes in a biological system. This process may result, for example, in the identification of *societal groupings* which are directly or indirectly responsible for the societal activity in question, or in appreciation of the relevance of certain *societal arrangements* (e.g. economic or political) and of the role of aspects of *culture* (e.g. societal values, aspirations, and belief) and of certain *artefacts* (e.g. machines).

THE HUMAN EXPERIENCE ORIENTATION

The second orientation focuses on the interrelationships between individual humans (or groups of humans) and the 'system as a whole'—that is, society and the biosphere.

The actual life experience of humans can usefully be viewed as being made up of three sets or clusters of variables. These are referred to respectively as the *biopsychic state*, the *immediate environment*, and the *behaviour pattern*. The relationships between these three different sets of variables and between them and the total environment are discussed at the beginning of chapter 3.

THE USE OF CHECK-LISTS AND THE IMPORTANCE OF INTANGIBLES

My colleagues and I have found *check-lists* to be an invaluable device for ensuring that, as far as possible, all important and relevant variables, concepts, and principles are taken into account in the biohistorical assessment of human situations. We have prepared check-lists for the different clusters of variables depicted in Figs. A.1 and A.2 of this Appendix and for Figs. 3.1 and 3.2 (chapter 3). The check-lists are used especially in the early stages of a study, when items deemed to be relevant to the situation are selected for detailed consideration in the project.

It is important to appreciate that the check-lists of variables do not provide any information on the nature of interrelationships in a system, nor do they give rise to predictions. They are no more than methodological devices to ensure that consideration is given to the full spectrum of variables and principles that may be pertinent.

Check-lists of variables

Biosphere

earth:

rock, soil
water content
nutrient content
content of other chemicals
content of living organic matter
content of decomposing organic matter
content of fossilized organic matter
depth of soil
depth of top-soil

water:

    in soil
    in atmosphere
    in waterways
    in oceans
    content of particulate matter
    content of living organic matter
    content of decomposing organic matter
    content of chemical compounds

atmosphere:

    its layers
    gas content (especially CO, CO^2, sulphur and nitrogen oxides, water)

biota:

    animals
    plants
    bacteria, fungi, protozoa, etc.
    viruses
    biomass

other considerations:

    pattern and level of bioproductivity
    soil gain or loss (and causes)
    changes in soil quality (e.g. chemical content, water content, organic component, texture, density)
    diversity of life forms
    characteristics of biogeochemical cycles
    inputs of wastes from society
    aesthetic and/or recreational value to humans

Biosphere—Society variables

    biometabolism of society
    technometabolism of society
    renewable resources—rates of extraction by society
    non-renewable resources—rates of extraction by society
    flow patterns between society and the biosphere of somatic energy and extrasomatic energy
    organic wastes—rate (and mode) of discharge by society
    technological (industrial) wastes—rate (and mode) of discharge by society
    impacts of societal activities on soil quantity, soil quality, atmosphere, water, biota.

Humans

life conditions:

    air quality

noise levels

radiation

ionization of atmosphere

diet—quantity and quality (energy, nutrients, contaminants, consistency)

water—availability and quality (and chemical and biotic contaminants)

psychotropic drugs—availability, kinds, use by others in environment

pharmaceutical drugs—availability, kinds, uses

pathogenic organisms—likelihood of contact, kinds

plants, animals, nature—contacts and interactions

goods—possessed by individual (and family); possessed by others in environment

machines in environment

built environment—general; proximity of buildings; size/scale of buildings; design of buildings (aesthetic and behavioural aspects)

dwellings—type; protection afforded; population density within; sanitation; ventilation; amenities (water supply, toilet, cooking, etc.); energy use

population density experienced

social interactions—number of interactions; number of different individuals; type of relationship

psychological support network—general

family experience—structure and size of family; degree and nature of inter
actions; extended family as source of support

community experience—community interaction; responsibility for local affairs; community spirit

work experience—interest value; variety versus monotony; sense of personal involvement (also see other life conditions items below); societal implications

job opportunities

co-operative small-group interaction—general

opportunities and incentives for initiative

opportunities and incentives for learning and practising manual skills

opportunities and incentives for creative behaviour

distance of work place from residence (travelling time, etc.)

consumer behaviour—general pattern

variety versus monotony in daily experience

interest value of personal environment

accessibility of information wanted

accessibility of information relevant to the understanding of the problems of society and of individuals

learning experience (in terms of spontaneity, relevance, and enjoyment)

educational opportunities

physical work pattern (level and kind)

sleeping and resting pattern

pattern of active recreation

pattern of passive recreation

opportunities for spontaneity in daily experience

physical violence in environment

other antisocial behaviours in environment

danger in environment

novelty value of environment

predictability/unpredictability in environment (likelihood of unexpected occurrences)
aesthetic quality of environment
opportunities for maintaining/improving status
goal-achievement cycles (short/long)
aspirations-fulfilment ratio

time budget—general
extent to which life conditions are conducive to sense of personal involvement, etc. (see list under *biopsychic state*, below)

other considerations:

personal freedom
membership of voluntary groupings
in-group, out-group experience
participation in decision-making
availability and quality of services—health, legal, educational, transport
availability and quality of consumer goods
mobility (geographical, social)
disparities in life conditions among the members of the community
in general—how people spend their time; *disparities* and *differentials* (the degree of variability among sections of the human population with respect to life conditions, as affected by economic status)

biopsychic state:

general variables:

stature
weight
age
colour of hair, skin, eyes

variables relevant to well-being:

nutritional state (and growth rate in children)
contagious disease
bacterial
protozoal
viral
'parasitic' disease—metazoal parasites
organ health
arteries and heart
lungs
gastro-intestinal tract
other
physical fitness
blood pressure
mental health

sense of personal involvement; challenge; belonging; comradeship and love; responsibility; interest; excitement; confidence; self-fulfilment; self-esteem; security; hope; being approved of; beauty; enjoyment; distress; deprivation; alienation; anomie; apathy; boredom; frustration; loneliness; inadequacy; insecurity; fear.

REFERENCES

Adena, M. A., Montesin, H. J., and Gibson, J. B. (1983). *Regional variation in mortality in Melbourne.* Department of Population Biology, Australian National University, Canberra.

Adovasio, J. M. and Carlisle, R. C. (1984). An Indian hunters' camp for 20 000 years. *Scientific American* **250**(5), pp. 104-8.

Albanese, A. A. and Oslo, Z. A. (1968). Proteins and amino-acids. In *Modern nutrition in health and disease* 4th edition (eds. M. G. Wohl, and R. S. Goodhart), pp. 95-155. Lea and Febiger, Philadelphia, PA.

Albury, W. R. (1980). Politics and rhetoric in the sociobiology debate. *Social Studies of Science* **10**, pp. 519-36.

Alexandersson, G. and Klevebring, B. I. (1978). *World resources: energy, metals, minerals.* Walter de Gruyter, Berlin.

Allaby, M. and Lovelock, J. (1980). Spray cans: the threat that never was. *New Scientist* **87**(1210), pp. 212-4.

Alland, A. (1968). War and disease: an anthropological perspective. In *War* (eds. M. Fried, M. Harris, and R. Murphy), pp. 65-75. Natural History Press, New York.

Alland, A. (1970). *Adaptation in cultural evolution: an approach to medical anthropology.* Columbia University Press, New York.

Allport, G. W. (1958). What units shall we employ. In *Assessment of human motives* (ed. G. Lindzey), pp. 239-60. Rinehart, New York.

Ambard, L. and Beaujard, E. (1904). Causes de l'hypertension arterielle. *Archives Generales de Medecine* **1**, pp. 520-33.

Anderson, R. and May, R. (1982). The logic of vaccination. *New Scientist* **96**(1332), pp. 410-15.

Antar, M. A., Ohlson, M., and Hodges, R. E. (1964). Changes in retail market food supplies in the United States in the last seventy years in relation to the incidence of coronary heart disease, with special reference to dietary carbohydrates and essential fatty acids. *American Journal Clinical Nutrition* **14**, pp. 169-78.

Ardrey, R. (1961). *African genesis.* Collins, London.

Ardrey, R. (1966). *The territorial imperative.* Collins, London.

Ardrey, R. (1970). *The social contract.* Collins, London.

Ardrey, R. (1976). *The hunting hypothesis.* Atheneum, New York.

Arndt, H. W. (1978). *The rise and fall of economic growth: a study in contemporary thought.* Longman Cheshire, Melbourne.

Arriaga, E. E. (1968). *New Life tables for Latin American populations in the nineteenth and twentieth centuries—Population Monograph no. 3.* Institute of International Studies, University of California, Berkeley, CA.

Australian Bureau of Mineral Resources. (1984). *Australian Mineral Industry Annual Review 1984.* Australian Bureau of Mineral Resources, Canberra.

Australian Department of Health. (1982). *Annual Report of the Director-General of Health 1981-82.* Australian Government Publishing Service, Canberra.

Australian National Heart Foundation. (1983). *A profile of Australians: a summary of the National Heart Foundation risk factor prevalence study.* National Heart Foundation, Canberra.

Bacon, F. (1905). *Novum Organum* (English, translated by R. Ellis and J. Spedding, with preface and notes). G. Routledge, London.

Baker, H. (1970). *Plants and civilization* (2nd edn). Macmillan, London.

Baklien, A. (1981). Chemicals, cancer and birth defects. *Search* 12(1-2), pp. 30-6.

Baldassare, M. and Fischer, C. S. (1977). The relevance of crowding experiments to urban studies. In *Perspectives on environment and behavior: theory, research and applications* (ed. D. Stokols), pp. 273-86. Plenum Press, New York.

Banks, A. L. (1969). Catastrophes and restraints. In *Population and food supply: essays on human needs and agricultural prospects* (ed. J. Hutchinson), pp. 47-60. Cambridge University Press, London.

Banton, M. (ed.)(1961). *Darwinism and the study of society: a centenary symposium.* Tavistock Publications, London.

Barash, D. P. (1977). *Sociobiology and behavior.* Elsevier, New York.

Barbour, I. G. (1980). *Technology, environment and human values.* Praeger, New York.

Barnaby, F., Kristofferson, L., Rodhe, H., Rotblat, J., and Prawitz, J. (1982). Conclusions. *Ambio* 11(2), pp. 161-2.

Barnaby, F. and Rotblat, J. (1982). The effects of nuclear weapons. *Ambio* 11(2-3), pp. 84-93.

Barnard, A. (1983). Contemporary hunter-gatherers: current theoretical issues in ecology and social organisation. *Ann. Rev. Anthropol.* 12, pp. 193-214.

Barnes, F. (1970). The biology of preneolithic man. In *The impact of civilization on the biology of man* (ed. S. V. Boyden), pp. 1-18. Australian National University Press, Canberra.

Barnet, R. J. and Muller, R. E. (1974). *Global reach: the power of the multinational corporations.* Simon and Schuster, New York.

Barnett, H. J. and Morse, C. (1963). *Scarcity and growth: the economics of natural resource availability.* Johns Hopkins Press, Baltimore, MD.

Barnett, S. A. (1973). On the hazards of analogies. In *Man and aggression* (2nd edn) (ed. A. Montagu), pp. 75-83. Oxford University Press, New York.

Barnett, S. A. (1981). *Modern ethology: the science of animal behavior.* Oxford University Press, New York.

Barnicot, N. A., Bennett, F. J., Woodburn, J. C., Pilkington, T. R. E., and Antonis, A. (1972). Blood pressure and serum cholesterol in the Hadza of Tanzania. *Human Biology* 44(1), pp. 87-116.

Bates, D. G. and Lee, S. H. (1979). The myth of population regulation. In *Evolutionary biology and human social behaviour: an anthropological perspective* (eds. N. A. Chagnon and W. Irons), pp.273-89. Duxbury Press, North Scituate, MA.

Baum, A. and Epstein, Y. M. (eds.)(1978). *Human response to crowding.* Lawrence Erlbaum, Hillsdale, NJ.

Baum, A., Harpin, R. E., and Valins, S. (1975). The role of group phenomena in the experience of crowding. *Environment and Behavior* 7(2), pp. 185-99.

Beard, T. C. (1979). *Promoting health: prospects for better health throughout Australia.* Australian Government Publishing Service, Canberra.

Beard, T. C., Gray, W. R., Cooke, H. M., and Barge, R. (1982). Randomised controlled trial of a no-added-sodium diet for mild hypertension. *Lancet* 2, pp. 455-8.

Beckerman, W. (1974). *In defence of economic growth.* Jonathan Cape, London.

Benjamin, B. (1971). Bereavement and heart disease. *Journal Biosocial Science* 3, pp. 61-7.

Bennett, J. W. (1976). *The ecological transition: cultural anthropology and human adaptation.* Pergamon, New York.

Bennett, K. A. (1979). *Fundamentals of biological anthropology.* W. C. Brown, Dubuque, IO.

Berndt, R.M. and Berndt, C. H. (1964). *The world of the first Australians: an introduction to the traditional life of the Australian Aborigines.* Ure Smith, Sydney.

Bews, J.W. (1935). *Human ecology.* Oxford University Press, London.

Bexenius, L. (ed.)(1981). *Report 81: on the alcohol and drug situation in Sweden.* Swedish Council for Information on Alcohol and Other Drugs, Stockholm.

Birch, C. (1980). The sustainable and just global society. In *Australian politics* (eds. H. Mayer, and H. Nelson), pp. 3-16. Longman Cheshire, London.

Blum, R. H. (1969a). A background history of drugs. In *Society and drugs: social and cultural observations* (eds. R. H. Blum and associates), pp. 3-23. Jossey-Bass, San Francisco, CA.

Blum, R. H. (1969b). A history of alcohol. In *Society and drugs: social and cultural observations* (eds. R. H. Blum and associates), pp. 25-42. Jossey-Bass, San Francisco, CA.

Blum, R. H. (1969c). A history of opium. In *Society and drugs: social and cultural observations* (eds. R. H. Blum and associates), pp. 45-58. Jossey-Bass, San Francisco, CA.

Blurton-Jones, N. and Reynolds, V. (eds.)(1978). *Human behaviour and adaptation.* Taylor and Francis, London.

Bocquet, A. (1979). Lake-bottom archaeology. *Scientific American* 240(2), pp. 48-56.

Borgia, G. (1980). Human aggression as a biological adaptation. In *The evolution of human social behavior* (ed. J. Lockard), pp. 165-91. Elsevier, New York.

Borrie, W. D. (1972). Population, environment and society. *Sir Douglas Robb Lectures.* Auckland University Press, Auckland.

Boserup, E. (1965). *The conditions of agricultural growth: the economics of agrarian change under population pressure.* Aldine, Chicago, IL.

Boulding, K. E. (1966). The economics of coming spaceship Earth. In *Environmental quality: in a growing economy* (ed. H. Jarrett), pp. 3-14. Johns Hopkins Press, Baltimore, MD.

Boyden, S. (1972). Ecology in relation to urban population structure. In *The structure of human populations* (eds. G. A. Harrison and A. J. Boyce), pp. 411-41 Clarendon Press, Oxford.

Boyden, S. (1973). Evolution and health. *Ecologist* 3(8), pp. 304-9.

Boyden, S. (1979). *An integrative ecological approach to the study of human settlements.* UNESCO, Paris.

Boyden, S. (1980). The need for an holistic approach to human health and well-being. In *Changing disease patterns and human behaviour* (eds. N. F. Stanley and R. A. Joske), pp. 622-44. Academic Press, London.

Boyden, S. and Millar, S. (1978). Human ecology and the quality of life. *Urban Ecology* 3, pp. 263-87.

Boyden, S., Millar, S., Newcombe, K., and O'Neill, B. (1981). *The ecology of a city and its people: the case of Hong Kong.* Australian National University Press, Canberra.

Briggs, A. (ed.)(1962). *William Morris: selected writings and designs.* Penguin Books, Harmondsworth, Middx.

Brock, J. F. (1963). Sophisticated diets and man's health. In *Man and his future* (ed. G. Wolstenholme), pp. 36-56. J. & A. Churchill, London.

Brooke-Thomas, R. (1979). *Human adaptation to a high Andean energy flow system.* Department of Anthropology, Pennsylvannia State University, University Park, PA.

Broomfield, J. H. (1981). The lethal Meccano. *Australian Journal of Politics and History* 27(1), pp. 30-9.

Brosin, H. W. (1968). The psychology of appetite. In *Modern nutrition in health and disease* (eds. M.G. Wohl, and R.S. Goodhart), pp. 63-75. Lea and Febiger, Philadelphia, PA.

Brothwell, D. and Sandison, A. T. (eds.)(1967). *Diseases in antiquity: a survey of the disease, injuries and surgery of early populations.* Charles C. Thomas, Springfield, MA.

Brown, L. R. (1970). Human food production as a process in the biosphere. In *The biosphere* (compiler *Scientific American*), pp. 93-104. W.H. Freeman, San Francisco, CA.

Brown, L. R. (1978). *The twenty-ninth day: accommodating human needs and numbers to the Earth's resources.* W. W. Norton, New York.

Brown, L. R. (1981). *Building a sustainable society.* W. W. Norton, New York.

Brown, L. R., *et al.* (1984). *State of the world 1984: a Worldwatch Institute report on progress toward a sustainable society.* W. W. Norton, New York.

Brown, L. R., *et al.* (1985). *State of the world 1985: a Worldwatch Institute report on progress toward a sustainable society.* W. W. Norton, New York.

Brown, L. R., *et al.* (1986). *State of the world 1986: a Worldwatch Institute report on progress toward a sustainable society.* W. W. Norton, New York.

Brown, L. R. and Finsterbusch, G. W. (1972). *Man and his environment: food.* Harper and Row, New York.

Bruch, H. (1978). *The golden cage: the engine of anorexia nervosa.* Random House, New York.

Brunig, E. F. (1977). The tropical rain forest: a wasted asset or an essential biospheric resource? *Ambio* 6(4), pp. 187-91.

Bryce-Smith, D. and Waldron, H. A. (1974). Lead, behaviour and criminality. *The Ecologist* 4(10), pp.367-77.

Bull, D. (1982). *A growing problem: pesticides and the third world poor.* Oxfam, Oxford.

Bullock, J. and Lewis, W. M. (1968). The influence of traffic on atmospheric pollution: the High Street—Warwick. *Atmospheric Pollution* 2, pp. 517-34.

Bunney, S. (1984). Did Lucy stand alone. *New Scientist* 101(1395), p. 18.

Burkitt, D. P. (1982). Diet and diseases of affluence. In *Diseases and the environment* (eds. A. R. Rees, and H. J. Purcell), pp. 167-73. John Wiley, New York.

Burkitt, D. and Trowell, H. C. (eds.) (1975). *Refined carbohydrate foods and disease: some implications of dietary fibre.* Academic Press, London.

Burns, S. (1977). *The household economy: its shape, origins and futures.* Beacon Press, Boston, MA.

Burrows, D. J. and Lapides, F. R. (eds.)(1970). *Alienation: a casebook.* Thomas Y. Crowell, New York.

Bushnell, G. H. S. (1976). The beginning and growth of agriculture in Mexico. *Phil. Trans. R. Soc. Lond. B.* **275**, pp. 117–20.

Butterworth, C. E. (1972). Iron 'undercontamination'? *The Journal of the American Medical Association* **220**(4), pp. 581–2.

Caffrey, B. (1968). Reliability and validity of personality and behavioral measures in a study of coronary heart disease. *J. Chron. Dis.* **21**, pp. 191–201.

Cairns, J. (1975). The cancer problem. *Scientific American* **233**(9), pp. 64–78.

Calhoun, J. B. (1962). Population density and social pathology. *Scientific American* **206**(2), pp.139–48.

Campbell, B. (1971). Sources of order and disorder. *New Scientist and Science Journal* **49**(742), pp. 566–68.

Caplan, A. L. (ed.)(1978). *The sociobiology debate: readings on ethical and scientific issues.* Harper and Row, New York.

Carefoot, G. L. and Sprott, E. R. (1969). *Famine on the wind: plant diseases and human history.* McGill–Queens University Press, Montreal.

Carroll, K. K. (1975). Experimental evidence of dietary factors and hormone-dependent cancers. *Cancer Research* **35**, pp. 3374–83.

Carroll, K. K. and Hopkins, G. J. (1979). Dietary polyunsaturated fat versus saturated fat in relation to mammary carcinogenesis. *Lipids* **14**(2), pp. 155–8.

Carr-Saunders, A. M. (1965). *World population: past growth and present trends.* Barnes and Noble, New York.

Carruthers, J., Clark, J., De Smet, M., Freckleton, F., McClelland, R., Pike, G., and Roe, M. (1973). Historical aspects of energy use by mankind. In *Energy and how we live* (compiler Australian-National Committee for Man and the Biosphere), pp. 27–66. Australian-UNESCO Seminar, 16–18 May 1973, Flinders University of South Australia, Canberra. MAB Publication No. 4.

Carson, R. L. (1962). *Silent spring.* Houghton Mifflin, Boston, MA.

Carter, G. F. (1977). A hypothesis suggesting a single origin of agriculture. In *Origins of agriculture* (ed. C.A. Reed), pp. 89–133. Mouton, The Hague.

Cary, M. (1962). *A history of Rome down to the reign of Constantine.* Macmillan, London.

Cashel, K. (1981*a*). Apparent consumption of foodstuffs and nutrients in Australia. *Proc. Nutri. Soc. Aust.* **6**, pp. 112–16.

Cashel, K. (1981*b*). National statistics: apparent consumption of foodstuffs and nutrients in Australia. *Menzies Foundation* **3**, pp. 203–19.

Cassel, J. (1976). The contribution of the social environment to host resistance. *American Journal of Epidemiology* **104**(2), pp. 107–23.

Cavalli-Sforza, L. L. and Bodmer, W. F. (1971). *The genetics of human populations.* W. H. Freeman, San Francisco, CA.

Chagnon, N. A. and Irons, W. (eds.)(1979). *Evolutionary biology and human social behaviour: an anthropological perspective.* Duxbury Press, North Scituate, MA.

Chambers, L. A. (1976). Classification and extent of air pollution problems. In *Air pollution: air pollutants, their transformation and transport* (3rd edn) (ed. A. C. Stern) Vol. 1, pp. 3-22. Academic Press, New York.

Chambers dictionary of science and technology (1974). Chambers, Edinburgh.

Chandler, T. and Fox, G. (1974). *3000 years of urban growth*. Academic Press, New York.

Charney, J., Stone, P. H., and Quirk, W. J. (1975). Drought in the Sahara: a biogeophysical feedback mechanism. *Science* **187**(4175), pp. 434-5.

Cherfas, J. (1983). Trees have made man upright. *New Scientist* **97**(1341), pp. 172-8.

Chick, H. (1958). Wheat and bread: a historical introduction. *Proceedings of the Nutrition Society* **17**, pp. 1-7.

Choi, S. C., Mirjafari, A., and Weaver, H. B. (1976). The concept of crowding: a critical review and proposal of an alternative approach. *Environment and behaviour* **8**(3), pp. 345-62.

Christian, J. J. (1961). Phenomena associated with population density. *Proceedings of the National Academy of Science* **47**, pp. 428-44.

Christie, M. (1980*a*). *Changing consumer behaviour in Papua New Guinea: its social and ecological implications*. Centre for Resource and Environmental Studies, Australian National University, for UNESCO, Canberra.

Christie, M. (1980*b*). *Consumer behaviour in Papua New Guinea*. PhD thesis, Centre for Resource and Environmental Studies, Australian National University, Canberra.

Christopher, A. J. (1980). Environmental lead questions: a point of view from Victoria. *Medical Journal of Australia* **2**, pp. 300-4.

Clark, C. and Haswell, M. (1970). *The economics of subsistence agriculture* (4th edn). Macmillan, London.

Clark, C. G. (1977). *Population growth and land use* (2nd edn). Macmillan, London.

Clark, J. G. D. (1952). *Prehistoric Europe: the economic basis*. Stanford University Press, Stanford, CA.

Clark, J. G. D. (1961). *World prehistory: an outline*. Cambridge University Press, Cambridge.

Clark, K. (1969). *Civilisation: a personal view*. BBC, London.

Clarke, R. and Hindley, G. (1975). *The challenge of the primitives*. Jonathan Cape, London.

Clarke, W. C. (1971). *Place and people: an ecology of a New Guinean community*. Australian National University Press, Canberra.

Clarke, W. C. (1973). The dilemma of development. In *The Pacific in transition: geographical perspectives on adaptation and change* (ed. H. C. Brookfield), pp. 275-98. Australian National University Press, Canberra.

Cleave, T. L. and Campbell, G. R. (1966). *Diabetes, coronary thrombosis and the saccharine disease*. John Wright, Bristol.

Cobb, S. (1976). Social support as a moderator of life stress. *Psychosomatic Medicine* **38**(5), pp. 301-14.

Cockburn, T. A. (1971). Infectious diseases in ancient populations. *Current Anthropology* **12**(1), pp. 45-62.

Coffin, D. L. and Knelson, J. H. (1976). Effects of sulfur dioxide and sulfate aerosol particles on human health. *Ambio* **5**(5-6), pp. 239-42.

Cohn, C. and Joseph, D. (1962). Influence of body weight and body fat on appetite of 'normal' lean and obese rats. *Yale Journal of Biology and Medicine* **34**, pp. 598–607.

Cole, H. S. D., Freeman, C., Jahoda, M., and Pavitt, K. L. R. (eds.)(1973). *Thinking about the future: a critique of the limits to growth*. Chatto and Windus, London.

Cole, L. C. (1971). Thermal pollution. In *Pollution* (eds. R. S. Leisner and E. J. Kormondy), pp. 63–7. Wm. C. Brown, Dubuque, IO.

Cook, E. (1971). The flow of energy in an industrial society. *Scientific American* **224**(3), pp. 135–44.

Cook, P. S. (1977). Diet and the hyperkinetic syndrome children. In *The impact of environment and lifestyle on human health* (eds. M. Diesendorf and B. Furnass), pp. 85–93. Society for Social Responsibility in Science (ACT), Canberra.

Cook, P. S. and Woodhill, J. M. (1976). The Feingold dietary treatment of the hyperkinetic syndrome. *Medical Journal of Australia* **2**(3), pp. 86–90.

Coombs, H. C. (1980). *Economic change and political strategy*. CRES: Working Paper HCC/WP19. Centre for Resource and Environmental Studies, Australian National University, Canberra.

Coon, C. S. (1962). *The origin of races*. Alfred A. Knopf, New York.

Council on Environmental Quality (1980). *The Global 2000 Report to the President*. Council on Environmental Quality, Washington DC.

Cowhig, J. (1972). Is there more to obesity than food? *New Scientist* **55**(803), pp. 31–3.

Crawford, M. A. (1968). Fatty-acid ratios in free living and domestic animals: possible implications for atheroma. *Lancet* **1**, pp. 1329–33.

Crossley, D. (1979). Energy and society: an annotated bibliography. *Social Alternatives* **1**(5), pp. 92–5.

Crutzen, P. J. and Birks, J. W. (1982). The atmosphere after a nuclear war: twilight at noon. *Ambio* **11**, pp. 115–25.

Dahlberg, F. (ed.)(1981). *Woman the gatherer*. Yale University Press, New Haven, CT.

Daly, H. E. (1972). The stationary-state economy. *The Ecologist* **2**(7), pp. 4–12.

Daly, H. E. (1973). How to stabilise the economy. *The Ecologist* **3**(3), pp. 90–6.

Daly, H. E. (1975). A model for a steady-state economy. In *Beyond growth: essays on alternative futures* (eds. W. R. Burch and F. H. Bormann), pp. 127–45. School of Forestry and Environmental Studies, Yale University, New Haven, CT.

Daly, H. E. (1977). *Steady-state economics: the economics of biophysical equilibrium and moral growth*. W. H. Freeman, San Francisco, CA.

Daly, H. E. (ed.)(1980). *Economics, ecology, ethics: essays toward a steady state economy*. W. H. Freeman, San Francisco, CA.

Darby, H. C. (1956). The clearing of the woodland in Europe. In *Man's role in changing the face of the Earth* (ed. W. C. Thomas), pp. 183–216. University of Chicago Press, Chicago, IL.

Darmstadter, J. (1971). *Energy in the world economy: a statistical review of trends in output trade and consumption since 1925*. Johns Hopkins Press, Baltimore, MD.

Dart, R. A. (1957). *The osteodontokeratic culture of Australopithecus Prometheus.* Transvaal Museum, Pretoria.

Darwin, C. (1873). *The expression of the emotions in man and animals.* J. Murray, London.

Davies, N. B. and Krebs, J. R. (1978). Introduction: ecology, natural selection and social behavior. In *Behavioural ecology: an evolutionary approach* (eds. J. R. Krebs and N. B. Davies), pp. 1-18. Blackwell Scientific Publications, Oxford.

Davis, K. (1965). The urbanisation of the human population. *Scientific American* **213**(3), pp. 41-53.

Davison, I. (1977). *Values, ends, and society.* University of Queensland Press, St. Lucia, Queensland.

Dawkins, R. (1976). *The selfish gene.* Oxford University Press, Oxford.

Day, L. H. and Day, A. T. (1964). *Too many Americans.* Delta, New York.

de Beer, G. (1975). Review of mankind evolving: the evolution of the human species by Theodosius Dobzhansky. In *Biological anthropology* (compiler S. H. Katz) Readings from Scientific American, pp. 415-17. W. H. Freeman, San Francisco, CA.

De Jouvenel, B. (1972). Utopia for practical purposes. In *Human identity in the urban environment* (eds. G. Bell and J. Tyrwhitt), pp. 164-70. Penguin, Harmondsworth, Middx.

DeLuca, J. R. (ed.)(1981). *Alcohol and health: fourth report to the US Congress.* Department of Health and Human Services, Rockville, MD.

de Lumley, H. (1969). A paleolithic camp at Nice. *Scientific American* **220**(5), pp. 42-50.

Derry, T. K. and Williams, T. I. (1960). *A short history of technology: from the earliest times to AD 1900.* Clarendon Press, Oxford.

Dimbleby, G. W. (1972). Impact of early man on his environment. In *Population and pollution* (eds. P. R. Cox and J. Peel), pp. 7-13. Academic Press, London.

Dimbleby, G. W. (1976). Climate, soil, and man. *Phil. Trans. R. Soc. Lond. B.* **275**, pp. 197-208.

Divale, W. T. (1972). Systematic population control in the middle and upper paleolithic: inferences based on contemporary hunter gatherers. *World Archaeology* **4**(2), pp. 222-43.

Dixon, P. W. (1979). A Neolithic and Iron Age site on a hilltop in Southern England. *Scientific American* **241**(5), pp. 143-50.

Dobzhansky, T. (1958). Evolution at work. *Science* **127**(3306), pp. 1091-8.

Dobzhansky, T. (1962). *Mankind evolving: the evolution of the human species.* Yale University Press, New Haven, CT.

Dobzhansky, T. (1963). Books: a debatable account of the origin of races. (Book review of 'The origin of races' by C. S. Coon). *Scientific American* **208**(2), pp. 169-72.

Doke, C. M. 1937. *Games, play and dances of the Khomani bushmen.* Witwatersrand Press, Johannesburg.

Doll, R. (1977). Strategy for detection of cancer hazards to man. *Nature* **265**, pp. 589-96.

Doll, R. and Armstrong, B. (1981). Cancer. In *Western diseases: their emergence and prevention* (eds. H. C. Trowell and D. P. Burkitt), pp. 93-112. Edward Arnold, London.

Dowie, J. A. (1972). Economic history: time for some anti-growth. *Australia Economic History Review* **12**, pp. 161–73.

Draper, P. (1975). !Kung Women: contrast in sexual egalitarianism in foraging and sedentary contexts. In *Towards an anthropology of women* (ed. R. R. Reiter), pp. 77–109. Monthly Review Press, New York.

Drummond, J. C. and Wilbraham, A. (1958). *The Englishman's food: a history of five centuries of English diet* (2nd edn). Jonathan Cape, London.

Dubos, R. (1961). *The dreams of reason: science and utopia.* Columbia University Press, New York.

Dubos, R. (1965). *Man adapting.* Yale University Press, New Haven, CT.

Dubos, R. (1970). The social environment. In *Environmental psychology: man his physical setting* (eds. H. M. Proshansky, W. H. Ittelson, and L.G. Rivlin), pp. 202–8. Holt, Rinehart and Winston, New York.

Dubos, R. and Dubos, J. (1953). *The white plague: tuberculosis, man and society.* Victor Gollancz, London.

Dunn, F. L. (1978). Epidemiological factors: health and disease in hunter-gatherers. In *Health and the human condition: perspectives on medical anthropology* (eds. M. H. Logan and E. E. Hunt), pp. 107–18. Duxbury Press, North Scituate, MA.

Dunphy, D. C. (1972). *The primary group: a handbook for analysis and field research.* Appleton-Century-Crofts, New York.

Durham, W. H. (1978). The coevolution of human biology and culture. In *Human behaviour and adaptation* (eds. N. Blurton Jones and V. Reynolds), pp. 11-32. Taylor and Francis, London.

Durham, W. H. (1979). Toward a coevolutionary theory of human biology and culture. In *Evolutionary biology and human social behavior: an anthropological perspective* (eds. N. A. Chagnon and W. Irons), pp. 39-59. Duxbury Press, North Scituate, MA.

Durkheim, E. (1951). *Suicide.* The Free Press, New York.

Eckholm, E. P. (1976). *Losing ground: environmental stress and world food prospects.* W. W. Norton, New York.

Eckholm, E. P. (1977). *The picture of health: environmental sources of disease.* W. W. Norton, New York.

Eckholm, E. P. (1982). *Down to earth: environmental and human needs.* W.W. Norton, New York.

Edwards, S. W. (1978). Nonutilitarian activities in the Lower Paleolithic: a look at the two kinds of evidence. *Current Anthropology* **10**(1), pp. 134–7.

Ehrlich, P. R. (1968). *The population bomb.* Ballantine, New York.

Ehrlich, P. R., Ehrlich, A. E., and Holdren, J. P. (1977). *Ecoscience: population, resources, environment* (2nd edn). W. H. Freeman, San Francisco, CA.

Eibl-Eibesfeldt, I. (1970). *Ethology: the biology of behavior.* Holt, Rinehart, and Winston, New York.

Eibl-Eibesfeldt, I. (1971). *Love and hate: on the natural history of basic behaviour patterns.* Methuen, London.

Eibl-Eibesfeldt, I. (1975). Aggression in !Ko-Bushmen. In *War, its causes and correlates* (eds. M. A. Nettleship, R. Dalegivens, and A. Nettleship), pp. 281–96. Mouton, The Hague.

Elliott, M. (1977). Conclusion: stopping the atomic juggernaut. In *Ground for concern: Australia's uranium and human survival* (ed. M. Elliott), pp. 209-16. Penguin, Harmondsworth, Middx.

Epstein, F. H. (1965). The epidemiology of coronary heart disease. *J. Chron. Dis.* **18**, pp. 735-74.

Epstein, F. H. (1979). Predicting, explaining and preventing coronary heart disease: an epidemiological view. *Modern Concepts of Cardiovascular Disease* **98**(2), pp. 7-12.

Etzioni, A. (ed.)(1971). *Alienation: from Marx to modern sociology.* Allyn and Bacon, Boston.

Evelyn, J. (1661). *Fumifugium: or the inconvenience of the aer and smoake of London dissipated* pp. 7, 11, 30, 37. National Society for Clean Air (1961), London.

Eyre, S. R. (1978). *The real wealth of nations.* Edward Arnold, London.

Falk, D. (1975). Comparative anatomy of the larynx in man and the chimpanzee: implications for language in Neanderthal. *Amer. J. Phys. Anth.* **43**(1), pp. 123-32.

Feingold, B. F. (1974). *Why your child is hyperactive.* Random House, New York.

Fenner, F. (1970). The effects of changing social organisation on the infectious diseases of man. In *The impact of civilisation on the biology of man* (ed. S. V. Boyden), pp. 48-76. Australian National University Press, Canberra.

Fenner, F., Henderson, D. A., Arita, I., Jezek, Z., and Laynyi, I. D. (1987, in press). *Smallpox and its eradication.* World Health Organization, Geneva.

Ferguson, A. (1767). *An essay on the history of civil society* (1966 edn) (ed. D. Forbes). Edinburgh University Press, Edinburgh.

Finlay-Jones, R. A. and McComish, M. J. (1972). Prevalence of diabetes mellitus in aboriginal lepers: the Derby Survey. *Medical Journal of Australia* **2**(3), pp. 135-7.

Finn, R., McConnochie, K., and Green J. R. (1982). The role of salt in hypertension. In *Disease and the environment* (eds. A. R. Rees and H. J. Purcell), pp. 109-13. John Wiley, New York.

Flinn, M. W. (ed.)(1965). *Report on the sanitary conditions of the labouring population of Britain (1842).* Author E. Chadwick (1842). Edinburgh University Press, Edinburgh.

Foley, R. (ed.)(1984). *Hominid evolution and community ecology: prehistoric human adaptation in biological perspective.* Academic Press, London.

Frankfort, H. (1951). *The birth of civilization in the near east.* Indiana University Press, Bloomington, IN.

Fraser, P. J., Khalil, M. A. K., Rasmussen, R. A., and Steele, L. P. (1984). Tropospheric methane in the mid-latitudes of the southern hemisphere. *Journal of Atmospheric Chemistry* **1**, pp. 125-35.

Frazer, W. M. (1950). *A history of English public health 1834-1939.* Balliere, Tindall and Cox, London.

Freedman, R. and Berelson, B. (1974). The human population. *Scientific American* **231**(3), pp. 31-9.

Freeman, C. (1974). The luxury of despair: a reply to Robert Heilbroner's Human Prospect. *Futures* **6**(6), pp. 450-62.

Freeman, D. (1964). Human aggression in anthropological perspective. In *The human history of aggression* (eds. J. D. Carthy and F. J. Ehling), pp. 109-19. Academic Press, London.

Fromm, E. (1968). *The revolution of hope: toward a humanized technology.* Harper and Row, New York.

Furnass, B. (1976). Changing patterns of health and disease. In *The magic bullet,* pp. 5–32. Society for Social Responsibility in Science ACT, Canberra.

Gabor, D., Colombo, U., King, A., and Galli, R. (1978). *Beyond the age of waste: a report to the Club of Rome.* Pergamon Press, Oxford.

Galbraith, J. K. (1958). *The affluent society.* Houghton Mifflin, Boston, MA.

Galbraith, J. K. (1971). *The new industrial state* (2nd edn). Houghton Mifflin, Boston, MA.

Galbraith, J. K. (1975). *Money: whence it came, where it went.* Pelican, Harmondsworth, Middx.

Galle, O. R., Gove, W. R., and McPherson, J. M. (1972). Population density and pathology: what are the relations for man? *Science* **176**(4030), pp. 23–30.

Gans, H. J. (1968). *People and plans: essays on urban problems and solutions.* Basic Books, New York.

Gans, H. J. (1976). Planning and social life: friendship and neighbor relations in suburban communities. In *Environmental psychology: people and their physical settings* (2nd edn) (eds. H. M. Proshansky, W. H. Ittelson, and L. G. Rivlin), pp. 564–5. Holt, Rinehart, and Winston, New York.

Garn, S. M. (1961). *Human races.* Thomas, Springfield, MA.

Garn, S. M. (ed.)(1964). *Culture and the direction of human evolution.* Wayne State University, Detroit, MI.

Garrels, R. M. and Lerman, A. (1977). The exogenic cycle: reservoirs, fluxes, and problems. In *Global chemical cycles and their alterations by man* (ed. W. Stumm), pp. 23–31. Dahlem Konferenzen, Berlin.

Garrison, F. H. (1929). *An introduction to the history of medicine.* Saunders, Philadelphia, PA.

Gerrard, J. W. (1974). Breast-feeding: second thoughts. *Pediatrics* **54**(6), pp. 757–64.

Gerrard, J. W. (1975a). Breast milk: a neglected asset. *Canadian Medical Association Journal* **112**, pp. 1281–2.

Gerrard, J. W. (1975b). Breast feeding: should it be recommended? *Canadian Medical Association Journal* **113**(2), pp. 138–9.

Gibson, J. (1977). Health, lifestyle and stress. In *The impact of environment and lifestyle on human health* (eds. M. Diesendorf and B. Furnass), pp. 52–5. Society for Social Responsibility in Science (ACT), Canberra.

Gifford, R. M. (1982). Global photosynthesis in relation to our food and energy needs. In *Photosynthesis. Vol 2: Development, carbon metabolism and plant productivity* (ed. D. Govindjee), pp. 459–95. Academic Press, New York.

Gifford, R. M. and Millington, R. J. (1975). *Energetics of agriculture and food production.* Commonwealth Scientific and Industrial Research Organisation, Canberra.

Glacken, C. J. (1967). *Traces on the Rhodian shore.* University of California Press, Berkeley, CA.

Glick, P. C. (ed.)(1975). *The population of the United States of America.* CICRED World population series 1974.

Goeller, H. E. and Weinberg, A. M. (1976). The age of substitutability: what do we do when the mercury runs out? *Science* **191**(4228), pp. 683–9.

Goldsmith, E., Allen, R., and Allaby, M. (eds.)(1972). *A blueprint for survival.* Tom Stacey, London.

Gollan, A. (1978). *The tradition of Australian cooking.* Australian National University Press, Canberra.

Golley, F. B. (1972). Energy flux in ecosystems. In *Ecosystem structure and function* (ed. J. A. Wiens), pp. 69–90. Oregon State University Press, Corvallis, OR.

Gottschalk, L., MacKinney, L. C., and Pritchard, E. H. (1969). *The foundations of the modern world 1300–1775: part one. History of mankind: cultural and scientific development* Vol. 4. George Allen & Unwin, London.

Gould, D. (1979). Coronary confusion. *New Scientist* **81**(1142), p. 458.

Gould, S. J. (1977). *Ever since Darwin: reflections in natural history.* W. W. Norton, New York.

Gowlett, J. A. J., Harris, J. W. K., Walton, D., and Wood, B. A. (1981). Early archaeological sites, hominid remains and traces of fire from Chesowanja, Kenya. *Nature* **294**(5837), pp. 125–9.

Graham, F. (1970). *Since Silent Spring.* Houghton Mifflin, Boston, MA.

Grainger, A. (1981). Will the world run out of timber? Reforesting Britain: a special report. *The Ecologist* **11**(2), pp.64–7.

Greenberg, S. B. (1974). *Politics and poverty: modernization and response in five poor neighbourhoods.* John Wiley, London.

Gregor, B. (1970). Denudation of the continents. *Nature* **228**(5268), pp. 273–5.

Gregory, M. S., Silvers, A., and Sutch, D. (eds.)(1978). *Sociobiology and human nature: an interdisciplinary critique and defense.* Jossey-Bass, San Francisco, CA.

Gribbin, J. (1979). Woodman, spare that tree. *New Scientist* **81**(1148), pp. 1016–18.

Grigg, D. B. (1974). *The agricultural systems of the world: an evolutionary approach.* Cambridge University Press, London.

Grigg, D. B. (1980). *Population growth and agrarian change: an historical perspective.* Cambridge University Press, Cambridge.

Gyorgy, P. (1971). The uniqueness of human milk: biochemical aspects. *American Journal of Clinical Nutrition* **24**, pp. 970–5.

Hall, E. T. (1966). *The hidden dimension.* Doubleday, New York.

Hall, R. H. (1981). Cancer and nutrition. *Ecologist* **11**, pp. 93–9.

Hamburg, D. A. (1964). Emotions in the perspective of human evolution. In *Expressions of the emotions in man* (ed. P. H. Knapp), pp. 300–17. International Universities Press, New York.

Hamburg, D. A. and Brodie, H. K. H. (1973). Psychological research on human aggression. *Impact of Science on Society* **23**, pp. 181–93.

Hamilton, W. D. (1964). The genetic theory of social behaviour. *J. Theor. Biol.* **7**, pp. 1–52.

Hammond, J. (1953). *Effects of artificial lighting on the reproductive and pelt cycles of mink.* W. Heffer & Sons, Cambridge.

Hare, R. (1954). *Pomp and pestilence: infectious disease, its origins and conquest.* Victor Gollancz, London.

Harries, J. M., Hubbard, A. W., Alder, F. E., Kay, M., and Williams, D. R. (1968). Studies on the composition of food: 3. The nutritive value of beef from intensively reared animals. *British Journal of Nutrition* **22**, pp. 21–31.

Harris, D. R. (1969). Agricultural systems, ecosystems and the origins of agriculture. In *The domestication of plants and animals* (eds. P. J. Ucko, and G. W. Dimbleby), pp. 3-15. Gerald Duckworth, London.

Harris, M. (1978). *Cannibals and kings: the origins of culture*. Collins, London.

Harris, M. (1979). *Cultural materialism: the struggle for a science of culture*. Random House, New York.

Harrison, G. A., Palmer, C. D., Jenner, D. A., and Reynolds, V. (1981). Association between rates of urinary catecholamine excretion and aspects of lifestyle among adult women in some Oxfordshire villages. *Human Biology* **53**(4), pp. 617-33.

Harrison, G. A., Weiner, J. S., Tanner, J. M., Barnicot, N. A., and Reynolds, V. (1977). *An introduction to human evolution, variation, growth, and ecology* (2nd edn). Oxford University Press, London.

Harwell, M. A. (1984). *Nuclear winter: the human and environmental consequences of nuclear war*. Springer-Verlag, New York.

Hawkes, J. (1954). *Man on earth*. Cresset Press, London.

Hawkes, J. (1963). Part one: prehistory. In *Prehistory and the beginnings of civilization* (eds. J. Hawkes and L. Woolley) Vol. 1, pp. 3-356. George Allen & Unwin, London.

Hawkes, J. G. (1969). The ecological background of plant domestication. In *The domestication of plants and animals* (eds. P. J. Ucko and G. W. Dimbleby), pp. 17-29. Gerald Duckworth, London.

Hay, R. L. and Leakey, M. D. (1982). The fossil footprints of Laetoli. *Scientific American* **246**(2), pp. 38-45.

Hayden, B. (1981). Subsistence and ecological adaptations of modern hunter/gatherers. In *Omnivorous primates: gathering and hunting in human evolution* (eds. R. S. O. Harding and G. Teleki), pp. 344-421. Columbia University Press, New York.

Hayes, D. (1977). *Rays of hope: the transition to a post-petroleum world*. W. W. Norton, New York.

Healy, P. (1975). Use of psychotropic drugs in Australia. *Informed Opinion* **14** Published by Mental and Drug Education Programme. NSW Health Commission, Sydney. pp. 1-23.

Heaton, K. (1981). Gallstones. In *Western Diseases: their emergence and prevention* (eds. H. C. Trowell and D. P. Burkitt), pp. 47-59. Edward Arnold, London.

Heichelheim, F. M. (1956). Effects of classical antiquity on the land. In *Man's role in changing the face of the Earth* (ed. W. C. Thomas), pp. 165-82. University of Chicago Press, Chicago, IL.

Heilbroner, R. L. (1970). *Between capitalism and socialism: essays in political economics*. Vintage Books, New York.

Heiser, C. B. (1973). *Seed to civilisation: the story of man's food*. W. H. Freeman, San Francisco, CA.

Helbaek, H. (1960). Ecological effects of irrigation in ancient Mesopotamia. *Iraq* **22**, pp. 186-96.

Henderson, H. (1981). *The politics of the solar age: alternatives to economics*. Anchor Press/Doubleday, New York.

Henderson, S. (1981). Social relationships, adversity and neurosis: an analysis of prospective observations. *British Journal of Psychiatry* **138**, pp. 391-8.

Henriques, F. (1959). *Love in action: the sociology of sex.* MacGibbon and Kee, London.

Herlihy, D. J. (1980). Attitudes towards the environment in medieval society. In *Historical ecology: essays in environment and change* (ed. L. J. Bilsky), pp. 100-16. Kennikat Press, Port Washington.

Herodotus. (1921). *The histories* (translated by A. de Selincourt). Heinemann, London.

Hewes, G. W. (1973). Primate communication and the gestural origin of language. *Current Anthropology* **14**(1-2), pp. 5-11.

Hibbert, C. (1963). *The roots of evil: a social history of crime and punishment.* Greenwood Press, Westport.

Higginson, J. (1976). A hazardous society? Individual versus community responsibility in cancer prevention. *Am. J. Public Health* **66**, pp. 359-66.

Hill, D. J. and Gray, N. J. (1982). Patterns of tobacco smoking in Australia. *Medical Journal of Australia* **1**(11), pp. 23-5.

Hinde, R. A. (1974). *Biological bases of human social behaviour.* McGraw-Hill, New York.

Hirsch, F. (1977). *The social limits to growth.* Routledge Kegan Paul, London.

Holdgate, M. W. (1983). *Trends in the world environment during the 1970's.* CRES Working Paper 1983/25. Centre for Resource and Environmental Studies, Australian National University, Canberra.

Holdgate, M. W., Kassas, M., and White, G. F. (eds.)(1982). *The world environment 1972-1982: a report by the United Nations Environment Programme* (Natural Resources and the Environment Series) Vol. 8. Tycooly International, Dublin.

Holdgate, M. W. and White, G. F. (eds.)(1977). *Environmental issues.* John Wiley & Sons, London.

Holland, S. (1975). *The socialist challenge.* Quartet Books, New York.

Holmberg, A. R. (1950). *Nomads of the long bow: the Siriono of Eastern Bolivia.* US Government Printing Office, Washington DC.

Hopkins, D. R. (1983). *Princes and peasants: smallpox in history.* University of Chicago Press, Chicago, IL.

Hordern, A. (1976). *Tranquility denied: stress and its impact today.* Rigby, Sydney.

Horton, J. (1964). The dehumanization of anomie and alienation: a problem of ideology of sociology. *British Journal of Sociology* **15**, pp. 283-300.

Howard, R. and Perley, M. (1982). *Acid rain: the devastating impact on North America.* McGraw-Hill, New York.

Hughes, J. D. (1975). *Ecology in ancient civilizations.* University of New Mexico Press, Albuquerque.

Hughes, J. D. and Thirgood, J. V. (1982). Deforestation in ancient Greece and Rome: a cause of collapse. *Ecologist* **12**, pp. 196-209.

Hughes, R. and Brewin, R. (1979). *The tranquillizing of America: pill popping and the American way of life.* Harcourt Brace Jovanovich, New York.

Hunn, E. S. (1981). On the relative contribution of men and women to subsistence among hunter gatherers of the Columbian plateau: a comparison with ethnolographic atlas summaries. *Journal of Ethnobiology* **1**(1), pp. 124-34.

Hunt, P. S. and Sali, A. (1979). The prevention of colorectal cancer. *Medical Journal of Australia* **1**, pp. 613-16.

Hunter, D. (1971). *The diseases of occupations* (4th edn). English University Press, London.

Illich, I. (1974). *Deschooling society*. Calder and Boyars, London.

Illich, I. (1975). *Tools for conviviality*. Collins, Glasgow.

Illich, I. (1978). *The right to useful unemployment: and its professional enemies*. Marion Boyars, London.

Insel, P. M. and Moos, R. H. (1974). The social environment. In *Health and the social environment* (eds. P. M. Insel and R. H. Moos), pp. 3–12. Lexington Books, Lexington, MA.

International Engergy Agency. (1981). *Energy policies and programmes of IEA countries*. OECD, Paris.

IUCN, UNEP, and WWF (International Union for Conservation of Nature and Natural Resources, United Nations Environment Programme, and the World Wildlife Fund) (1980). *World conservation strategy: living resource conservation for sustainable development*. IUCN, Gland, Switzerland.

Irons, W. (1979). Natural selection, adaptation, and human social behaviour. In *Evolutionary biology and human social behaviour* (eds. N. A. Chagnon and W. Irons), pp. 4–39. Duxbury Press, North Scituate, MA.

Isaac, G. (1978). The food-sharing behaviour of protohuman hominids. *Scientific American* **238**(4), pp. 90–108.

Isaksen, I. A. and Stordal, F. (1981). The influence of man on the ozone layer: readjusting the estimates. *Ambio* **10**, pp. 9–16.

Jacobs, J. (1969). *The economy of cities*. Random House, New York.

Jacobsen, T. and Adams, R. M. (1958). Salt and silt in Ancient Mesopotamian agriculture. *Science* **128**(3334), pp. 1251–8.

Jacobson, B. (1981). Women: smoking's new victims. *New Scientist* **90**(1254), pp. 506–8.

James, W. (1983). *The principles of psychology*. Harvard University Press, Cambridge, MA.

Janick, J, Schery, R. W., Woods, F. W., and Ruttan, V. W. (eds.)(1969). *Plant Science: an introduction to world crops*. W. H. Freeman, San Francisco, CA.

Jelliffe, D. B. and Jelliffe, E. F. P. (eds.) (1978). *Human milk in the modern world: psychosocial, nutritional and economic significance*. Oxford University Press, Oxford.

Jenkins, D. (1974). *Job power: blue and white collar democracy*. Heinemann, London.

Johanson, D. C. and Edey, M. A. (1981). *Lucy: the beginning of humankind*. Granada, London.

Johnston, L. D., Bachman, J. G., and O'Malley, P. M. (1982). *Student drug use, attitudes and beliefs: national trends 1975–1982*. United States Department of Health and Human Services, Washington DC.

Jones, G. (1980). *Social Darwinism and English thought: the interaction between biological and social theory*. John Spiers and Margaret A. Boden, Sussex.

Jones, R. (1969). Fire-stick farming. *Australian Natural History* **16**, pp. 224–8.

Jones, R. (1975). The neolithic palaeolithic and the hunting gardeners: man and land in the antipodes. In *Quaternary Studies* (eds. R. P. Suggate and M. M. Cresswell), Bulletin 13, pp. 21–34. Royal Society of New Zealand, Wellington.

Jones, R. (1978). Why did the Tasmanians stop eating. fish? In *Explorations in ethnoarchaeology* (ed. R. A. Gould), pp. 11-47. University of New Mexico Press, Albuquerque.

Joyce, C. (1980). Hyperactivity debate gives food for thought. *New Scientist* **88**(1225), p. 279.

Joyce, C. (1984). Toxic fumigant banned for US grain harvest. *New Scientist* **101**(1396), p. 7.

Julienne, A., Smith, L., Thomson, N., and Gray, A. (1983). *Summary of Aboriginal mortality in New South Wales country regions, 1980-1981.* Department of Health, Sydney, New South Wales.

Kahn, H. and Wiener, A. J. (1967). *The year 2000: a framework for speculation on the next thirty-three years.* Macmillan, New York.

Kaminer, B. and Lutz, W. P. W. (1960). Blood pressure in Bushmen of the Kalahari Desert. *Circulation* **22** (Part 2, August), pp. 289-95.

Kannel, W. B. (1976). Coronary risk factors: recent highlights from the Framingham study. *Australian and New Zealand Journal of Medicine* **6** (Supplement 4).

Kannel, W. B., Dawber, T. R., Kagan, A., Revotskie, N., and Stokes, J. (1961). Factors of risk in the development of coronary heart disease—six year follow-up experience: The Framingham study. *Annals of Internal Medicine* **55**(1), pp. 33-50.

Kaufmann, W. (1973). *Without guilt and justice.* Peter A. Wyden, New York.

Kee, R. (1980). *Ireland a history.* Weidenfeld & Nicolson, London.

Keen, H., Thomas, B. J., Jarrett, R. J., and Fuller, J. H. (1978). Nutritional factors in diabetes mellitus. In *Diet of man: needs and wants* (ed. J. Yudkin), pp. 89-108. Applied Science Publishers, London.

Keeling, C. D. (1973). Industrial production of carbon dioxide from fossil fuels and limestone. *Tellus* **25**(2), pp. 174-98.

Keesing, R. M. (1976). *Cultural anthropology: a contemporary perspective.* Holt, Rinehart and Winston, New York.

Keith, A. (1946). *Essays on human evolution.* Watts, London.

Keller, S. (1972). Neighbourhood concepts in sociological perspective. In *Human identity in the urban environment* (eds. G. Bell and J. Tyrwhitt), pp. 276-89. Penguin, Harmondsworth, Middx.

Kermode, G. O. (1972). Food additives. *Scientific American* **226**(3), pp. 15-21.

Keyfitz, N. and Flieger, W. (1968). *World population: an analysis of vital data.* University of Chicago Press, Chicago, IL.

Keys, A. (1975). Coronary heart disease: the global picture. *Atherosclerosis* **22**, pp. 149-92.

Klein, R. G. (1973). *Ice-age hunters of the Ukraine.* University of Chicago Press, Chicago, IL.

Kormondy, E. J. (1969). *Concepts of ecology.* Prentice-Hall, Englewood Cliffs, NJ.

Kormondy, E. J. (1974). Natural and human ecosystems. In Human ecology (ed. F. Sargent), pp. 27-43. North Holland Publishing Co., The Hague.

Kramer, S. N. (1967). *The Sumerians.* University of Chicago Press, Chicago, IL.

Kroeber, A. L. and Kluckhohn, C. (1952). *Culture: a critical review of concepts and definitions.* Peabody Museum Papers 47. Harvard University Press, Cambridge, MA.

Kropotkin, P. (1904). *Mutual aid: a factor of evolution* (2nd edn). William Heinemann, London.

Lancet (editorial) (1970). Ambition and disease: the chicken or the egg? *Lancet*, 19 December, pp. 1291-2.

346   *References*

Langer, W. L. (1964). The black death. *Scientific American* **210**(2), pp. 114–21.

Laslett, P. (1971). *The world we have lost*. Methuen, London.

Latham, M. C. (1975). Nutrition and infection in national development. *Science* **188**(4188), pp. 561–5.

Leach, G. (1976). *Energy and food production*. IPC, London.

Leakey, R. E. (1973). Evidence for an advanced Plio-Pleistocene hominid from East Rudolf, Kenya. *Nature* **242**(5398), pp. 447–50.

Leakey, R. E. (1981). *The making of mankind*. Michael Joseph, London.

Leakey, R. and Lewin, R. (1979). The origins of human language. *New Scientist* **83**(1173), pp. 894–7.

Leavesley, J. (1984). The impact of disease on civilization. *The Medical Journal of Australia* **141**(6), pp. 377–9.

Lee, A. M. (1972). An obituary for 'alienation'. *Social Problems* **20**(1), pp. 121–6.

Lee, R. B. (1969). !Kung Bushmen subsistence: an input–output analysis. In *Environment and cultural behaviour: ecological studies in cultural anthropology* (ed. A. P. Vayda), pp. 47–79. The Natural History Press, New York.

Lee, R. B. and DeVore, I. (eds.)(1968). *Man the hunter*. Aldine, Chicago, IL.

Leiss, W. (1978). *The limits to satisfaction: on needs and commodities*. Marion Boyars, London.

Lenneberg, E. H. (1967). *Biological foundations of language*. John Wiley & Sons, New York.

Levine, S. (1971). Stress and behavior. *Scientific American* **224**(1), pp. 26–31.

Lewin, R. (1979). An ancient cultural revolution. *New Scientist* **83**(1166), pp. 352–5.

Lewin, R. (1984). Practice catches theory in kin recognition. *Science* **223**(4460), pp. 1049–51.

Lewis, W. A. (1955). *The theory of economic growth*. George Allen & Unwin, London.

Lewontin, R. C. (1977). Biological determinism as a social weapon. In *Biology as a social weapon* (ed. Ann Arbor Sciences for the People Editorial Committee), pp. 6–18. Burgess, Minneapolis, MN.

Lieberman, P., Crelin, E. S., and Klatt, D. H. (1972). Phonetic ability and related anatomy of the newborn and adult human, neanderthal man, and the chimpanzee. *American Anthropologist* **74**(3), pp. 287–307.

Lieth, H. and Whittaker, R. H. (eds.)(1975). *Primary productivity in the biosphere*. Ecological Studies 14. Springer-Verlag, Berlin.

Lipid Research Clinics Program (1984*a*). The Lipid Research Clinics coronary primary prevention trial results. 1 Reduction in incidence of coronary heart disease. *Journal of the American Medical Association* **251**, pp. 351–64.

Lipid Research Clinics Program (1984*b*). The Lipid Research Clinics coronary primary prevention trial results. 2 The relationship of reduction in incidence of coronary heart disease to cholesterol lowering. *Journal of the American Medical Association* **251**, pp. 365–73.

Lipman-Blumen, J. (1972). How ideology shapes women's lives. *Scientific American* **226**(1), pp. 34–42.

Livingstone, F. B. (1962). Population genetics and population ecology. *American Anthropologist* **64**, pp. 44–53.

Lloyd, A. (1979). Environmental ministers come clean. *New Scientist* **82**(1155), p. 524.

Lloyd, G. E. R. (ed.)(1978). *Hippocratic writing*. Translated by J. Chadwick, W. N. Mann, I. M. Lonie and E. T. Wittington. Penguin, Harmondsworth, Middx.

Lonnroth, M., Johansson, T. B., and Steen, P. (1980). *Solar versus nuclear: choosing energy futures*. Pergamon Press, Oxford.

Lorenz, K. Z. (1966). *On aggression*. Methuen, London.

Lorenz, K. Z. (1974). *Civilised man's eight deadly sins*. Methuen, London.

Lovins, A. (1977). *Soft energy paths: towards a durable peace*. Ballinger, Cambridge, MA.

Lovins, A. B. (1975). *World energy strategies: facts issues and options*. Friends of the Earth International, San Francisco, CA.

Lovins, A. B. and Price, J. H. (1975). *Non-nuclear futures: the case for an ethical energy strategy*. Friends of the Earth International, San Francisco, CA.

Lumsden, C. J. and Wilson, E. O. (1981). *Genes, mind and culture: the coevolutionary process*. Harvard University Press, Cambridge, MA.

MacNeish, R. S. (1964). The origins of the New World civilisation. *Scientific American* **211**(5), pp. 29–37.

MacNeish, R. S. (1965). The origins of American agriculture. *Antiquity* **39**, pp. 87–94.

Maddi, S. R. (1968). *Personality theories: a comparative analysis*. Doresy Press, Homewood, IL.

Maddocks, I. (1961). Possible absence of essential hypertension in two complex island populations. *Lancet* 19 August, pp.396–9.

Maddox, J. (1972). *The doomsday syndrome*. Macmillan, London.

Madsen, K. B. (1959). *Theories of motivation: a comparative study of modern theories of motivation*. Munksgaard, Copenhagen.

Malinowski, B. (1944). *A scientific theory of culture and other essays*. University of North Carolina Press, Chapel Hill, NC.

Marcuse, H. (1972). *Counter-revolution and revolt*. Allen Lane, London.

Marshall, L. (1976). Sharing, talking, and giving. In *Kalahari hunter-gatherers: studies of the !Kung San and their neighbours* (eds. R. B. Lee and I. De Vore), pp. 349–71. Harvard University Press, Cambridge, MA.

Martin, P. S. (1967). Prehistoric overkill. In *Pleistocene extinctions: the search for a cause* (eds. P. S. Martin and H. E. Wright), pp. 75-120. Yale University Press, New Haven, CT.

Maslow, A. H. (1954). *Motivation and personality*. Harper, New York.

May, J. M. (1965). *The ecology of malnutrition in Africa*. Harper, New York.

Mayer, J. (1968). *Overweight: causes, cost and control*. Prentice-Hall, NJ.

Maynard Smith, J. (1964). Group selection and kin selection. *Nature* **201**(4924), pp. 1145-7.

McCance, R. A. and Widdowson, E. M. (1956). *Breads, white and brown: their place in thought and social history*. Pitman Medical Publishing Company, London.

McKeown, T. and Brown, R. G. (1955). Medical evidence related to English population changes in the Eighteenth Century. *Population Studies* **9**(2), pp. 119–41.

McKeown, T. and Record, R. G. (1962). Reasons for the decline of mortality in England and Wales during the Nineteenth Century. *Population Studies* **16**(2), pp. 94–122.

McLeod, M. (1983). World-wide pollution. *Canberra Times*, May 11, p. 33.

McMichael, A. J., McCall, M. G., Hartshorne, J. M., and Woodings, T. L. (1980). Patterns of gastro-intestinal cancer in European migrants to Australia: the role of dietary change. *International Journal of Cancer* **25**, pp. 431–7.

McMichael, J. (1979). Facts and atheroma: an inquest. *British Medical Journal* **1**, pp. 173–5.

McNeill, W. H. (1976). *Plagues and peoples*. Anchor Doubleday, New York.

Meadows, D. H., Meadows, D. L., Randers, J., and Behrens, W. W. (1972). *The limits to growth*. Universal Books, New York.

Medawar, P. (1982). *Pluto's republic*. Oxford University Press, Oxford.

Meggers, B. J. (1954). Environmental limitation on the development of culture. *American Anthropologist* **56**(5, part 1), pp. 801–24.

Mellaart, J. (1964). A neolithic city in Turkey. *Scientific American* **210**(4), pp. 94–104.

Mellaart, J. (1975). The origins and development of cities in the near east. In *Janus: essays in ancient and modern studies* (ed. L. L. Orlin), pp. 5–22. Center for Coordination of Ancient and Modern Studies, Ann Arbor, MI.

Merton, R. K. (1957). *Social theory and social structure*. Free Press, New York.

Mesarovic, M. and Pestel, E. (1975). *Mankind at the turning point: the second report of the Club of Rome*. Hutchinson, London.

Midgley, M. (1978). *Beast and man: the roots of human nature*. Cornell University Press, Ithaca, NY.

Mikellides, B. (ed.) (1980). *Architecture for people: explorations in a new humane environment*. Holt, Rinehart and Winston, New York.

Milbrath, L. W. (1981). *General report: U.S. component of a comparative study of environmental beliefs and values*. Environmental Studies Center, State University of New York, New York.

Milgram, S. (1970). The experience of living in cities. *Science* **167**(3924), pp. 1461–8.

Millar, S. (1979). *The biosocial survey in Hong Kong*. Centre for Resource and Environmental Studies, Australian National University, Canberra.

Miller, H. P. (1964). *Rich man, poor man*. Thomas Y. Crowell, New York.

Millstone, E. (1984). Food additives: a technology out of control? *New Scientist* **104**(1426), pp.20–4.

Minc, S. (1960). The civilised pattern of human activity and coronary heart disease. *Medical Journal of Australia* **2** 16 July, pp. 87–91.

Minc, S. (1967). Emotions and ischemic heart disease. *American Heart Journal* **73**(5), pp.713–16.

Mishan, E. J. (1967). *The cost of economic growth*. Penguin, Harmondsworth, Middx.

Monckeberg, F. (1969). Malnutrition and mental behaviour. *Nutrition Reviews* **27**(7), pp. 191–3.

Montagu, A. (ed.)(1973). *Man and aggression* (2nd edn). Oxford University Press, New York.

Montagu, A. (1976). *The nature of human aggression*. Oxford University Press, New York.

Montagu, M. F. A. (ed.)(1968). *Man and aggression*. Oxford University Press, New York.

Moore, A. M. T. (1975). The excavation of Tell Abu Hureyra in Syria: a preliminary report. *Proceedings of the Prehistory Society* **41**, pp. 50–69.

Moore, A. M. T. (1979). A pre-Neolithic farmers' village on the Euphrates. *Scientific American* **241**(2), pp. 50–58.

Moos, R. H. (1976). *The human context: environmental determinants of behaviour*. John Wiley, New York.

Morgan, E. (1982). *The aquatic ape*. Souvenir Press, London.

Morgan, E. (1984). The aquatic hypothesis. *New Scientist* **102**(1405), pp. 11–13.

Morris, J. N., Adam, C., Chave, S. P. W., Sirey, C., and Epstein, L. (1973). Vigorous exercise in leisure-time and the incidence of coronary heart disease. *Lancet* **1**, pp. 333–9.

Morris, J. N., Marr, J. W., and Clayton, D. G. (1977). Diet and heart: a postscript. *British Medical Journal* 19 November, pp. 1307–17.

Mumford, L. (1966). *The city in history*. Penguin, Harmondsworth, Middx.

Mumford, L. (1971). *The pentagon of power*. Secker and Warburg, London.

Murton, R. K. and Kear, J. (1975). The role of daylength in regulating the breeding seasons and distribution of wildfowl. In *Light as an ecological factor: II* (eds. G.C. Evans, R. Bainbridge, and O. Rackham), pp. 337–60. Blackwell Scientific Publications, Oxford.

Myers, N. (1981). The hamburger connection: how Central America's forests became North America's hamburgers. *Ambio* **10**(1), pp. 3–8.

Myers, N. (ed.) (1985). *The Gaia atlas of planet management for today's caretakers of tomorrow's world*. Pan Books, London.

Myers, R. H. (Chairman) (1980). *Report of the Committee of Inquiry into Technological Change in Australia: Vol. 1. Technological change and its consequences*. Australian Government Publishing Service, Canberra.

Narr, K. J. (1956). Early food-producing populations. In *Man's role in changing the face of the Earth* (ed. W. L. Thomas), pp. 134–151. University of Chicago Press, Chicago, IL.

National Air Pollution Control Administration (1970). *Air quality criteria for carbon monoxide*. National Air Pollution Control Administration Publication No AP-62. US Department for Health, Education, and Welfare, Washington DC.

National Clearinghouse for Alcohol Information (1981). *Fact sheet: selected statistics on alcoholism*. National Clearinghouse for Alcohol Information, Rockville, MD.

National Information Service (Canberra) (1982). Death and drug abuse in Australia, 1969 to 1980. *Technical Information Bulletin* **69** October, pp. 1–29.

Naughton, J. and Hellerstein, H. K. (eds.) (1973). *Exercise testing and exercise training in coronary heart disease*. Academic Press, New York.

Needleman, H. L., Gunnolleviton, A., Reed, R., Peresie, H., and Maher, C. (1975). Duodenal ulcer and diet. In *Refined carbohydrate foods and disease: some implications of dietary fibre* (eds. D. P. Burkitt and H. C. Trowell), pp. 279–310. Academic Press, London.

Neel, J. V. (1962). Diabetes mellitus: a 'thrifty' genotype rendered detrimental by 'progress'? *American Journal of Human Genetics* **14**, pp. 353-62.

Neel, J. V., Fajans, S. S., Conn, J. W., and Davidson, R. T. (1965). Diabetes Mellitus. In *Genetics, and the epidemiology of chronic diseases* (eds. J. V. Neel, M. W. Shaw, and W. J. Schull), pp. 105-32. US Department of Health, Education and Wellfare, Washington DC.

Nelkin, D. and Fallows, S. (1978). The evolution of the nuclear debate: the role of public participation. *Annual Review of Energy* **3**, pp. 275-312.

Newcombe, K. (1976). *A brief history of concepts of energy and the rise of energy use by humankind.* Centre for Resource and Environmental Studies, Australian National University, Canberra.

Newman, O. (1972). *Defensible space: people and design in the violent city.* Architectural Press, London.

Newman, P. and Hogan, T. (1981). A review of urban models: towards a resolution of the conflict between populace and planner.*Human Ecology* **9**(3), pp. 269-304.

Newton, M. (1971). Mammary effects. *American Journal of Clinical Nutrition* **24**, pp. 987-91.

Nicol, H. (1967). *The limits of man.* Constable, London.

Nisbet, R. A. (1970). *The social bond: an introduction to the study of society.* Alfred A. Knopf, New York.

Nixon, R. M. (1970). The state of the Union: the President's message delivered before a joint session of the Congress, January 22, 1970, *Weekly Compilation of Presidential Documents* **6**(4), 26 January, pp. 58-66. US Government Information Service, Washington DC.

Ohlson, M. A. (1969). Dietary pattern and effect on nutrient intake. *World Review of Nutrition and Dietetics* **10**, pp.13-43.

Oke, T.R. (1978). *Boundary layer climates.* Methuen, London.

Oppacher, F. (1977). Philosophical dimensions of a conserver society: quality of life in a conserver society. In *Conserver Society project: report on phase II. Vol. 4: Values and the conserver society.* (Project director, K. Valastakis). Study no. 14. GAMMA, University of Montreal/McGill University, Montreal.

OECD, (1979). *Interfutures: facing the future.* Organization for Economic Cooperation and Development, Paris.

O'Riordan, T. (1976). *Environmentalism.* Pion, London.

Paddayya, K. (1979). On evidence for non-utilitarian activities in Lower Palaeolithic man. *Current Anthropology* **20**, p. 417.

Passel, P.and Ross, L. (1972). Don't knock the $2-trillion economy. *New York Times Magazine* 5 March, p. 14-15, 64, 68-70.

Passmore, J. (1970). *The perfectibility of man.* Duckworth, London.

Passmore, J. (1980). *Man's responsibility for nature: ecological problems and Western traditions.* Duckworth, London.

Patterson, W. C. (1976). *Nuclear power.* Penguin, Harmondsworth, Middx.

Pearce, F. (1983). The cartography of cancer. *New Scientist* **99**(1374), pp. 682-3.

Pearson, D. (1970). *The chemical analysis of foods* (6th edn). J. & A. Churchill, London.

Peel, J. D. Y. (1971). *Herbert Spencer: the evolution of a sociologist.* Heinemann Educational Books, London.

Perkins, H. C. (1974). *Air pollution.* McGraw-Hill, New York.

Petersen, W. (1969). *Population* (2nd edn). Collier-Macmillan, London.

Peterson, N. (1971). Open sites and the ethnographic approach to the archaeology of hunter-gatherers. In *Aboriginal man and environment in Australia* (eds. D. J. Mulvaney and J. Golson), pp. 239-48. Australian National University Press, Canberra.

Peterson, N. (1978). The traditional pattern of subsistence to 1975. In *Nutrition of Aborigines in relation to the ecosystem of central Australia* (eds. B. S. Hetzel and H. J. Frith), pp. 25-37. Commonwealth Scientific and Industrial Research Organization, Canberra.

Phillips, C. S. (1971). The revival of cultural evolution in social science theory. *Journal of Developing Areas* **5** (April), pp. 337-70.

Phillips, E. D. (1973). *Greek Medicine.* Thames and Hudson, London.

Pimentel, D. (1977). World food, energy, man and environment. In *Energy, agriculture and waste management* (ed. W. J. Jewell), pp. 5-16. Ann Arbor Science, Ann Arbor, MI.

Pollard, S. (1968). *The idea of progress: history and society.* C. A. Watts, London.

Postel, S. (1985). Protecting forests from air pollution and acid rain. In *State of the world: a Worldwatch Institute report on progress toward a sustainable society* (eds. L. R. Brown *et al*), pp. 97-123. W. W. Norton, New York.

Pounds, N. J. G. (1979). *An historical geography of Europe: 1500-1840.* Cambridge University Press, Cambridge.

Powles, J. (1973). On the limitation of modern medicine. *Science, Medicine and Man* **1**(1), pp.1-30.

Preston, S. H., Keyfitz, N., and Schoen, R. (1972). *Cause of death: life tables for national populations.* Seminar Press, New York.

Prosser, C. L. (ed.) (1958). *Physiological adaptation.* American Physiological Society, Washington.

Pyke, M. (1968). *Food and society.* John Murray, London.

Raab, W. (1966). Emotional and sensory stress factors in myocardial pathology: neurogenic and hormonal mechanisms in pathogenesis, theory and prevention. *American Heart Journal* **72**(4), pp. 437-576.

Raab, W. (1970). *Preventive myocardiology: fundamentals and targets.* American Lecture Services, Springfield, MA.

Rada, J. (1980). *The impact of micro-electronics: a tentative appraisal of information technology.* International Labour Office, Geneva.

Ramanathan, V., Cicerone, R. J., Singh, H. B., and Kiehl, J. T. (1985). Trace gas trends and their potential role in climate change. *Journal of Geophysical Research* **90**(D3), pp. 5547-66.

Rapoport, A. (1975). Natural history of aggression, edited by J. D. Carthy and F. T. Ebling: review article. In *Biological anthropology:* (compiler S. H. Katz). Readings from Scientific American, pp. 418-20. W.H. Freeman, San Francisco, C. A.

Raskall, P. (1978). Who's got what in Australia: the statistics of wealth. *Journal Australian Political Economy* **2** (June), pp. 3-16.

Rasmussen, W. D. (1982). The mechanization of agriculture. *Scientific American* **247**(3), pp. 49-61.

Reader, J. (1981). *Missing links: the hunt for earliest man.* Collins, London.

Redfield, R. (1953). *The primitive world and its transformations.* Cornell University Press, Ithaca, NY.

Redman, C. L. (1978). *The rise of civilization: from early farmers to urban society in the Ancient Near East.* W.H. Freeman, San Francisco, CA.

Reed, C. A. (1977*a*). A model for the origin of agriculture in the Near East. In *Origins of agriculture* (ed. C. A. Reed), pp. 543–67. Mouton, The Hague.

Reed, C. A. (1977*b*). Origins of agriculture: discussion and some conclusions. In *Origins of agriculture* (ed. C. A. Reed), pp. 879–953. Mouton, The Hague.

Reed, C. A. (ed.) (1977*c*). *Origins of agriculture.* Mouton, The Hague.

Rendel, J. M. (1970). The time scale of genetic change. In *The impact of civilisation on the biology of man* (ed. S. V. Boyden), pp. 27–47. Australian National University Press, Canberra.

Rescher, N. (1969). *Introduction to value theory.* Prentice-Hall, Englewood Cliffs, NJ.

Reynolds, V. (1962). Kinship and the family in monkeys, apes and man. *Man* 3, pp. 209–23.

Reynolds, V. (1966). Open groups in human evolution. *Man* 1, pp. 441–52.

Reynolds, V. (1980). *The biology of human action* (2nd edn). W. H. Freeman, Oxford.

Ridker, R. G. and Watson, W. D. (1980). *To choose a future: resource and enviromental consequences of the alternative growth paths.* Published for Resources for the Future by Johns Hopkins University Press, Baltimore, MD.

Riesman, D. (1950). *The lonely crowd.* Yale University Press, New Haven, CT.

Roberts, A. (1979). *The self-managing environment.* Allison and Busby, London.

Robertson, J. (1979). *The sane alternative: a choice of futures.* River Basin Publishing Company, St. Paul, MN.

Robinson, J. (1972). The second crisis of economic theory. *The American Economic Review* **62**(2), pp. 1–10.

Rohrlich-Leavitt, R., Sykes, B., and Weatherford, E. (1975). Aboriginal women: male and female anthropological perspectives. In *Towards an anthropology of women* (ed. R. R. Reiter), pp. 110–27. Monthly Review Press, New York.

Rokeach, M. (1973). *The nature of human values.* Free Press, New York.

Roper, M. K. (1975). Evidence of warfare in the Near East from 10000–4300 B.C. In *War, its causes and correlates* (eds. M. A. Nettleship, R. Dalegivens, and A. Nettleship), pp. 299–343. Mouton, The Hague.

Rosenman, R. H. and Friedman, M. (1974). The central nervous system and coronary heart disease. In *Health and the social environment* (eds. P. M. Insel and R. H. Moos), pp. 93–106. Lexington Books, Lexington, MA.

Roszak, T. (1970). *The making of a counter culture: reflections on the technocratic society and its youthful opposition.* Faber & Faber, London.

Roszak, T. (1979). *Person/planet: the creative disintegration of industrial society.* Victor Gollancz, London.

Rotblat, J. (1979). Nuclear energy and nuclear weapon proliferation. In *Nuclear energy and nuclear weapon proliferation* (eds. F. Barnaby, J. Goldblat, B. Jasani, and J. Rotblat), pp. 373–435. Taylor and Francis, London.

Rotondo, H. (1965). Adaptability of human behavior. In *Environmental determinants of community well-being: proceedings of the special session held during the third meeting of the PAHO Advisory Committee on Medical Research*, (*1964, 17 June*), pp. 37–40. Pan American Health Organization, Washington DC.

Rotty, R. M. (1973). Commentary on and extension of calculative procedure for CO_2 production. *Tellus* **25**, pp. 508–17.

Rotty, R. M. (1979). Energy demand and global climate change. In *Man's impact on climate* (eds. W. Bach, J. Pankrath, and W. Kellogg), pp. 269–83. Elsevier Scientific Publishing Company, Amsterdam.

Rotty, R. M. and Weinberg, A. M. (1977). How long is coal's future? *Climatic Change* **1**, pp.45–57.

Ruse, M. (1979). *Sociobiology: sense or nonsense*. D. Reidel, Dordrecht, Holland.

Russell, B. (1946). *History of western philosophy*. George Allen & Unwin, London.

Russell, W. M. S. (1968). The slash-and-burn technique. *Natural History* **77**(3), pp. 58–65.

Rutter, M. and Jones, R. R. (eds.)(1983). *Lead versus health: sources and effects of low level lead exposure*. John Wiley, Chichester.

Sackett, D. L., Gibson, R. W., Bross, I. D. J., and Pickren, J. W. (1968). Relation between aortic atherosclerosis and the use of cigarettes and alcohol: an autopsy study. *New England Journal of Medicine* **279** (December 26), pp. 1413–20.

Sagan, C., Toon, O. B., and Pollack, J. B. (1979). Anthropogenic albedo changes and the earth's climate. *Science* **206**(4425), pp. 1363–8.

Sahlins, M. (1972). *Stone age economics*. Aldine Publishing Co, Chicago, IL.

Salaman, R. N. (1949). *The history and social influence of the potato*. Cambridge University Press, Cambridge.

Sauer, C. O. (1956). The agency of man on the earth. In *Man's role in changing the face of the earth* Vol. 1, (ed. W. L. Thomas), pp. 49–69. University of Chicago Press, Chicago, IL.

Sauer, C. O. (1969). *Agricultural origins and dispersals: the domestication of animals and foodstuffs* (2nd edn). MIT Press, Cambridge, MA.

Schild, R. (1976). The final paleolithic settlement of the European plain. *Scientific American* **234**(2), pp. 88–99.

Schmandt-Besserat, D. (1978). The earliest precursor of writing. *Scientific American* **238**(6), pp. 38–47.

Schumacher, E. F. (1973). *Small is beautiful: a study of economics as if people mattered*. Blond and Briggs, London.

Scitovsky, T. (1976). *The joyless economy: an inquiry into human satisfaction and consumer dissatisfaction*. Oxford University Press, London.

Selinger, B. (1976). Food additives: the Australian scene. In *The magic bullet* (ed. M. Diesendorf), pp. 70–92. Society for Social Responsibility in Science (ACT), Canberra.

Selye, H. (1976). *Stress in health and disease*. Butterworths, Boston.

Service, E. R. (1962). *Primitive social organisation: an evolutionary perspective* (2nd edn). Random House, New York.

Seymour, J. (1980). Soil erosion: can we dam the flood. *Ecos* **25**, pp. 3–11.

Shekelle, R. B., Shyrock, A. M., Paul, O., Lepper, M., and Stamler, J. (1981). Diet, serum cholesterol and death from coronary heart disease: the Western Electric study. *New England Journal of Medicine* **304**(2), pp. 65–70.

Shulman, M. (1975). *The ravenous eye*. Coronet Books, London.

Silberbauer, G. B. (1965). *Report to the Government of Bechuanaland on the Bushman*

Survey. Bechuanaland Government, Gaberones.

Simmel, G. (1969). The metropolis and mental life. In *Classic essays on the culture of cities* (ed. R. Sennett), pp. 47-60. Appleton-Century-Crofts, New York.

Simmons, L. W. (1945). *The role of the aged in primitive society.* Yale University Press, New Haven, CT.

Simon, J. L. (1981). *The ultimate resource.* Princeton University Press, Princeton, NJ.

Simpson, G. G. (1950). *The meaning of evolution.* Oxford University Press, London.

Singer, P. (1981). *The expanding circle: ethics and sociobiology.* Clarendon Press, Oxford.

Singer, S. F. (1970). Human energy production as a process in the biosphere. In *The biosphere* (compiler *Scientific American*), pp. 105-14. W. H. Freeman, San Francisco, CA.

Singh, G. and Geissler, E. A. (1985). Late Cainozoic history of vegetation, fire, lake-levels and climate at Lake George: New South Wales, Australia. In *Philosophical Transactions of the Royal Society.* pp. 379-447. London.

Singh, N. (1978). *Economics and the crisis of ecology* (2nd edn). Oxford University Press, Delhi.

Sinnett, P. F. (1975). *The people of Murapin.* E.W. Classey, Farrington, Oxfordshire.

Sinnett, P. and Whyte, M. (1977). Papua New Guinea: a comparison of health status and disease patterns. In *The impact of environment and lifestyle on human health* (eds. M. Diesendorf and B. Furnass), pp. 9-20. Society for Social Responsibility in Science (ACT), Canberra.

Sinnett, P. and Whyte, M. (1981). Papua New Guinea. In *Western diseases: their emergence and prevention* (eds. H. C. Trowell and D. P. Burkitt), pp. 171-87. Edward Arnold, London.

Sjoberg, G. (1965). The origin and evolution of cities. *Scientific American* **213**(3), pp. 55-63.

Skogerboe, G. V. (1974). Agricultural systems. In *Human ecology* (ed. F. Sargent), pp. 127-45. North Holland, Amsterdam.

Smith, A. (1910). *The wealth of nations* Vols. 1 and 2. J. M. Dent & Sons Ltd, London.

Smith, J. M. (1975). *The theory of evolution* (3rd edn). Penguin, Harmondsworth, Middx.

Solheim, W. G. (1972). An earlier agricultural revolution. *Scientific American* **226**(4), pp. 34-41.

Sors, A. (1980). Assessing the health risks of global pollution. *Ambio* **9**(2), pp. 89-96.

South Africa. Medical Officer of Health. (1965). *Report on the health of Johannesburg in 1965.*

Spooner, B. (ed.)(1972). *Population growth: anthropological implications.* MIT Press, Cambridge, MA.

Spuhler, J. N. (ed.)(1959). *The evolution of man's capacity for culture.* Wayne State University Press, Detroit, MI.

Spuhler, J. N. (1968). Assortative mating with respect to physical characteristics. *Eugenics Quarterly* **15**(2), pp. 128-40.

Starr, C. (1979). Energy and power. In *Energy* (compiler *Scientific American*). W. H. Freeman, San Francisco, CA.

Steinberg, A. G. (1965). Genetics and diabetes. *Fifth Congress of the International Diabetes Federation, July 1965, Toronto*. Excerpta Medica International Congress Series No. 84.

Steinhart, J. S. and Steinhart, C. E. (1974). Energy in the US food system. *Science* **184**(4134), pp. 307-16.

Stokinger, H. E. and Coffin, D. L. (1968). Biologic effects of air pollutants. In *Air pollution* Vol. 1 (2nd edn) (ed. A. C. Stern), pp. 445-546. Academic Press, New York.

Stone, E. (1763). An account of the success of the bark of the willow in the cure of agues. *Philosophical Transactions of the Royal Society* **53**, pp. 195-200.

Strahler, A. N. (1956). The nature of induced erosion and aggradation. In *Man's role in changing the face of the earth* (ed. W. L. Thomas), pp.621-38. Chicago University Press, Chicago, IL.

Stringer, C. (1984). Human evolution and biological adaptation in the pleistocene. In *Hominid evolution and community ecology: prehistoric human adaptation in biological perspective* (ed. R. Foley), pp. 55-83. Academic Press, London.

Surrey, J. and Huggett, C. (1976). Opposition to nuclear power: a review of international experience. *Energy Policy* **4**(4), pp. 286-307.

Susser, M. W. and Watson, W. (1971). *Sociology in medicine*. Oxford University Press, London.

Taher, A. H. (1982). *Energy a global outlook: the case for effective international co-operation*. Pergamon Press, Oxford.

Taylor, R. (1979). *Medicine out of control: the anatomy of a malignant technology*. Sun Books, Melbourne.

Theophrastus, (1961). *Enquiry into plants, and minor works on odours and weather signs* Vol. II. (Engl. transl. Sir Arthur Hort.) William Heinemann, London.

Thomas, E. M. (1959). *The harmless people*. Secker and Warburg, London.

Thomas, W. L. (ed.)(1956). *Man's role in changing the face of the Earth*. Vol. 1. University of Chicago Press, Chicago, IL.

Thomis, M. I. (ed.)(1972). *Luddism in Nottinghamshire*. Phillimore, London.

Thoresen, C. E., Friedman, M., Gill, J. K., and Ulmer, D. K. (1932). The recurrent coronary prevention project: some preliminary findings. *Acta Med. Scand.* (Suppl.) **660**, pp. 172-92.

Tiger, L. (1969). *Men in groups*. Random Press, New York.

Tiger, L. and Fox, R. (1972). *The imperial animal*. Secker and Warburg, London.

Tinbergen, N. (1972). Functional ethology and the human sciences. *Proc. R. Soc. Lond. B*. **182**, pp. 385-410.

Tinbergen, N. and the editors of *Life* (1966). *Animal behavior*. Time-Life International, Amsterdam.

Tobias, P. V. (1964). Bushman huntergatherers: a study in human ecology. In *Ecological studies in Southern Africa* (ed. D. H. S. Davis), pp. 67-86. W. Junk, The Hague.

Tobias, P. V. (1969). Commentary on new discoveries and interpretations of early African fossil hominids. *Yearbook of Physical Anthropology 1967* **15**, pp. 24-30.

Tobias, P. V. (1975). Long or short hominid phylogenies: paleontological and molecular evidences. In *The role of natural selection in human evolution*(ed. F.

M. Salzano), pp. 89-118. North Holland Publishing Co, Amsterdam.

Toffler, A. (1970). *Future shock*. Random House, New York.

Toffler, A. (1981). *The third wave*. Pan Books, London.

Tonnies, F. (1955). *Community and association*. Translated by C. P. Loomis. Routledge and Kegan Paul, London.

Torrens, I. M. (1984). What goes up must come down: the acid rain problem. *OECD Observer* **129**, pp. 9-15.

Trevelyan, G. M. (1950). *Illustrated English social history* (2nd edn). Longmans, London.

Trevelyan, G. M. (1951). *Illustrated English social history: Chaucer's England and the early Tudors* Vol. 1. Longmans, Green and Co, London.

Trivers, R. L. (1971). The evolution of reciprocal altruism. *Quarterly Review of Biology* **46**, pp. 35-57.

Trowell, H. (1975). Some historical aspects of milling cereals and refining sugar. In *Refined carbohydrate foods and disease: some implications of dietary fibre* (eds. D. P. Burkitt and H. C. Trowell), pp. 43-6. Academic Press, London.

Trowell, H. (1981). Hypertension, obesity, diabetes mellitus and coronary heart disease. In *Western diseases: their emergence and prevention* (eds. H. C. Trowell and D. P. Burkitt), pp. 3-32. Edward Arnold, London.

Trowell, H. C. and Burkitt, D. P. (eds.)(1981). *Western diseases: their emergence and prevention*. Edward Arnold, London.

Tudge, C. (1977). Bread in the head. *New Scientist* **76**(1083), pp. 812-13.

Turco, R. P., *et al.* (1983). Nuclear winter: global consequences of multiple nuclear explosions. *Science* **222**, pp. 1283-92.

Turnbull, C. (1961). *The forest people: a study of the Pygmies of the Congo*. Simon & Schuster, New York.

Turnbull, C. (1965). *Wayward servants: the two worlds of the African Pygmies*. Natural History Press, Garden City, NJ.

Turnbull, C. (1972). *The mountain people*. Simon & Schuster, New York.

Tuve, G. L. (1976). *Energy, environment, populations and food: our four interdependent crises*. John Wiley, New York.

Ucko, P. J. and Dimbleby, G. W. (1969). Introduction: context and development of studies of domestication. In *The domestication of plants and animals* (eds. P. J. Ucko and G. W. Dimbleby), pp. xvii-xxvi. Gerald Duckworth, London.

UNCTAD. (1978). *Transfer of technology: its implications for development and environment*. Compiler United Nations Conference on Trade and Development Secretariat. United Nations, New York.

Unesco (1975). *Population education in Asia: a source book*. United Nations Educational, Scientific, and Cultural Organization Regional Office for Education in Asia, Bangkok, Thailand.

United Nations (1968). *1967 demographic yearbook*. United Nations, New York.

United Nations (1978). *Transfer of technology: its implications for development and environment*. United Nations, New York.

United Nations (1981*a*). *United Nations 1979/1980 Statistical Yearbook*. United Nations, New York.

United Nations (1981*b*). *1979 yearbook of world energy statistics*. United Nations, New York.

United Nations (1982). *1980 demographic yearbook*. United Nations, New York.
United Nations Department of International Economic and Social Affairs (1985). *Demographic yearbook 1983* (35th edn). United Nations, New York.
US Department of Commerce (1976). *World population 1975*. US Department of Commerce and US Bureau of the Census, Washington DC.
US Department of Commerce (1983). *Statistical abstract of the United States: national data book and guide to sources* (103rd edn). US Department of Commerce, Washington DC.
US Department of Health and Human Services (1981). *Third Annual Report from the Secretary of the Department of Health and Human Services to the President and Congress of the United States: drug abuse prevention, treatment and rehabilitation in fiscal 1980*. Department of Health and Human Services, Rockville, MD.
Valaskakis, K. (project director) (1977). *Conserver society project: report on phase II. Vol. 1: The selective conserver society*. GAMMA, University of Montreal / McGill University, Montreal.
Vickers, G. (1970). *Freedom in a rocking boat: changing values in an unstable society*. Allen Lane, London.
Vines, G. (1984). Does fit mean healthy? *Scientific American* 101(1398), pp. 14–21.
Waldron, I. and Ricklefs, R. E. (1973). *Environment and population: problems and solutions*. Holt, Rinehart and Winston, New York.
Ward, B. and Dubos, R. (1972). *Only one earth: the care and maintenance of a small planet*. Penguin, Harmondsworth, Middx.
Ware, C. F., Panikkar, K. M., and Romein, J. M. (1966). *The twentieth century. Vol. 6: History of mankind: cultural and scientific development*. George Allen and Unwin, London.
Washburn, S. L. (1960). Tools and human evolution. *Scientific American* 203(3), pp. 62–75.
Webster, D. (1981). Late pleistocene extinction and human predation: a critical overview. In *Omnivorous primates: gathering and hunting in human evolution* (eds. R. S. O. Harding and G. Teleki), pp. 556–94. Columbia University Press, New York.
Wells, H. G. (1961). *The outline of history*. Cassell, London.
Wendorf, F. *et al.* (1979). Use of barley in the Egyptian Late Paleolithic. *Science* 205(4413), pp. 1341–7.
Wetstone, G. S. and Rosencranz, A. (1983). *Acid rain in Europe and North America: national responses to an international problem*. Environmental Law Institute, Washington DC.
Weyer, E. M. (1932). *The Eskimos: their environment and folkways*. Archon Books, Hamden, CN.
Weyer, E. (1959). *Primitive peoples today*. Hamish Hamilton, London.
Wheelwright, E. L. (1978). *Capitalism, socialism or barbarism?The Australian predicament: essays in contemporary political economy*. Australia and New Zealand Book Co, Sydney.
White, L. A. (1949). *The science of culture: a study of man and civilization*. Farrar, Straus and Company, New York.
Whitehouse, R. (1977). *The first cities*. Phaidon, Oxford.
Whittaker, R. H. and Likens, G. E. (1975). The biosphere and man. In *Primary*

productivity of the biosphere (eds. H. Lieth and R. H. Whittaker), pp. 305-28. Springer-Verlag, Berlin.

Wiet, G., Elisseeff, V., Wolff, P., and Navdov, J. (1975). *The great Medieval civilizations. Vol. 3: History of mankind: cultural and scientific development.* George Allen & Unwin, London.

Wilkinson, R. G. (1973). *Poverty and progress.* Methuen, London.

Williams, E. S. (1980). *Australian Royal Commission of Inquiry into Drugs.* Australian Government Publishing Service, Canberra.

Williams, R. M. (1978). *British population* (2nd edn). Heinemann Educational Books, London.

Wills, C. (1970). Genetic load. *Scientific American* **222**(3), pp. 98-107.

Wilson, E. O. (1975). *Sociobiology: the new synthesis.* Belknap Press, Cambridge.

Wilson. E. O. (1978). *On human nature.* Harvard University Press, Cambridge, MA.

Wilson, E. O. (1978). Introduction: what is sociobiology? In *Sociobiology and human nature* (eds. M. S. Gregory, A. Silvers, and D. Sutch), pp. 1-12. Jossey-Bass, San Francisco, CA.

Wittfogel, K. A. (1956). The hydraulic civilisation. In *Man's role in changing the face of the Earth* (ed. W. C. Thomas), pp. 152-64. University of Chicago Press, Chicago, IL.

Wohlwill, J. F. (1974). Human adaptation to levels of environmental stimulation. *Human Ecology* **2**(2), pp. 127-47.

Wolff, K. H. (ed.)(1950). *The sociology of Georg Simmel.* Free Press, New York.

Wood, B. A. (1978). *Human evolution.* John Wiley, New York.

Wood, C. (1981). The psyche and the heart. *New Scientist* **92**(1278), pp. 366-9.

Woodburn, J. (1968*a*). An introduction to Hadza ecology. In *Man the hunter* (eds. R. B. Lee and I. DeVore), pp. 49-55. Aldine, Chicago, IL.

Woodburn, J. (1968*b*). Stability and flexibility in Hadza residential groupings. In *Man the hunter* (eds. R. B. Lee and I. DeVore), pp. 103-10. Aldine, Chicago, IL.

Woodburn, J. (1980). Hunters and gatherers today and reconstruction of the past. In *Soviet and Western Anthropology* (ed. E. Gellner), pp. 95-117. Duckworth, London.

Woodhill, J., Palmer, J., and Blacket, R. (1969). Dietary habits and their modification in coronary prevention programme for Australians. *Food Technology in Australia* **21**(6), pp. 264-71.

Woods, L. E. (1984). *Land degradation in Australia.* Australian Government Publishing Service, Canberra.

Woolley, L. (1963). The beginnings of civilisation: part 2. In *Prehistory and the beginnings of civilization. Vol. 1: History of mankind: cultural and scientific development* (eds. J. Hawkes and L. Woolley), pp. 357-854. George Allen and Unwin, London.

Worchel, S. (1978). The experience of crowding: an attributional analysis. In *Human responses to crowding* (eds. A. Baum and Y. M. Epstein), pp. 328-56. Lawrence Erlbaum Associates, New York.

World Health Organization (1947). Development and Constitution of the W.H.O. *Chronicle of the World Health Organization* **1**(1-2).

World Health Organization (1972). *Health hazards of the human environment.* World Health Organization, Geneva.

Young, G. L. (1974). Human ecology as an interdisciplinary concept: a critical inquiry. In *Advances in ecological research* (ed. A. MacFadyen), pp. 1-105. Academic Press, London.

Young, J. Z. (1971). *An introduction to the study of man.* Clarendon Press, Oxford.

Yudkin, J. (1964). The need for change. In *Changing food habits* (eds. J. Yudkin and J. C. McKenzie), pp. 15-24. MacGibbon and Kee, London.

Yudkin, J. (1967). Sugar and coronary thrombosis. *New Scientist* **33**(536), pp. 542-3.

Ziegler, P. (1969). *The black death.* Collins, London.

Zihlman, A. and Tanner, N. (1978). Gathering and the hominid adaptation. In *Female hierarchies* (eds. L. Tiger and H. T. Fowler), pp. 163-94. Beresford Book Service, Chicago, IL.

Zinsser, H. (1935). *Rats, lice, and history.* George Routledge, London.

Zohary, D. (1969). The progenitors of wheat and barley in relation to domestication and agricultural dispersal in the Old World. In *The domestication of plants and animals* (eds. P. J. Ucko and G. W. Dimbleby), pp. 47-66. Gerald Duckworth, London.

Zolotas, X. (1981). *Economic growth and declining social welfare.* Bank of Greece, Athens.

INDEX